SUNSPOTS

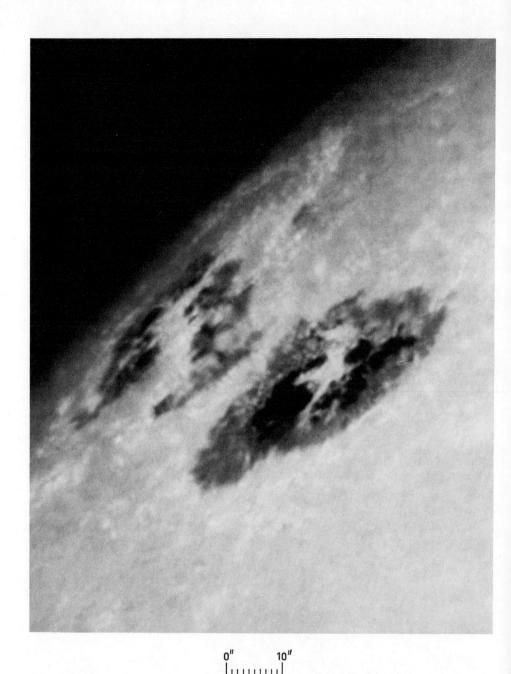

0″   10″

Frontispiece. Large sunspot group near east limb (June 9, 1957). Note the intrusion of facular material into spot nearest limb (cf. Chapter 7, Section 7.2).

SUNSPOTS

R. J. Bray and R. E. Loughhead

Commonwealth Scientific and Industrial Research Organization
National Measurement Laboratory, Sydney, Australia

DOVER PUBLICATIONS, INC., NEW YORK

OTHER BOOKS BY R. J. BRAY
AND R. E. LOUGHHEAD

The Solar Granulation (Chapman and Hall, 1967)

The Solar Chromosphere (Chapman and Hall, 1974)

Published in Canada by General Publishing Company,
Ltd., 30 Lesmill Road, Don Mills, Toronto, Ontario.
Published in the United Kingdom by Constable and
Company, Ltd., 10 Orange Street, London WC2H 7EG.

This Dover edition, first published in 1979, is a
republication of the work originally published by Chapman and Hall Ltd., London, and John Wiley & Sons,
Inc., New York, in 1964. Printing errors on pages 3, 251
and 295 have been corrected in this Dover edition. Only a
brief Editor's Note to the International Astrophysics
Series has been omitted.

International Standard Book Number: 0-486-63731-X
Library of Congress Catalog Card Number: 78-73064

Manufactured in the United States of America
Dover Publications, Inc.
180 Varick Street
New York, N.Y. 10014

Acknowledgements

The authors wish to thank Miss Joan M. Cook, the Reference Librarian of the National Standards Laboratory, for her untiring assistance in procuring out-of-the-way literature. We also thank Miss M. McKechnie, the Chief Librarian, for her interest and support, and Mr P. A. Kazakov for his careful translation of many Russian papers. We are indebted to Mr R. A. H. Partridge for valuable advice in the preparation of the Figures, and Mrs S. Williams for her care in carrying out this work. Miss Jennifer Pilgrim made a substantial contribution to the reduction of the observations described in Section 7.3.4, and also prepared many of the Plates. Additional help in photographic processing was provided by Mr H. R. Gillett. For other assistance we are indebted to Miss Marie McCabe, Mr D. G. Norton, and other colleagues too numerous to mention.

For supplying illustrative material we thank the following persons and organizations:

Drs V. Bumba and J. Kleczek, Ondrejov Observatory (Plate 7.3).

Dr R. E. Danielson and Prof. and Mrs M. Schwarzschild, Princeton University (Plates 2.2, 2.3, 3.1a).

Mr W. C. Miller, Mt Wilson and Palomar Observatories (Plate 4.1).

Perkin-Elmer Corp. (Plate 2.2).

Public Library of New South Wales (Plates 1.1, 1.2).

Science Museum, London (Plate 1.3).

Prof A. B. Severny, Crimean Astrophysical Observatory (Plates 5.1, 5.2).

For permission to reproduce copyright material we are indebted to Mr Stillman Drake, Prof. M. Waldmeier, Geest and Portig K.-G. (Leipzig), Doubleday and Co., McGraw-Hill Book Co., Penguin Books Ltd., the University of Chicago Press, the Carnegie Institution of Washington, the International Astronomical Union, and the Editors of the following journals: *Annales d'Astrophysique, Astrophysical Journal, Australian Journal of Physics, Monthly Notices of the Royal Astronomical Society, The Observatory,* and *Zeitschrift für Astrophysik.*

Finally, we have to thank the C.S.I.R.O. Executive and Dr R. G. Giovanelli, Chief of the Division of Physics, for permission to publish this book.

Commonwealth Scientific and Industrial
Research Organization
National Measurement Laboratory
Sydney, Australia

R. J. Bray
R. E. Loughhead

Authors' Preface

Mankind has recently embarked upon a great scientific adventure—the exploration of the Earth's environment by means of rockets, satellites, and space probes. The Earth's environment is largely dominated by influences arising from transient active phenomena occurring on the Sun. It is not surprising, therefore, that a large part of the current endeavour in the new science is being devoted to the measurement of particle fluxes, magnetic fields, and electromagnetic radiation of all kinds in the space between the Earth and the Sun, and to their correlation with visible phenomena on the solar disk. On the other hand, we are still far from an understanding of the nature of solar activity itself, or of the complex inter-relationships between the various phenomena which comprise it. Our aim in writing this book has been to give a detailed account of just one aspect of solar activity, namely sunspots. It is appropriate that one of the first monographs dealing with a single aspect of solar activity should be devoted to sunspots: not only have they been studied longer than any other solar feature but, as we shall see below, they are the most fundamental manifestation of solar activity.

The transient active phenomena on the Sun occur within localized regions known as *activity centres*. Besides sunspots, the phenomena associated with activity centres include chromospheric flares, prominences, photospheric and chromospheric faculae, and coronal condensations. In addition, activity centres are responsible for the emission of enhanced solar radio noise and X-ray radiation and also, perhaps, solar cosmic rays. It is now generally accepted that these diverse and often spectacular phenomena owe their origin to the presence of *magnetic fields*; for example, the energy liberated in chromospheric flares is so large that the only possible source seems to be the magnetic field associated with neighbouring sunspots. The field configuration in an activity centre can be pictured in an idealized way as a system of gigantic loops of magnetic flux whose ends are anchored to sunspots of opposite polarity. The spots thus play a unique and important role in this complex field system: not only do they serve as anchorage points for the magnetic flux loops, but nowhere else on the Sun does the field strength attain such high values. In a certain sense, therefore, the magnetic field of the centre can be said to 'belong' to the associated group of spots.

A second great scientific project which is currently engaging the attention of physicists all over the world is the magnetic confinement of hot plasmas, with the ultimate aim of exploiting the almost limitless energy of thermonuclear fusion. Compared with other astrophysical systems or with laboratory plasmas the material in the visible layers of a sunspot is relatively lowly ionized. Nevertheless, calculation shows that magnetic forces will generally predominate over purely hydrodynamic forces, and it follows that the study of the dynamical behaviour of sunspots falls within the domain of *magnetohydrodynamics*. In this connection the study of sunspots—and, indeed, of all the active phenomena on the Sun—deserves the serious attention of plasma physicists, for here we have naturally occurring ionized or partially ionized systems which provide challenging testing grounds for magnetohydrodynamic theories. In fact, it is worth recalling that the rapid development of magnetohydrodynamics in the early 1940's was largely stimulated by Hannes Alfvén's pioneering work on the theory of the origin of sunspots.

The physicist usually has the opportunity of making direct physical measurements in the laboratory on the systems in which he is interested, but the astronomer or solar physicist has to rely purely on observation and inference. It may surprise the physicist, therefore, to learn just how much detailed knowledge is now available about physical conditions and processes in sunspots. It has been our aim in writing this book to provide a comprehensive description of this knowledge and, where possible, to outline the theoretical explanation of the various features which observational study has uncovered. Our basic method of attack has been to proceed from the particular to the general, the point of view being continually widened throughout the course of the book. Thus we start with the individual sunspot, dealing successively with its morphology (Chapter 3), physical conditions (Chapter 4), and magnetic field (Chapter 5). In Chapter 6 we turn to a description of the properties of sunspot groups, while in Chapter 7 we enlarge our point of view even further to consider the role played by sunspots in activity centres. Finally, in Chapter 8 we discuss modern magnetohydrodynamic theories of the origin of sunspots and of the solar cycle as a whole, ending the chapter with a description of magnetic activity on stars other than the Sun.

On several occasions we have felt it necessary to include discussions of certain topics which, at first sight, might be thought to lie outside the scope of a book on sunspots. One such topic is the *photospheric granulation* (Chapter 3, Section 3.2), a knowledge of which is an essential pre-requisite to an understanding of sunspots. In fact, the presence of granulation in the umbrae of sunspots shows that the basic convective processes responsible for the photospheric granulation also operate in sunspots, although until recently it was believed that the sunspot magnetic field would suppress the convection. However, a discussion of the photospheric granulation which includes the important advances made in recent years has hitherto been lacking from the astronomical literature.

A second such topic is *magnetic stars* (Chapter 8, Section 8.5). The inclusion of a brief account of this topic is justified on the grounds that magnetic activity is by no means a phenomenon unique to the Sun. In fact, although magnetic activity on the solar scale would be undetectable on other stars owing to their greater distance, magnetic activity on a much more lavish scale has been detected on stars covering a wide range of spectral types. The question thus arises of whether the solar cycle is related to some general form of magnetic activity occurring on a faster and more spectacular scale on certain other stars. If so, it follows that the study of sunspots and other magnetic phenomena on the Sun is really part of the wider study of stellar magnetism—a study which, in view of its possible significance to the problem of stellar evolution, constitutes one of the most important and exciting problems confronting present-day astrophysics.

Finally, the inclusion of several lengthy discussions of various *observing methods* hardly requires justification: nearly all of our knowledge of sunspots is derived from observation, and each new improvement in technique usually results in a significant increase in our knowledge. Thus the introduction of modern high-resolution observing methods, which are described in Chapter 2, has yielded much new information about the fine structure of sunspots, while the application of new methods of measuring and mapping solar magnetic fields (Chapter 5) is currently leading to a substantial revision of earlier ideas about the field configuration in sunspots. In describing modern observing methods, the authors have been at pains to point out limitations in existing techniques and to indicate, when possible, the direction in which future improvements may lie.

Modern methods of observing the Sun are not only extremely sophisticated in themselves but, in addition, often require extensive background knowledge not readily accessible from published sources. For example, in order to understand some of the problems involved in the techniques of high-resolution solar photography, it is essential to have some knowledge of the properties and origin of atmospheric seeing (Chapter 2, Section 2.2). Similarly, to provide a proper basis for the quantitative interpretation of records obtained with solar magnetic field measuring instruments of various types, a knowledge is required of the theory of formation of Fraunhofer lines in the presence of a magnetic field (Chapter 5, Sections 5.2 and 5.3). Some understanding of topics such as these is essential to the serious student of sunspots; the authors hope that their inclusion in this book, together with descriptions of some of the actual instruments employed in sunspot observations, will also prove helpful to solar physicists studying other aspects of the Sun.

R.J.B.
R.E.L.

Contents

Plates

SUNSPOTS

CHAPTER 1

Historical Introduction

1.1 Pre-Telescopic Observations of Sunspots

The earliest known reference to a sunspot, dating back to the middle of the
fourth century B.C., is attributed to Theophrastus of Athens (c. 370–290 B.C.),
a pupil of Aristotle. However, for the earliest *systematic* records of naked-eye
spots we must turn from Greece to China. During the period 28 B.C. to 1638 A.D.
there are no less than 112 descriptions of outstanding sunspots in the official
Chinese histories, not counting numerous references in a variety of other sources.
Early records of sunspots also appear in the annals of Japan and Korea.

In the western literature of the same period, on the other hand, references to
naked-eye observations of sunspots are very rare and fragmentary. It has been
suggested that this is a consequence of the great respect paid to the teachings
of Aristotle, who maintained that the Sun was a perfect body without blemish—
a belief which actually became part of orthodox Christian theology during the
Middle Ages. The first reference to a sunspot after Theophrastus appears to be
that in Einhard's *Life of Charlemagne*, describing a spot seen around 807 A.D.
but interpreted as a transit of Mercury. Abū-l-Fadl Ja'far ibn al-Muktafī
(906–977 A.D.) is said to have recorded that the philosopher al-Kindī observed
a sunspot in May 840 A.D. and attributed it to a transit of Venus. Russian
chronicles of the fourteenth century give imaginative descriptions of spots on
the Sun seen in 1365 and 1371 A.D. through the haze of forest fires. Other
observations are also known, dating from about 1200 A.D. (Ibn Rushd) and
about 1450 A.D. (Guido Carrara and his son Giovanni). Nevertheless, the widespread belief in the 'perfection' of the Sun persisted to the dawn of the telescopic era: in fact, even Johann Kepler (1571–1630 A.D.) mistakenly attributed
a spot seen on May 18, 1607 to a transit of Mercury.

1.2 The First Telescopic Observations of Sunspots: Fabricius, Galileo, Scheiner, and Harriot

The study of solar physics—and, indeed, of astrophysics as a whole—began in
the year 1611, when sunspots were observed for the first time through the
telescope. The honour of this discovery is shared by four men: Johann Goldsmid (1587–1616) in Holland, Galileo Galilei (1564–1642) in Italy, Christopher
Scheiner (1575–1650) in Germany, and Thomas Harriot (1560–1621) in
England. It is not certain which of these four actually made the first sunspot

observations, but the priority of publication undoubtedly belongs to the Dutchman Goldsmid, who is generally known by his Latinized name Fabricius. He was the son of the pastor in the little Dutch village of Osteel. The father was himself a keen observer of the heavens; he was acquainted with the Danish astronomer Tycho Brahe (1546–1601), and was a friend of Johann Kepler. The son was therefore brought up in an astronomical environment. Although his equipment was probably inferior to Galileo's and Scheiner's, Fabricius inferred from his observations that the Sun must rotate. However, he does not appear to have appreciated the importance of this conclusion and pursued his investigations no further.

·On the other hand, the second of the four men, Galileo, went about his researches into sunspots with all his characteristic sagacity. He disposed once and for all of the suggestion that spots might really be small planets revolving around the Sun, pointing out, for example, that this hypothesis was quite incompatible with the observed changes in their size and shape. Instead, he showed that the spots belonged to the Sun and he inferred from his observations that the Sun revolves around a fixed axis with a period of about one lunar month. He noticed that spots frequently occurred in groups and that the spots within a single group moved relatively to one another. Finally, he found that spots were confined to two relatively narrow belts adjacent to the solar equator. It was to be more than 150 years before any important addition was to be made to this basic knowledge of sunspots. Some of Galileo's original sunspot drawings in his book *Istoria e Dimostrazioni intorno alle Macchie Solari* have been reproduced by Berry (1898: see Fig. 55). It is interesting to note that in some cases these show a clear distinction between the umbra and the penumbra.

The third of the four co-discoverers of sunspots, Christopher Scheiner, was born in Swabia, Southern Germany, and became a Jesuit priest at the age of twenty. He taught Hebrew and mathematics at the University of Ingolstadt and later became a professor of mathematics at Rome. When he first observed the spots he suspected some defect in his telescope, but he soon became convinced of their actual existence. He reported his discovery to his ecclesiastical superior who, however, absolutely refused to believe in the reality of the spots. In fact, Scheiner was not allowed to publish his observations under his own name. Instead, he announced his discovery in three anonymous letters addressed to Mark Welser, a wealthy merchant of Augsburg and a friend of Galileo. Welser sent Scheiner's letters to Galileo who replied in three famous letters giving, for the first time, a detailed account of his own researches into sunspots.[1] In the following year, 1613, Galileo's letters were published by the Lincean Academy in the book *Macchie Solari* mentioned above.

[1] The reason for Galileo's delay in publishing his own results is explained in his first letter to Welser. We quote from Stillman Drake's (1957) translation: 'And I, indeed, must be more cautious and circumspect than most other people in pronouncing upon anything new. As Your Excellency well knows, certain recent discoveries that depart from common and popular opinions have been noisily denied and impugned, obliging me to hide in silence every new idea of mine until I have more than proved it.'

At first Scheiner believed that sunspots must be small planets revolving around the Sun. However, this view was crushingly refuted by Galileo in his letters and eventually Scheiner was led by his own observations to realize that the Sun rotates with a period of about 27 days. Like Galileo, Scheiner found that sunspots occurred only in a narrow belt, which he termed the 'royal' zone, extending some 30° on either side of the equator (see Section 6.3.5). However, whereas Galileo's work on sunspots occupied only a couple of years, Scheiner assiduously observed the Sun over a long period of time. His collected observations were published in 1630 in a volume entitled *Rosa Ursina sive Sol*, dedicated to the Duke of Orsini. (The title of the book derives from the badge of the Orsini family, which was a rose and a bear.) The collection of sunspot drawings contained in *Rosa Ursina*, mostly covering the period 1625 to 1627, represents Scheiner's most important contribution as they have helped the course of the sunspot cycle to be traced back to the time of the first telescopic observations of sunspots (cf. Section 6.3.1). Plate 1.1 is a reproduction of one of the original drawings, showing the apparent paths of two spots across the solar disk at different times of the year. In both spots the umbra and penumbra are clearly distinguished. Plate 1.2 is a reproduction of a drawing in *Rosa Ursina*, showing Scheiner's telescope. The telescope was, in effect, equatorially mounted and the solar image was viewed by projection.

The controversy between Galileo and Scheiner concerning the nature of sunspots unfortunately developed into a quarrel regarding their personal claims to the discovery. This controversy made Scheiner a bitter enemy of Galileo and probably contributed substantially to the hostility with which Galileo was henceforth regarded by the Jesuits. In his book *Macchie Solari* Galileo for the first time publicly declared his adherence to the Copernican theory of the motion of the earth. In due course, he was secretly denounced to the Inquisition and, in 1616, he was summoned before Cardinal Bellarmine on the order of the Pope and warned neither to defend nor to hold certain Copernican views. But it was the publication in 1632 of Galileo's great astronomical treatise, *A Dialogue on the Two Principal Systems of the World, the Ptolemaic and Copernican*, which finally led to his trial before the Inquisition at Rome on the grave charge of heresy. There followed one of the saddest scenes in the history of Christianity when on June 22, 1633, in the great Hall of the Inquisition in the Dominican Convent of Santa Maria sopra Minerva, Galileo was made to kneel down and publicly recant his heresies. The fight between Galileo and the Church was ostensibly waged over the Copernican doctrine, but the real issue was the right of a scientist to teach and defend his scientific beliefs. In this conflict the advantage gained by the Church was only temporary.

Less is known about the work of the fourth pioneer sunspot observer, Thomas Harriot. At one time mathematical tutor to Sir Walter Raleigh, Harriot accompanied Sir Richard Grenville to Virginia as a surveyor. On his return, he was introduced by Raleigh to the Earl of Northumberland, under whose patronage he devoted the rest of his life to scientific studies.

1.3 Discovery of the Wilson Effect in Sunspots

The first memorable period in the telescopic observation of sunspots ended with the publication of Scheiner's *Rosa Ursina* in 1630. Thereafter, interest seems to have lapsed and it was not until 1769 that any further significant advance was made. In that year the attention of Alexander Wilson (1714–1786), professor of astronomy in the University of Glasgow, was attracted to a very large spot nearing the west limb of the Sun. When first seen, the spot had a penumbra of uniform width but, as it moved closer to the limb, Wilson noticed that the penumbra on the side remote from the limb gradually contracted and finally disappeared. Moreover, when the spot reappeared at the east limb some two weeks later, the same behaviour was shown by the penumbra on the opposite side of the spot, now the one remote from the limb (cf. Chapter 3, Plate 3.23). The phenomenon was subsequently observed by Wilson in other spots and is now known as the 'Wilson effect'. To explain his observations Wilson advanced the hypothesis that spots represent saucer-shaped depressions in the Sun's surface, formed by the partial removal of the luminous matter which he believed to cover the supposedly dark solid interior of the Sun.

However, the true explanation of the Wilson effect lies in the higher transparency of the spot material compared with the photosphere (cf. Section 3.8). Wilson's work accordingly ranks as the first *physical* investigation into the properties of the individual sunspot. This is in sharp contrast to the efforts not only of his predecessors but also of his successors for upwards of a century afterwards, which were directed towards statistical studies of the occurrence and distribution of spots. Wilson's remarkable discovery was therefore far ahead of its time. Indeed, other physical discoveries of comparable importance were not forthcoming until 1906, when Hale and his collaborators at Mt Wilson began their fundamental investigations into the nature of the sunspot spectrum (cf. Section 1.7).

1.4 Discovery of the Sunspot Periodicity and the Latitude Drift: Schwabe, Carrington, Wolf, and Spörer

The law of sunspot periodicity was discovered by Heinrich Schwabe (1789–1875) of Dessau in Germany. The son of an official, Schwabe studied science at the University of Berlin and then returned to Dessau to commence business as an apothecary. His early career left little time for diversion but, in 1826, he procured a small telescope from Munich and begun to observe the Sun. His object was to search for a possible planet inside the orbit of Mercury. Believing that such a body must sooner or later betray its presence by crossing in front of the solar disk, Schwabe carefully recorded the occurrence of sunspots for 43 years. His first announcement of a probable 10-year periodicity in the occurrence of spots, made in 1843, unfortunately attracted little attention.

Nevertheless, Schwabe continued his patient observations and finally, in 1851, a table giving his sunspot statistics from 1826 onwards, clearly demonstrating the periodicity, was published by Humboldt in the third volume of his famous treatise *Kosmos*. In this way the world suddenly became aware of the existence of a phenomenon which had escaped the notice of telescopic observers for over 200 years. In 1857 Schwabe was awarded the Gold Medal of the Royal Astronomical Society for his achievement.

News of Schwabe's discovery profoundly interested an Englishman, Richard Christopher Carrington (1826–1875), who was just then in the process of building his own private observatory at Redhill in Surrey. The son of a wealthy brewer, Carrington was originally intended for the ministry but while at Trinity College, Cambridge, his attention was diverted to astronomy. To gain experience he first worked for three years as an assistant at Durham Observatory and then built his own observatory. His telescope was an equatorially mounted 4·5-inch refractor driven by clockwork, which was arranged to project an image of the Sun onto a white screen. During the years 1853–1861 he made a vast number of sunspot observations which were published in 1863 under the auspices of the Royal Society in a classic monograph entitled *Observations of the Spots on the Sun*. One of Carrington's most important contributions to solar physics was his discovery that the average latitude of the spots decreases steadily from the beginning to the end of each solar cycle (cf. Section 6.3.5). Owing to the death of his father in 1858, the care of the family brewery devolved upon Carrington and in 1861 he was forced to terminate his sunspot observations. This marked the effective end of Carrington's scientific career.

Confirmation of both Schwabe's and Carrington's discoveries was quickly forthcoming. Rudolf Wolf (1816–1893) of Berne searched through all available records of sunspot observations back to the time of Galileo and Scheiner and was able to derive the much more accurate estimate of 11 years for the average duration of the sunspot cycle. In 1848 Wolf introduced the now well-known *relative sunspot number R* as a measure of sunspot activity (cf. Section 6.3.1). In 1855 he was appointed director of the new Zürich Observatory and there instituted a programme for the daily determination of R. This work has been continued ever since and the Zürich Observatory has remained the world centre of information regarding sunspot numbers.

The latitude drift discovered by Carrington was investigated in more detail by Gustav Spörer (1822–1896) and is, in fact, quite frequently referred to as 'Spörer's law'. Spörer was born in Berlin and began to observe sunspots in 1860 with the object of determining the law of solar rotation. With the aid of a government endowment he organized a small solar observatory at Anclam in Pomerania and undertook sunspot observations there until the end of 1873. Thereafter he continued his work at the Potsdam Astrophysical Observatory, which is now well-known for its interest in the measurement of sunspot magnetic fields.

1.5 The Beginning of Sunspot Photography: Warren de la Rue

The first successful photograph of the Sun was a daguerreotype taken by the physicists H. Fizeau and L. Foucault in Paris on April 2, 1845. The exposure time was 1/60 second and two large sunspot groups were recorded. Great difficulty was experienced in obtaining an exposure time short enough to avoid burning-out the image and the experiment was not repeated. However, in 1857 the challenge was taken up by a remarkable Englishman, Warren de la Rue (1815–1889), who was asked by the Royal Society to design a telescope for solar photography for the Kew Observatory. This instrument, which became the prototype for a number of later photoheliographs, was subsequently transferred to the Royal Observatory, where it was used to begin the Greenwich series of sunspot photographs in April, 1874.

Born in Guernsey, Warren de la Rue was educated at the École Sainte-Barbe in Paris and subsequently made a fortune as a paper manufacturer in England. His interest in observational astronomy was first aroused in 1840 when he visited the engineer James Nasmyth to consult him about appliances for making white lead. Nasmyth was well experienced in the technique of casting and polishing speculum mirrors and was, in fact, engaged in the casting of a 13-inch speculum at the time of de la Rue's visit. With the aid of Nasmyth and the telescope-maker William Lassell, de la Rue proceeded to master the art of speculum making himself. His largest reflector was a 13-inch equatorial of 10-foot focal length which he erected first in his garden at Canonbury, London, and later at his new private observatory at Cranford in Middlesex. Using the newly invented collodion wet-plate process, de la Rue in 1853 obtained the first lunar photographs of substantial value. He continued his photographic observations of the Moon at Cranford and eventually received the Gold Medal of the Royal Astronomical Society in recognition of his lunar studies.

The man responsible for persuading the Royal Society to provide the funds for the construction of de la Rue's photoheliograph was the English astronomer Sir John Herschel (1792–1871), who from 1847 onwards had repeatedly advocated the construction of an instrument to obtain a daily photographic record of sunspots. The photoheliograph was built by the optician A. Ross under the supervision of de la Rue. A photograph of the instrument, which is now in the Science Museum at South Kensington, London, is shown in Plate 1.3. The objective is a 3·5-inch achromat of 50 inches focal length, corrected for the photographic (violet) region of the spectrum. An enlarging lens is placed behind the primary focus to form a 4-inch image of the Sun on the photographic plate. Rapid exposures were achieved by means of a spring-loaded, roller-blind shutter operating in the focal plane. The photoheliograph was installed in a dome at the Kew Observatory in 1858. At first progress was slow and it was two years before most of the initial difficulties were overcome. In 1860 de la Rue took the photoheliograph on an eclipse expedition to Spain and, on his return, re-erected it at Cranford. However, in 1861 it was again transferred to

Kew and placed in the hands of the new observatory superintendent, Balfour Stewart. Under the care of de la Rue and Stewart the daily record of sunspots recommenced and continued unbroken until 1872. By this time the observations extended over a complete solar cycle but it was felt that the work, however desirable, could no longer be continued at Kew.

At this juncture the Royal Observatory at Greenwich wisely decided to step into the breach and in 1873 the Kew photoheliograph was transferred to Greenwich for an experimental period. The Greenwich series of sunspot photographs was commenced in April, 1874. In the same year the firm of Dallmeyer completed the construction of several photoheliographs in connection with plans to observe the transit of Venus in December, 1874. In 1875 one of these instruments was brought into operation at Greenwich to maintain the daily sunspot records, while the others were sent to Mauritius, Dehra Dun (India), and the Royal Observatory at the Cape of Good Hope. The Dallmeyer photoheliograph is still in routine use at the Royal Observatory's new site at Herstmonceux Castle in Sussex. In its original form it had a 4-inch objective which, with an enlarging lens, formed an image of the Sun 4 inches in diameter. The enlarging system was altered to give an image size of nearly 8 inches in 1884, the original objective was replaced by a new 4-inch lens in 1910, and the enlarging system was again modified in 1926. The photoheliograph was moved to Herstmonceux in 1949 and is now attached to the 6-inch Newbegin refractor. The Greenwich series of sunspot photographs has continued without any essential interruption from 1874 to the present day.

1.6 Pioneers in High-Resolution Photography of Sunspots: Janssen, Hansky, and Chevalier

The spatial resolution of the photographs obtained with the Kew photoheliograph and its successors is inadequate to reveal the detailed fine structure of sunspots and the solar photosphere which, as we shall see in Chapter 3, has a scale of only 1–2″ of arc (750–1500 km). The necessary improvement in technique was first accomplished by the French astronomer, Pierre Jules Janssen (1824–1907). The son of an eminent musician, he was educated at the University of Paris, where he subsequently obtained the post of professor of physics. He began to study the solar spectrum in the early 1860's and achieved fame by his discovery in 1868 that with the aid of a spectroscope it was possible to see prominences outside of an eclipse.[2] As a result of this work, he was appointed director of a new astrophysical observatory set up at Meudon, in the vicinity of Paris. There, using a refractor of aperture 13·5 cm, Janssen brought the art of high-resolution photography of sunspots and the solar photosphere to a high degree of perfection. His collected observations, extending over a period of some 20 years, together with reproductions of some of his original photographs, were published in 1896 (cf. Chapter 2, Table 2.1).

[2] The same discovery was made independently by Sir Norman Lockyer.

Another pioneer in the field of high-resolution solar photography was the young Russian astronomer Alexis Hansky (1870–1908). After returning from an eclipse expedition to the Novaya Zemlya Islands in 1896, he spent some time at the Meudon Observatory in France. In 1905 he was appointed an assistant astronomer at the Pulkovo Observatory near Leningrad where, using a conventional astrograph and an enlarging camera, he turned his attention to high-resolution photography of the solar disk. His main interest lay in the photospheric granulation, but during the course of this work he obtained some very high-quality photographs of sunspots (cf. Chapter 2, Table 2.1); these were published posthumously in 1908.

A third pioneer in the art of high-resolution solar photography was Father Stanislas Chevalier (1852–1930), who for many years was the director of the Zô-Sè Observatory in China. This observatory was founded as a branch of an older Jesuit observatory situated at Zi-Ka-Wei on the outskirts of Shanghai. The Zô-Sè Observatory was built on a low hill some 15 miles from Shanghai and was equipped with twin 40 cm refractors of 7 m focal length carried on the same equatorial mounting, one designed for visual work and the other for photography. Although the instrument was brought into operation in 1901, observations of the Sun did not start until 1904. Thereafter, Chevalier obtained a large number of sunspot photographs, some of which were reproduced in an atlas published in 1916 (cf. Section 3.7.1).

Although the photographs obtained by Janssen, Hansky, and Chevalier clearly revealed the wealth of fine detail present in sunspots and the solar photosphere, all three men failed to exploit the full potentialities of high-resolution photography. Moreover, they did not realize just how much physical information could be gained from observations of this kind. In fact, nearly half a century was to elapse before the new technique, in an improved form, was fully utilized as a research tool for the study of sunspots and the solar photosphere.

1.7 Beginning of the Mt Wilson Era: George Ellery Hale

The founder of the Mt Wilson Observatory was George Ellery Hale (1868–1938). He was born in Chicago and graduated in engineering at the Massachusetts Institute of Technology. However, his interest in astronomy had been stimulated at an early age and, upon graduation, he commenced solar observations at a small private observatory at Kenwood, Chicago, financed by his father. Here, aided by F. Ellerman, he perfected the spectroheliograph and with it obtained over 3000 photographs of solar phenomena.[3] Hale's intense interest in sunspots and, in particular, in their spectra dated from his early days at Kenwood. Visual observations of the remarkable phenomena of widening and changes of intensity shown by sunspot lines (see Section 4.3.1) made him realize

[3] The principle of the spectroheliograph was actually worked out independently by three men: Hale, Deslandres, and Evershed.

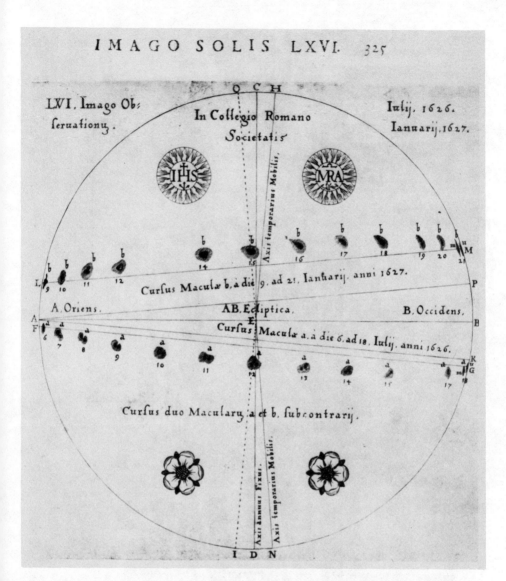

PLATE I.I. Sunspot drawing from Scheiner's *Rosa Ursina*, showing the apparent paths of two spots across the solar disk at different times of the year. In both spots the umbra and penumbra are clearly distinguished.

Maculæ et Faculæ ex uariis obseruandj modis, stabiliuntur.

Constitutione Refractoria composita.

Daniel Widmann Sculpsit.

PLATE 1.2. Scheiner's telescope, from a drawing in *Rosa Ursina*. The solar image was viewed by projection.

PLATE 1.3. The original Kew photoheliograph. (*By courtesy of the Science Museum, London.*)

the importance of obtaining photographs of sunspot spectra with much higher dispersion than had hitherto been employed. To achieve this end, Hale decided to build a fixed horizontal telescope to feed a long-focus spectrograph. Thanks to the generosity of a Miss Helen Snow, this plan culminated in the construction of the Snow horizontal telescope at the Yerkes Observatory in 1903. Unfortunately, the mirrors of the Snow telescope had to be made of glass and, from the outset, difficulty was experienced in obtaining good definition owing not only to distortion of the mirrors by the Sun's radiation but also to the presence of currents of heated air in the horizontal optical path.

Hale spent the winter of 1903 in southern California and here his attention was drawn to the possibilities of Mt Wilson as the site for a mountain observatory, devoted primarily to solar research but also equipped with a large reflector for stellar work. On his return to Yerkes Observatory Hale applied to the newly-established Carnegie Institution of Washington for a grant to transfer the Snow telescope to Mt Wilson. So it was that during the years 1904–5 the instrument was carried in sections by mules up the mountainside and re-erected at the south-east edge of the 5700-foot mountain. A spectroheliograph of 5-foot focal length and a solar spectrograph of 18-foot focal length were then installed and their use soon led to two discoveries of fundamental importance concerning the nature of sunspots.

The first of these came as a result of a photographic study of sunspot spectra. For this work light was allowed to enter the spectrograph slit only from the umbra, while comparison spectra were obtained by covering the centre of the slit and allowing light from the photosphere to enter on each side. As a working hypothesis Hale assumed that the temperature within spots is lower than in the photosphere and evidence to confirm this was soon obtained. A spectroscopic laboratory was set up on Mt Wilson and a study was made of the dependence of line intensities on temperature. The combined observational and laboratory studies left no doubt that spots are cooler than the photosphere.

The second discovery arose indirectly from observations made with the spectroheliograph. Using specially sensitized plates Hale for the first time succeeded in obtaining photographs of the Sun in Hα light. On some of these photographs he found that the chromosphere around spots showed distinct evidence of a vortex structure (see Section 7.3.2). Although the occurrence of a well-marked vortex pattern is now known to be an exceedingly rare phenomenon, it was nevertheless this observation that led Hale to look for magnetic fields in sunspots. The intuitive reasoning that led Hale to this discovery is not valid in the light of present day knowledge (cf. Section 8.2), but for historical interest it is well worth quoting Hale's own words: 'Without entering at present into further details, a single suggestion relating to the possible existence of magnetic fields on the Sun may perhaps be offered. We know from the investigations of Rowland that the rapid revolution of electrically charged bodies will produce a magnetic field, in which the lines of force are at right angles to the plane of revolution. Corpuscles emitted by the photosphere may perhaps be drawn into the vortices, or a preponderance of positive or negative ions may result from some other cause. When observed along the lines of force, many of the lines in the spot spectrum should be double, if they are produced in a strong magnetic field. Double lines, which look like reversals, have recently been photographed in spot spectra with the 30-foot spectrograph of the tower telescope, confirming the visual observations of Young and Mitchell. It should be determined whether the components of these double lines are circularly polarized in opposite directions, or, if not, whether other less obvious indications of a magnetic field are present. I shall attempt the necessary observations as soon as a suitable spot appears on the Sun'.

The necessary observations were obtained on June 25, 1908, with the aid of the large spectrograph of the 60-foot solar tower. (This telescope was erected at Mt Wilson in 1907 according to plans drawn up by Hale and C. G. Abbot and incorporated several novel features designed to overcome the difficulties experienced with the Snow telescope.) To carry out the magnetic field tests a Fresnel rhomb (acting as a quarter-wave plate) and a Nicol prism were mounted above the spectrograph slit. The spot spectrum was photographed in the region $\lambda\lambda 6000$–6200 and it was found that the relative intensities of the components of the doublet lines were reversed when the Nicol was rotated through 90°. This provided conclusive proof of the presence of the longitudinal Zeeman effect, and hence of the presence of a magnetic field (see Section 5.2).

The definition obtained with the 60-foot solar tower and the vertical 30-foot spectrograph was found to be markedly superior to that of the horizontal Snow telescope. Nevertheless, Hale found that neither the diameter of the solar image (17 cm) nor the dispersion of the spectrograph were large enough for the best work on sunspot spectra. Accordingly, funds were secured from the Carnegie Institution for the construction of the now world-famous 150-foot solar tower and a 75-foot vertical spectrograph, which were brought into operation in 1912. Judged by the quality of some of the observations subsequently obtained, the 150-foot tower stands out as the finest solar telescope of this type yet built. With it Hale and his collaborators undertook many fundamental investigations into the nature of sunspots and their magnetic fields, the results of which are described in later chapters.

The advent in 1912 of the 150-foot solar tower with its powerful spectrograph marks the beginning of the modern era of sunspot research. Hale himself, worn out by years of overwork, was forced to resign the directorship of the Mt Wilson Observatory in 1923. However, his influence and enthusiasm persisted there for many years. So important, indeed, were the contributions made by Hale and his chief collaborators, W. S. Adams, F. Ellerman, C. E. St John, H. D. Babcock, S. B. Nicholson, and E. Pettit, that the period between 1905 and 1930 may justly be referred to as the Mt Wilson era in solar physics. Under Hale the art of solar observation reached a peak which has seldom been surpassed. Although primarily an observationalist and experimentalist, Hale from the outset sought to bring his acute physical insight to bear on astronomical problems. He saw solar physics as an integral and fundamentally important part of the broader realm of astrophysics, and planned his investigations accordingly. It was no accident, therefore, that Mt Wilson became a leading centre for both stellar and solar research.

1.8 Chronological Summary

c. 350 B.C... Earliest reference to a sunspot, attributed to Theophrastus of Athens.

28 ..... Start of systematic sunspot records in official Chinese histories.

c. A.D. 807 . Sunspot seen around this time recorded in Einhard's *Life of Charlemagne*.

840 ..... Sunspot reported to have been seen by the philosopher al-Kindī.

c. 1200 ..... Observation of a sunspot by Ibn Rushd.

1365⎱ .... Russian chronicles describe sunspots seen through the haze of
1371⎰   forest fires.

c. 1450 ..... Observation of a sunspot by Guido Carrara and his son Giovanni.

1607 ..... Sunspot seen in this year mistakenly attributed by Kepler to a transit of Mercury.

1611 ..... First telescopic observations of sunspots by Fabricius, Galileo, Scheiner, and Harriot.

1613 ..... Publication of Galileo's *Istoria e Dimostrazioni intorno alle Macchie Solari*.

1630 ..... Publication of Scheiner's *Rosa Ursina sive Sol*.

1769 ..... Discovery of the Wilson effect.

1843 ..... First announcement of the existence of the sunspot cycle made by Schwabe.

1845 ..... Fizeau and Foucault obtain successful daguerreotype of Sun.

1848 ..... R. Wolf introduces the relative sunspot number.

1851 ..... Publication of Schwabe's sunspot statistics in the third volume of Humboldt's *Kosmos*.

1857 ..... De la Rue asked by the Royal Society to design a photohelio-graph for the Kew Observatory.

1858 ..... Announcement of Carrington's discovery of the latitude drift of sunspots.

1861 ..... Systematic photography of sunspots commences at Kew.

1873 ..... Kew photoheliograph transferred to the Royal Observatory at Greenwich.

1874 ..... Start of Greenwich sunspot records.

1875 ..... Dallmeyer photoheliograph brought into operation at Greenwich.

1896 ..... Publication of Janssen's collected observations of sunspots and the solar photosphere.

1903 ..... Construction of the Snow horizontal telescope at Yerkes Observatory.
1904 ..... Start of sunspot observations at Zô-Sè.
1905 ..... Snow telescope re-erected at Mt Wilson.
1906 ..... Mt Wilson studies of the sunspot spectrum show that spots are cooler than the photosphere.
1907 ..... Construction of the 60-foot solar tower telescope at Mt Wilson.
1908 ..... Posthumous publication of Hansky's sunspot photographs.
1908 ..... Hale discovers magnetic fields in sunspots.
1909 ..... Motions in sunspot penumbrae discovered spectroscopically by J. Evershed (cf. Section 4.4.2).
1912 ..... Completion of the 150-foot solar tower telescope at Mt Wilson.

REFERENCES

ADAMS, W. S. [1955] 'Early solar research at Mt Wilson', *Vistas in Astronomy*, ed. A. BEER, vol. 1, p. 619. London, Pergamon.
ARMITAGE, A. [1950] *A Century of Astronomy*. London, Sampson Low.
BALL, R. S. [1906] *Great Astronomers*, 2nd ed. London, Pitman.
BERRY, A. [1898] *A Short History of Astronomy*. London, John Murray.
CLERKE, A. M. [1887] *A Popular History of Astronomy During the Nineteenth Century*, 2nd ed. Edinburgh, A. and C. Black.
DRAKE, S. [1957] *Discoveries and Opinions of Galileo*. New York, Doubleday.
GOLDBERG, L. [1953] *Introduction to 'The Sun'*, ed. G. KUIPER. Univ. Chicago Press.
GRANT, R. [1852] *History of Physical Astronomy*. London, Robert Baldwin.
HALE, G. E., and NICHOLSON, S. B. [1938] 'Magnetic observations of sunspots, 1917–1924, Part I', *Pub. Carnegie Inst.*, No. 498.
KING, H. C. [1955] *The History of the Telescope*. London, Griffin.
MACPHERSON, H. [1933] *Makers of Astronomy*. Oxford Univ. Press.
NEEDHAM, J. [1959] *Science and Civilisation in China*, vol. 3, p. 434. Camb. Univ. Press.
NEWTON, H. W. [1958] *The Face of the Sun*. London, Penguin Books.
SARTON, G. [1947] 'Early observations of the sunspots', *Isis* 37, 69.
SCHOVE, D. J. [1950] 'The earliest dated sunspot', *J. Brit. Astron. Assoc.* 61, 22.
VAUCOULEURS, G. DE [1961] *Astronomical Photography*. London, Faber and Faber.

CHAPTER 2

High-Resolution Observing Methods

2.1 Introduction

In the preceding chapter we have described the progress in understanding the nature of sunspots resulting from visual observations and, later, direct photography, during the 'historical' era 1611–1912. The year 1907 saw the completion of the 60-foot tower telescope at Mt Wilson, and other solar telescopes of this type soon followed. These telescopes were designed to supply large prism or grating spectrographs with a bright, stationary solar image, so that features on the Sun could be studied with a much higher spectroscopic dispersion than had hitherto been employed. As a result of the widespread use of instruments of this type the next few decades saw a great increase in our knowledge of the spectroscopic and magnetic properties of sunspots. However, the same period saw little or no increase in our knowledge of the fine detail in sunspots and, in fact, the techniques developed by Janssen, Hansky, and Chevalier around the turn of the century remained until quite recently unsurpassed. This lack of progress was due chiefly to the difficulty of photographing the fine detail in sunspots under normal conditions of daytime seeing. As we shall see in the next chapter, many of the most interesting and important features of sunspots have a scale of only 1–2″ of arc (750–1500 km). Recent observations indicate that the magnetic fields of sunspot umbrae also have a fine structure with a scale of this order; the same is probably true for sunspot velocity fields. In addition, the fine detail of sunspots is relatively long-lived: some of the features have lifetimes of several hours. Consequently, the *systematic* observation of such detail poses an observational problem of considerable magnitude.

The poor quality of daytime seeing can be attributed to two major causes: (*a*) temperature inhomogeneities within the telescope and its dome caused by the heating effect of the Sun, and (*b*) thermal convection currents in the lower atmosphere set up as a consequence of ground heating. Recent advances in the techniques of high-resolution direct photography have stemmed from attempts to overcome both these problems.

The first major advance was made at the Pic-du-Midi Observatory in the early 1940's when the late Bernard Lyot pioneered the application of the cinematographic technique to the photography of sunspots and the solar photosphere. By taking a large number of photographs at intervals short compared with the time-scale of the phenomenon under study, he was able to obtain a few comparatively unaffected by seeing. In addition, a water-cooled focal-plane diaphragm was used to eliminate at least one source of telescope heating. Although Lyot himself went no further than to demonstrate the value of the cinematographic technique, the work at the Pic-du-Midi has in recent years been successfully carried on by J. Rösch.

The next advance was made in Australia in 1957 when the Physics Division of the Commonwealth Scientific and Industrial Research Organization constructed a 5-inch photoheliograph specifically designed for high-resolution cinematography of sunspots and the solar photosphere. A number of novel features were incorporated in this telescope, and for the first time careful precautions were taken to eliminate all sources of telescope heating. It has since been extensively used by the present authors; many of the observations described in the next chapter were obtained with this instrument.

Examination of films taken with the 5-inch photoheliograph demonstrated in a very striking manner the *intermittency* of daytime seeing: sudden moments of comparative calm were found to occur even on days of generally poor seeing. In 1959, in order to take better advantage of this effect, the authors, in collaboration with D. G. Norton, designed and constructed a 'seeing monitor' to provide a continuous measure of the seeing and, more important, to trigger the camera shutters when the good moments occurred. The use of this device has since resulted not only in improved observations but also in a better understanding of the nature and origin of daytime seeing.

Parallel to the development of improved methods of observing the Sun from the ground, a radically different approach to the problem was being pioneered in England and France by Blackwell, Dewhirst, and Dollfus. These workers constructed a solar telescope for operation from a manned balloon, and carried out several flights to heights of 20,000 feet. Shortly afterwards, Schwarzschild and his co-workers in the U.S.A. sent an automatic telescope in an unmanned balloon to a height of 80,000 feet, and obtained photographs of the photospheric granulation of unsurpassed definition. More recently, Schwarzschild has obtained high-quality photographs of sunspots from flights made with improved methods of guiding and focusing the telescope.

Considerable experience has now been gained in the techniques of high-resolution photography at ground level and far more is now known about daytime seeing than was the case only a few years ago. In the light of this experience substantial improvements should now be possible in the design of solar installations. In particular, it should be possible to construct a telescope–spectrograph combination capable of supplying spectroscopic and magnetic observations of the fine detail in sunspots, which are urgently needed to supplement the information already derived by direct photography. (In view of the bulk and complexity of the equipment required for observations of this type, balloon and satellite techniques will probably make little contribution to this field for many years to come.) In the opinion of the authors, a radical approach to the design of solar telescopes, and probably of spectrographs and magnetographs, will be necessary if such observations are to be systematically obtained.

In this chapter we shall attempt to give a systematic account of modern methods of high-resolution sunspot photography. To understand the problems involved a knowledge of the properties and origin of atmospheric seeing is an essential pre-requisite; this topic is discussed in Section 2.2. Night-time seeing, which has been far more widely studied than daytime seeing, is included in this account because a knowledge of night-time seeing and its origin throws valuable light on the corresponding daytime problem. The solar seeing monitor, together with some of the results derived with its aid, is described in Section 2.2.4, while the origin of daytime or solar seeing is discussed in Section 2.2.5.

Modern methods of sunspot cinematography are described in Section 2.3, including the recent technique of seeing-monitor control (Section 2.3.2). The techniques of solar photography from manned and unmanned balloons are described in Sections 2.3.3 and 2.3.4 respectively.

Finally, in Section 2.4 we discuss some of the factors affecting the performance of solar telescopes, and indicate how the experience gained in modern high-resolution techniques can serve as a guide to possible improvements in their design. Solar installations designed for spectroscopic observations of the fine detail in sunspots are included in this discussion. The following topics are considered: location of observatory site (Section 2.4.2); height of telescope above ground level (Section 2.4.3); the advantages and disadvantages of a tower-type reflector on the one hand, and an equatorial refractor on the other (Section 2.4.4); shielding from wind and prevention of heating effects (Section 2.4.5); and choice of aperture, image diameter, and emulsion (Section 2.4.6). No final answer to some of the questions raised can be given at the present time; however, the experience and knowledge now available indicate, in some cases, the way to further progress.

2.2 Atmospheric Seeing and its Origin

2.2.1   NIGHT-TIME OR 'STELLAR' SEEING

Atmospheric seeing can cause the telescopic image of a star to show three effects: brightness variations ('scintillation'); rapid image motions, usually accompanied by pulsation and image degradation; and, occasionally, slow image motions. Some astronomers restrict the word 'seeing' to image motion and degradation. However, since it is convenient to have a blanket term to cover all the effects, we shall use the word in its broader sense.

Scintillation has been more widely observed and measured than any of the other seeing effects, mainly owing to the comparative ease of the observations. Records of brightness fluctuations of stars obtained with oscillographs show frequencies in the range 1·0 to 500 c.p.s. (Mikesell, Hoag, and Hall, 1951; Ellison and Seddon, 1952; Gifford, Johnson, and Wilson, 1955; Elsässer, 1960). The amplitude of the fluctuations usually has its maximum value at a frequency in the neighbourhood of 10 c.p.s. At large zenith distances the high frequency components (i.e., of the order of several hundred cycles per second) are usually absent, but the low frequency components have a greater amplitude than at the zenith. The amplitude of the variations may be as great as several tenths of the mean intensity and is strongly dependent on the aperture of the telescope, increasing markedly as the aperture is reduced and attaining its maximum value at an aperture of a few inches (Ellison and Seddon, 1952; Ellison, 1956). An extended object, such as a planet, shows a smaller scintillation amplitude than a star, since the brightness variations of neighbouring points are uncorrelated.

Rapid image motions are found to be closely correlated with *image degradation*. The frequencies of the image motions lie mainly in the range 0·1 to 10 c.p.s. (Elsässer, 1960; Mayer, 1960). Mayer (1960) found an r.m.s. image excursion varying on different nights from 0".4 to 1" of arc, while Danjon and Couder (1935: cf. p. 94) quote a value exceeding 0".3 as typical for a large part of France; the value becomes lower than 0".1 only very exceptionally, perhaps for only a few hours per year. Using a rapidly-trailing star image technique, Gaviola (1948) found that occasionally for periods of the order of a hundredth of a second the image diameter approached its theoretical value, although a strong diffuse background was still present (an extended object would have continued to show degradation). Most of the time, however, the diameter of the actual star image ('tremor disk') greatly exceeds the diameter of the central peak of the theoretical diffraction pattern corresponding to the aperture used. Ellison (1956) found an average tremor disk diameter of 2–5" of arc for the Edinburgh 36-inch reflector (theoretical resolution 0".1), while Elsässer (1960) reports typical values of the r.m.s. image excursion at various observatories ranging from 0".3 to 0".5. The widths of the trails on Gaviola's photographs are of the order of 1–2". Meinel (1960b) has published a histogram showing the distribution of tremor disk diameters for 150 nights of observation at Kitt Peak, Arizona; the most frequent value was 1".4. Finally, van Biesbroeck (cf. Keller, 1953) has visually resolved double stars as close as 0".1 with the 82-inch reflector of the McDonald Observatory, Texas, under exceptionally good conditions; even then, however, a diffuse background and dancing of the pattern were evident.

Summarizing, tremor disk diameters of 0·1–0".2 of arc correspond to exceptionally good seeing conditions and occur only rarely even at a good site. 0·5–2".0 are typically occurring average values, while values exceeding 2", which correspond to rather poor stellar seeing, occur far more frequently than the astronomer would wish. The best planetary and lunar *photographs* have a resolution of about 0".4, corresponding to the theoretical resolution of a 12-inch aperture. As we shall see in Section 2.2.3, the values corresponding to daytime seeing are considerably worse than the values quoted above.

The dependence of image motion on zenith distance has been investigated by Mayer (1960); the resulting curve has the character of a *saturation* curve, the saturation value being substantially attained at $Z = 76°$.

Finally, *slow image motions* (called 'accidental refractions' by Danjon and

OBJECTIVE
COVER

EXPOSURE
CONTROLLER

PHOTOELECTRIC
GUIDER

SEEING
MONITOR

PLATE 2.1. The Sydney 5-inch photoheliograph and its mounting. The photoheliograph is mounted on the top face of a 10-foot 'equatorial spar', which also carries the exposure controller, photoelectric guider, and seeing monitor. In the background, portion of C.S.I.R.O.'s 64-element 'Chris Cross' interferometer can be seen. The optical layout of the photoheliograph is shown in Fig. 2.3.

PLATE 2.2. 12-inch balloon-borne stratospheric telescope. The 12-inch primary mirror is located inside the left end of the tube, the small rotating secondary near the right. The housing for the magnifying lens is visible just above the secondary. (*Photo by courtesy of Perkin–Elmer Corp.*)

Couder, 1935: cf. p. 86) have a period ranging from a few seconds up to half a minute or so, with amplitudes of the same order of magnitude as the rapid motions. They affect neighbouring objects identically, whereas the more rapid motions are uncorrelated; they are attributed by Land (1954) to disturbances in the immediate neighbourhood of the telescope objective.

Little is yet known about the dependence of daytime or night-time seeing on location or climate. The western and southwestern regions of the U.S.A., including Texas, Arizona, and California, seem to be particularly well-favoured, as do certain regions of France (Dimitroff and Baker, 1945: cf. p. 76). However, the potentialities of many other countries, such as Australia, are largely unknown, owing to lesser astronomical activity. There is no real evidence that mountain observatories experience better seeing than those at sea level. The choice of a mountain site is usually governed by the desire for freedom from haze, smoke, and city lights. Mt Wilson Observatory (California) is a notable exception, however: this site was chosen by G. E. Hale after extensive tests of the solar seeing (King, 1955: cf. p. 324). The question of selecting a site for a solar observatory is discussed in Section 2.4.2.

As we shall see in the next section, the various phenomena grouped under the term seeing exercise a certain degree of independence; this is due to the fact that different regions of the atmosphere are responsible for the different phenomena.

2.2.2   ORIGIN OF NIGHT-TIME SEEING

A comparison of theory and observation shows that scintillation is produced by disturbed layers high in the atmosphere, whereas image motion originates quite near the telescope. Thus Elsässer and Siedentopf (1959) conclude from a theoretical analysis that the main contribution to the scintillation comes from layers 8 km above sea level, while the most important contribution to image motion originates at sea level, the effect falling off exponentially with increasing height. They find that at sea level the r.m.s. fluctuations in refractive index lie between 0.3×10^{-8} and 1.3×10^{-8}, the corresponding fluctuations in temperature lying between 0.03 and $0.13°C$. Other authors have given the following estimates for the height of the scintillation layer: 3–4 km (Chandrasekhar, 1952); 5 km (Ellison and Seddon, 1952); 3–30 km, depending on the frequency components considered (Gifford, Johnson, and Wilson, 1955); and greater than 20 km (Fürth, 1956).

The size of the disturbing elements usually considered responsible for scintillation is of the order of 10 cm (Chandrasekhar, 1952; Ellison and Seddon, 1952; Elsässer, 1960). Image motions are thought to be due to somewhat larger elements.

There is a large body of evidence which shows, in agreement with the conclusions of Elsässer and Siedentopf, that image motion and degradation are produced by disturbed regions of the atmosphere *at much smaller heights* than those responsible for scintillation—on occasion in the immediate neighbourhood of the dome or even in the tube of the telescope itself. Thus while Mikesell (cf. Elsässer, 1960) found that only wind velocities at heights between 8 and 12 km showed any correlation with scintillation, Elsässer and Heynekamp found a very close correlation between image motion and *surface* wind. Similarly, indoor experiments carried out by Smith, Saunders, and Vatsia (1957) using a camera, a resolution test pattern, and an artificial source of turbulence revealed much greater degradation when the turbulence was located near the camera than when it was located near the target. Under the circumstances, therefore, it is easy to understand why image motion and scintillation are largely independent, so that image quality can vary markedly from night to night, or during the course of a single night, without any noticeable variation in the scintillation (Elsässer, 1960; Mayer, 1960).[1]

Hosfeld (1954) and Steavenson (1955) have investigated the effect of thermal disturbances in the immediate neighbourhood of the dome and telescope. Hosfeld found that when warm air was allowed to escape through the dome aperture marked image motion occurred but the scintillation remained unchanged. Steavenson found considerable degradation of a star image when there was a temperature difference of 10–12°C between the telescope mirror and the air in the tube; the degradation was much reduced when the difference was only 2–3°C. Realization of the deleterious effects caused by such disturbances has led several workers to suggest ways of eliminating them. The interested reader is referred to articles by Rösch (1955, 1956), Steavenson (1955), and Sisson (1960).

2.2.3 DAYTIME OR 'SOLAR' SEEING

An image resolution of 1″ of arc or better is required for effective observations of the fine detail in sunspots or the solar photosphere (Chapter 3); this figure corresponds to the theoretical limit of resolution of a 5-inch telescope. Yet in spite of the fact that most solar telescopes have apertures larger than 5 inches—in some cases much larger—this limit of resolution is only attained comparatively rarely, even in direct photography, where exposure times may be less than one-thousandth of a second. Seeing conditions which enable 1″ of arc resolution to be obtained are therefore regarded as excellent by solar astronomers, although this resolution is considerably worse than the best achieved at night. If this resolution could be systematically attained for all types of observations, including spectroscopic and magnetic observations (where effective exposures of several seconds may be necessary), a considerable improvement in our knowledge of sunspots would result.

It is of interest to inquire how often 1″ seeing occurs. Keenan (1953) remarks that it is the common experience of solar observers to work for several months without encountering more than one or two days of 1″ or better seeing. Salanave

---

[1] Mikesell, Hoag, and Hall (1951), however, do not agree that the two phenomena are uncorrelated: they found that the high frequency (i.e., greater than 100 c.p.s.) components of the scintillation were consistently absent, and the low frequency components much less pronounced, when the image degradation was small.

(1957), who observed the Sun with a 6·5-inch refractor at Junipero Serra Peak, California, estimates that 1″ seeing prevailed for one per cent of the time, corresponding to four hours of very good seeing during six weeks' observation. Our own experience, based on several seasons of sunspot photography with a 5-inch telescope specifically designed for high-resolution work (Section 2.3.1) supports these estimates. Even on the best days, only one per cent of the photographs—taken repeatedly at 5-second intervals throughout the day—remain nearly completely unaffected by seeing and show the full performance of which the telescope is capable.[2] The quality of successive photographs can vary to a very marked extent: an excellent photograph can be sandwiched between two photographs so poor that they barely reveal the granular structure of the photosphere. Since the solar seeing varies so rapidly from moment to moment, some form of photoelectric measurement is necessary to determine the frequency of occurrence and duration of the moments of good seeing. Some further information on the statistics of good seeing, derived with the aid of a solar seeing monitor, is given in the next section.

1″ seeing, as described above, implies conditions such that the photospheric granulation or any other similar fine detail can be distinguished both visually and on direct photographs taken with a short exposure time. It must be emphasized, however, that such seeing conditions—even if prolonged for several seconds or a minute—would not necessarily enable spectroscopic or magnetic observations (where effective exposure times of several seconds may be necessary) of equal resolution to be obtained. This is due to the fact that even when the image is little degraded, shifts and distortions amounting to several seconds of arc may be present: these shifts are readily apparent when high-quality photographs of the Sun, taken at a short enough time interval apart that no real change could occur, are carefully examined. In direct photography the exposure time is small enough to 'freeze' the image motion to some extent, but this is not the case for observations requiring exposure times amounting to several seconds; such observations would not necessarily show the same resolution as direct photographs taken during the same period.

2.2.4 A SOLAR SEEING MONITOR

As we shall see in Section 2.2.5, poor solar seeing owes its origin chiefly to telescope and dome heating on the one hand, and thermal convection currents in the lower atmosphere on the other. In the case of a telescope mounted in the open air and provided with appropriate cooling devices, such as the 5-inch photoheliograph described in Section 2.3.1, poor seeing due to the first of these causes is eliminated. There remains, however, the problem of poor seeing produced by convection currents arising from ground heating. Experience gained with the photoheliograph showed that not only is this residual seeing highly variable from moment to moment but, even on generally poor days,

[2] Rösch (1959a) gives a figure of two per cent for his best films, taken from the Pic-du-Midi. Rösch's techniques are described in Section 2.3.1.

moments of good seeing do occur. It is true that by employing the cinematographic technique with a relatively short interval between successive frames some of the good moments are automatically taken advantage of. Nevertheless, it seemed to the authors that further improvement would result if, instead of photographing at fixed intervals, exposures were triggered automatically at the moments of best seeing. In addition, this procedure would reduce film wastage: this can be quite considerable, since in order to study the evolution of the 1–2″ of arc detail in sunspots, good sequences lasting several hours are required. Accordingly, the authors, in collaboration with D. G. Norton,

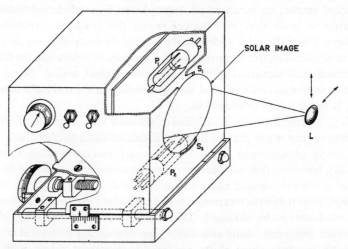

Fig. 2.1. A solar seeing monitor. L, magnifying lens; S_1 and S_2, slits; P_1 and P_2, photocells. The photocell box is provided with motion along the optical axis to compensate for variation in the solar image diameter throughout the year. The pre-amplifiers are located inside the box.

designed and constructed a photoelectric device to monitor the seeing continuously and to trigger the camera shutters at the good moments.

The seeing monitor (Bray, Loughhead, and Norton, 1959) is carried on the same mounting as the 5-inch photoheliograph (cf. Plate 2.1). A 3·5-inch objective and a magnifying lens L form a solar image, 109 mm in diameter, on a white plate (Fig. 2.1). To allow accurate centring of the image, L is provided with fine motions in two directions at right angles to the optical axis. Two narrow circular slits cut in the plate, S_1 and S_2, allow light from the east and west limbs (and the sky beyond) to fall on two vacuum photocells, P_1 and P_2. The distance between the inner edges of S_1 and S_2 is 108 mm, so that light reaches the cells from two arcs of the solar limb each 0·5 mm. (about 9″ of arc) in width. With perfect seeing the amount of light falling on each cell would remain constant. When the seeing is poor, however, the intensity recorded by each cell fluctuates irregularly; the corresponding photocell output then contains both a steady and a fluctuating component. The fluctuating component, which superficially resembles a random noise signal, is a measure of the seeing quality. After pre-amplification, the cell outputs are amplified and added in a second stage, and the fluctuating component of the combined signal is passed to a cathode-ray oscilloscope, a vacuum tube voltmeter, a chart recorder, and a sensitive moving-coil meter relay (see block diagram in Fig. 2.2).

The cathode-ray tube, which has a long-persistence phosphor, provides an instantaneous display of the seeing signal. The voltmeter is set on a peak-to-peak a.c. scale and gives a quantitative (though arbitrary) measure of the seeing,

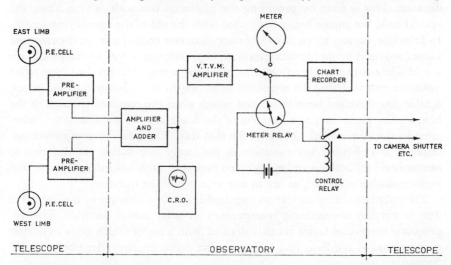

Fig. 2.2. Block diagram of seeing monitor and telescope triggering system. The seeing signal from the vacuum tube voltmeter is fed to a sensitive meter relay, whose pointer carries a contact at its lower end. This contact can close with a second contact attached to the case of the instrument. The latter can be manually adjusted so that the contacts close whenever the seeing signal falls to some pre-determined, acceptable value. This in turn closes a control relay which operates the shutters. Simultaneous tracings of the seeing signal can be obtained with a chart recorder.

averaged over a period comparable with the time constant of the meter. The chart recorder provides a permanent record of the seeing fluctuations, and will eventually enable a statistical analysis to be made of the frequency of occurrence and duration of the moments of good seeing. Finally, the meter relay controls the taking of exposures at the moments of best seeing; it can be set to trigger off whenever the mean amplitude of the fluctuating seeing signal falls below any pre-assigned level, the triggering level being adjustable. (An exposure is made within a small fraction of a second of the signal falling to a 'good' value; if the seeing remains good, exposures continue automatically at a rate of one every 2–3 seconds—the time required for the camera to wind on.)

The purpose of *adding* the two photocell outputs is to minimize the effect on the seeing signal of momentary guiding displacements or wind shake. To a first approximation, a momentary shift of the image—if sufficiently small—causes no alteration of the summed outputs. For this reason, arcs of the limb of finite width (9″ of arc), rather than a tangential image, are used. 'Matched' photocells having a sensitivity of 8·4 microamps/lumen are employed in the monitor.

Although the instrument measures only the amplitude of the *fluctuating* component of the light flux falling on the photocells, it is desirable to keep the diameter of the solar image—and hence the widths of the luminous arcs falling on the photocells—the same throughout the year despite the varying Earth–Sun distance. This is done by providing the photocell box with motion along the optical axis, the image being re-focused with the aid of the magnifying lens L. In principle, on any given day the image diameter could be set at the required value (109 mm) by simple measurement; in practice, however, such a procedure would allow only a crude adjustment, since the seeing enlarges the image to an unknown extent. Using the measured focal lengths of the image-forming lenses, a table has therefore been computed which gives the required position of the box, on any given day, as a function of the Sun's angular semidiameter (tabulated in the *Astronomical Ephemeris*) for that day. The total range of movement required is 17·6 mm. Any variation in the image size during the day due to atmospheric refraction can be neglected except at such low solar altitudes that high-resolution observations are in any case out of the question.

The only remaining correction required is for slow changes in the light flux due to varying atmospheric transparency or solar zenith distance. The appropriate correction factor is easily derived from a meter which gives an output proportional to the light flux falling on the exposure controller photocell (cf. Section 2.3.1).

Apart from the pre-amplifiers, all the electronics associated with the seeing monitor are located inside the telescope control hut. Two meters in the control hut monitor the *direct current* outputs from the two photocells and enable the observer to verify that the solar image is correctly centred on the photocell box.

The peak-to-peak voltage of the fluctuating seeing signal, measured by the vacuum tube voltmeter (Fig. 2.2), provides a quantitative measure of the seeing. Although this measure is on an arbitrary scale, which depends on the size and brightness of the solar image, the sensitivity of the photocells, and electronic circuit constants, it is of interest to indicate roughly the values which actually occur in practice, and to discuss how they correlate with (*a*) the visual appearance of the limb of the solar image on the seeing monitor box, and (*b*) the resolution shown by direct photographs of the granulation or sunspots taken at the same time.

A commonly occurring value for the seeing signal is 1–2 volts. At this level, the limb shows easily discernible distortions; simultaneous photographs of the granulation show considerable degradation, although the granulation may be fairly well resolved over localized portions. Worse values are not uncommon: on poor days, the signal may remain at 4–5 volts for a considerable portion of the time. Values as high as 10 volts do occur but are fairly exceptional; only on one occasion has a value as high as 15 volts been observed. If the seeing signal *continuously* exceeds 1–2 volts, high-resolution observations are out of the question.

Turning now to the happier side of the picture, a seeing level of 0·8 volts is regarded as 'promising' as far as 1–2" of arc direct photography is concerned. The limb then appears sharp and well-defined, and a fairly close visual inspection is required before any waviness can be perceived. About 10 per cent of the photographs taken at this level may be expected to show 1" resolution over substantially the whole of their areas; the remaining photographs would be mediocre, with an occasional downright bad one.[3] On a day of generally good seeing the number of occasions when the signal drops to 0·8 volts or below may be between, say, fifty and two hundred. Such periods usually last for only a second or so, occasionally for as long as a minute. When a quiet period occurs the seeing signal declines with great rapidity (e.g., it may drop from, say, 5 volts to only 0·5 volts in a few seconds or even in less than a second), and afterwards increases to a more average value equally quickly. On a day of generally poor seeing there may be only half-a-dozen such periods—indeed days do occur when there are no such good moments at all. Under the circumstances, the triggering level is usually set at 0·8 volts. Should the seeing signal decline to a lower value, exposures continue automatically; only if an embarrassingly large number occur is the triggering level manually reset to a lower value.

Finally, when the seeing signal declines to 0·3–0·5 volts, any flicker of the limb on the seeing monitor image becomes barely perceptible. A fair percentage of photographs taken at this level show the full theoretical resolution of the 5-inch lens (1" of arc). On no occasion has a seeing signal of less than 0·3 volts been observed.

Credit for constructing the first device to provide a quantitative measure of the solar seeing must go to H. Siedentopf (1939). Like the seeing monitor described above, Siedentopf's device measured photoelectrically the distortions of the solar limb due to seeing; instead of two tangential slits, a single *radial* slit was used.

2.2.5   ORIGIN OF SOLAR SEEING

In the case of night-time seeing, while scintillation is caused by disturbed layers at heights of several kilometres or more, image motion and degradation are due to temperature inhomogeneities in the first few hundred feet of the atmosphere and in the immediate neighbourhood of the telescope and dome (cf. Section 2.2.2). During a clear night the lower atmosphere is in a state of stable stratification: any refractive index fluctuations are the result of distortion and mixing of

---

[3] An exact correlation between the value of the seeing signal and the quality of the corresponding photograph is not to be expected, since the seeing monitor samples the seeing only at the east and west limbs: the seeing at other points of the image may sometimes be different.

the layers (of different refractive index) due to wind. During the day, on the other hand, the temperature fluctuations associated with rising thermal convection currents set up by ground heating (discussed below) are much greater than any fluctuations present in the lower atmosphere during the night. Moreover, direct heating of the telescope and dome by the Sun results in thermal inhomogeneities in the immediate neighbourhood of the telescope which are much more damaging than those at night. Therefore, it is easy to see why the best resolution achieved during the day is an order of magnitude worse than the best attained in night-time observations.

During the day, *heating of the telescope and dome* can be a most serious source of trouble; in fact, with some solar telescopes the seeing may be dominated by thermal currents due to purely local heating. This is particularly true for conventional tower telescopes with their long optical paths and mirror optics. Mirror heating can be avoided by confining their illumination to moments when actual exposures are being made; however, this expedient is inconvenient in some types of observing programmes. In cases where the seeing is dominated by purely local heating the best seeing would be expected to occur shortly after sunrise—i.e., before the telescope has had a chance to heat up. Many solar astronomers do in fact find that the best seeing occurs in the early morning (see, for example, King, 1955: p. 326). Our own experience is different, probably owing to the absence of a dome and the use of cooling devices: we find that although the early morning is sometimes associated with fair seeing, the very best moments usually occur when the Sun has attained a moderate altitude. Often very good seeing occurs between 11 a.m. and 1 p.m. The reduction of telescope and dome heating by appropriate design is discussed in some detail in Sections 2.3.1 and 2.4.5.

In the absence of telescope heating, the most serious source of bad solar seeing is the *thermal convection currents* set up as a consequence of ground heating. This is not to say that there is *no* contribution from layers at heights of several kilometres or more, which as we have seen in Section 2.2.2 are the ones responsible for stellar scintillation. However, during the day the lower atmosphere—i.e. the first few hundred feet—is certainly the region in which the temperature inhomogeneities have their greatest magnitude. And, as in the case of stellar seeing, it is mainly inhomogeneities in this region which are responsible for *degradation* of the image, although it must be remembered that in the case of an extended object scintillation is necessarily accompanied by some degradation, since the brightness variations of neighbouring points are uncorrelated.[4]

A considerable amount of work—both experimental and theoretical—has been carried out in investigating the physical conditions in the lower layers of the atmosphere. Although several pictures have been suggested, there is still

[4] Smith, Saunders, and Vatsia (1957), using a camera, a resolution test pattern, and a horizontal path in the open air found that while the resolution decreased rapidly as the separation increased up to 500 feet, beyond that distance it approached a limiting value. It follows by analogy that inhomogeneities a few hundred feet or more from a solar telescope can be expected to cause less harm to the image than those much nearer the ground.

some uncertainty concerning the exact form taken by the convection elements. According to Priestley (1959), the situation is as follows: there are two regimes, a *forced* convection regime close to the ground, where the motions are turbulent (the turbulence is caused by surface wind); and a *free* convection regime, where the motion which carries the heat is set up by buoyancy forces. There is a well-marked transition between the two regimes, the transition occurring typically at a height of 1·5 m when the average wind velocity is 4–5 m/sec. To quote Priestley: '. . . near the ground the mechanically generated turbulence can carry the heat away as rapidly as the situation requires. But its intensity declines with increasing height, and at some level, which will be lower the lighter the wind and the stronger the heat flux, it becomes inadequate for the task, and free convection must take over.' It is clear that unless a telescope is *very* close to the ground and the wind very strong, the line-of-sight from the telescope will not pass through the turbulent region; hence for our purposes we need only consider temperature inhomogeneities in the free convection region.

When the temperature at some point in this region is recorded as a function of time, the resulting traces (Priestley, 1959: Fig. 19) show well-marked alternating disturbed and quiet periods. During the disturbed periods the temperature fluctuations are of the order of 1 °C, whereas during the quiet periods any residual fluctuations still present seem to be less than 0·1 °C. Simultaneous velocity measurements show that the disturbed periods are associated with *ascending* air, and the quiet periods with *descending* air. There is no cessation of turbulence in the descending air: there is, however, a very marked decrease in the temperature fluctuations. It is these, of course, which are responsible for the refractive index inhomogeneities.

Finally, when simultaneous measurements of temperature are made at *different* heights ranging from 1·5 m to 32 m (Priestley, 1959: Fig. 20), the resulting traces show a well-marked correlation with, however, a decided 'shear' from one level to the next.

Priestley interprets these observations as indicating that the convective heat transfer of the boundary layer is dominated by continuing plumes, which originate at the boundary between the forced convection and free convection regimes and which move with the speed of the wind at some quite low level. A buoyant plume forms a very efficient convective mechanism; in addition, theory indicates that although some special circumstance is required to bring such a convective column into existence, it will be maintained without the continuation of any preferred heat supply (Priestley, 1959: p. 86). This provides additional reason for expecting that the convection will take place in the form of columns rather than short bursts or bubbles.[5] (There is no observational evidence which would suggest the existence of a regular *cellular* convection, e.g., of the Bénard type.)

---

[5] Observations which support an alternative theory, namely the Scorer–Ludlam bubble theory (Scorer, 1958), have been considered by Plank (1959). Woodward (1960) has exhaustively discussed the likely shapes of convective elements at various stages of their existence.

There seems little doubt that poor solar seeing occurs when the line-of-sight passes through one or more such convective columns, each containing temperature inhomogeneities of the order of 1 °C, whereas the sudden moments of good seeing occur when the line-of-sight momentarily passes through the more uniform regions surrounding the plumes (Bray and Loughhead, 1961). The existence of convective plumes thus provides a ready explanation for the *intermittency effect* in the solar seeing. The quiet periods in the temperature traces last for periods ranging from several seconds to a minute (Priestley, 1959: Fig. 19)— i.e. somewhat longer than the observed periods of good seeing described in Section 2.2.4. They also appear to be more frequent. These discrepancies can be understood, however, when it is realized that the telescope has to look through a forest of disturbing convective columns: good seeing occurs only when, momentarily, a 'clear' patch lies along the line-of-sight. For geometrical reasons, it is evident that the best chance of having a line-of-sight free from disturbances occurs near noon, assuming that the *number* of columns is independent of the time of day. Hence although the ground heating is then at a maximum, this should be the time of best seeing for a telescope in which *internal* seeing has been eliminated (see above).

The extent to which daytime observations are dominated by ground heating has been strikingly demonstrated in experiments carried out by Siedentopf and Wisshak (1948). These workers measured the variation with the time of the day of the *scintillation* amplitude of an artificial source along a *horizontal* path. They found that on a cloudless day there was a strong maximum around midday; two distinct minima, one shortly after sunrise, the other shortly before sunset; and a rather weak maximum after sunset. On days of intermittent cloud, there was a close correlation between scintillation amplitude and sunshine. These results were explained on the basis of the variation of the lapse rate with ground heating; they do not necessarily contradict our impression that good seeing often occurs around noon since, as pointed out above, geometrical factors enter the picture when the path is no longer horizontal.[6]

2.3 Modern Methods of Solar Photography

2.3.1 GROUND-LEVEL PHOTOHELIOGRAPHS

The application of the cinematographic technique to high-resolution photography of sunspots and the solar photosphere was pioneered by Lyot in the early 1940's. Although Lyot himself went no further than to demonstrate the

---

[6] We are indebted to Dr C. H. B. Priestley for drawing our attention to the existence of quiet and disturbed periods in temperature traces derived from measurements made in the boundary layer of the atmosphere, and for discussing their relevance to the problem of solar seeing. In a private letter, Priestley points out that the intermittency is found to be most marked on days of light wind, since the more intense, fine-scale turbulence which is mechanically generated in strong winds inhibits the occurrence of the more purely convective bursts. This remark is consistent with our own impressions concerning the correlation between solar seeing and surface wind velocity, although we find that it is the *cold* winds which are the most damaging.

value of the new technique, his work has since been successfully carried on by J. Rösch (1957, 1959a).

Rösch's telescope is a 23-cm (9-inch) refractor which, apart from the use of a water-cooled focal-plane diaphragm, is of conventional design; it is mounted in a dome at the Pic-du-Midi Observatory in the French Pyrenees. Exposures are made in groups spaced half a minute or a minute apart at the rate of 24 frames per second, each group of exposures lasting for a few seconds; the exposure time is of the order of 1/200 second. Recently Rösch has also used a 15-inch telescope with success, but no details have been published.

In 1958 Loughhead and Burgess published a description of a 5-inch photoheliograph specifically designed for high-resolution cinematography of sunspots and the solar photosphere in which, for the first time, an attempt was made to eliminate all sources of telescope heating. This telescope is mounted in the open air and incorporates a number of novel features, which are described below.

As shown in Plate 2.1, the photoheliograph is mounted in the open air in flat country (altitude 200 feet above sea level), some 30 miles west of Sydney, Australia. In order to protect the telescope from solar heating, the entire front area (excluding the 5-inch objective lens L_1 and the auxiliary guider telescope) is covered by a hollow aluminium shield D_1 (cf. Fig. 2.3). The front surface of D_1 contains numerous perforations through which air is continually drawn by a suction system. The temperature of the shield is thus kept close to the ambient value, and it can perform its function of keeping the Sun's rays off other parts of the telescope without itself giving rise to damaging currents of heated air. A similar shield, S_1, protects the objective lens; this shield, however, is automatically removed from the light path by means of a solenoid when an exposure is due. At the prime focus there is a diaphragm D_2, which allows only light from that portion of the image which is to be photographed to proceed further into the telescope. Like D_1 and S_1, D_2 is cooled by air suction.[7] The reduced pressure in the suction system is maintained by a $\frac{1}{2}$ h.p. electric forge-blower (working in reverse), which is situated some 15 feet from the telescope and connected to S_1, D_1, and D_2 by an underground pipe and flexible couplings. The suction system is also used to cool the motor which operates the rotating shutter and the solenoid which opens S_1.

The telescope is designed for time-lapse cinematography of any selected region of the solar disk, such as a sunspot group, on 35 mm film in 'white' light; a fixed interval between successive exposures varying from 5 to 120 seconds can be employed. Alternatively, as described in the next section, exposures can be automatically triggered by means of the seeing monitor whenever the quality

---

[7] The effectiveness of this method of cooling has been demonstrated in a laboratory experiment: interferograms of the air over a perforated metal surface heated by electric radiators to simulate solar radiation show marked distortions of the wavefront, but these disappear when air is sucked through the perforations (Loughhead and Burgess, 1958: Plate 2). The effectiveness of the system is due to the fact that any heated air in the neighbourhood of the surface is drawn off and replaced by air at the ambient temperature before it can collect and give rise to convection currents.

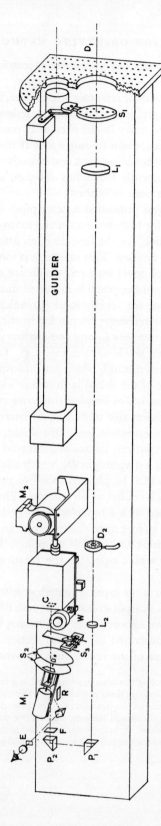

Fig. 2.3. The Sydney 5-inch photoheliograph: optical layout. C, camera; D_1, front shield; D_2, prime-focus diaphragm; L_1, objective; L_2, magnifying lens; M_1, shutter motor; M_2, camera motor; P_1, P_2, prisms; R, reflex mirror; E, eyepiece; S_1, objective shutter; S_2, sector-disk shutter; S_3, blade shutter; W, clock; F, green filter. The photoheliograph mounting is shown in Plate 2.1.

of the seeing momentarily reaches some pre-determined, acceptable value. The optical layout is shown in Fig. 2.3: L_1, the 5-inch objective, is a cemented doublet carefully mounted free of strain. L_1 forms a 16 mm solar image on the prime-focus diaphragm D_2, and the portion corresponding to the aperture in D_2 is magnified by a second lens L_2; the effective image diameter at the camera gate is 20 cm. The use of a magnifying lens provides a relatively large solar image without the long optical path usually associated with solar telescopes of the tower type. L_2 is provided with graduated screws giving motions in the north-south and east-west directions, so that any desired region of the solar image can be brought onto the camera gate. The optical system is designed to keep any off-axis aberrations introduced by shifting L_2 away from the optical axis of the objective within the Rayleigh limit. F is a glass filter passing 800Å of the spectrum centred on 5400Å. The telescope is focused by moving L_2 parallel to the optical axis, the solar image being examined with the aid of a reflex mirror R and an eyepiece E; as a consequence of the absence of heating effects within the telescope, the focus is found to be stable over a wide range of ambient temperatures.

The shutter unit, mounted in front of the camera, consists of a solenoid-operated blade shutter S_3 and a rotating sector-disk shutter S_2. The latter consists of two coaxial sector disks, which are driven by an electric motor M_1, one of the disks rotating at one-sixth the speed of the other. The high-speed disk controls the exposure time; with a sector angle of $4°$ the exposure time is of the order of a millisecond. The function of the slow-speed disk is to give the blade shutter time to close before a double exposure can occur; the blade shutter sequences the exposures.

In order to obtain exposures of uniform density regardless of varying sky transparency or solar zenith distance, it is essential to have some form of automatic exposure control. The exposure controller (cf. Plate 2.1) consists of a small auxiliary telescope and a photoelectric monitor, whose output is applied to the shutter motor in such a way as to keep the product of the exposure time and the light flux constant to within one per cent. The electronic circuit is designed to stabilize the speed of the shutter motor against changes either in mechanical load or in power supply voltage.[8]

All operations involved in taking exposures at the required times are sequenced automatically by an electronic programme controller situated inside the telescope control hut. The image of a clock (W in Fig. 2.3), together with the date, is recorded on a corner of each frame.

The film normally used is Recordak Micro-File, a fine-grained emulsion; it is developed for about 5 minutes in D-19 to give a gamma of about 3.

The telescope is kept directed towards the Sun with the aid of a photoelectric guider fed by an auxiliary 3·5-inch telescope (cf. Plate 2.1). This forms a solar image on an occulting disk behind which are placed four photoelectric cells.

[8] A somewhat similar rotating shutter for sunspot photography, not featuring, however, automatic exposure control, has been described by Brekke (1959).

Amplified signals from these cells are used to control small electric motors providing motion in hour angle and declination. Guiding to 1–2″ of arc can be achieved under wind-free conditions.[9]

In general, the 5-inch photoheliograph has proved to be a very successful tool for exploring the fine detail in sunspots and the solar photosphere; many of the observations described in Chapter 3 were obtained with this instrument. In addition, experience derived from its use can serve as a guide to possible improvements in the design of solar telescopes (this topic is discussed in Section 2.4).

2.3.2 USE OF A SEEING MONITOR TO AID SOLAR OBSERVATIONS

The 5-inch photoheliograph, described above, is designed to take photographs of any selected region of the solar disk at fixed intervals which can be varied from 5 to 120 seconds or, alternatively, exposures can be triggered off by the seeing monitor (Section 2.2.4) whenever the quality of the seeing reaches some pre-determined, acceptable level.

In setting the triggering level, the observer is guided by a knowledge derived from previous experience of the value of the seeing signal voltage corresponding to good seeing (cf. Section 2.2.4) and by the general level of the seeing quality in the period during which the equipment is being adjusted for the day's observations. Should he set the level at too optimistic a value, the result would be too few exposures; should he choose too pessimistic a value, on the other hand, the result would be an unnecessarily large number. The triggering circuit is so arranged (cf. Fig. 2.2) that should the seeing improve at any time to a point where it becomes *superior* to that corresponding to the chosen triggering level, exposures continue automatically at a rate of one every 2–3 seconds. The triggering level can easily be manually re-set to a new value without stopping the operation of the equipment, should the general quality of the seeing change markedly during the course of the day.

It remains now to discuss to what extent the use of the seeing monitor has improved the actual performance of the telescope. The chief improvement is, of course, in the number of good photographs that can be obtained in any given period. Our experience shows that even on days of generally poor seeing, when photography by the normal time-lapse method would be out of the question, the seeing monitor technique enables one to obtain a certain number of good photographs. And more important, on days of generally good seeing, a greater number of very good-quality photographs are obtained. This fact is of great importance in studying the evolution of the longer-lived fine structures in sunspots and photospheric faculae: for such observations, good-quality sequences lasting for several hours are required (cf. Bray and Loughhead, 1961). Such sequences are extremely difficult to obtain by the normal time-lapse method—

[9] The guider and the telescope mounting are based on drawings kindly supplied by Dr W. O. Roberts of the High Altitude Observatory, Climax, Colorado, U.S.A. Dubov (1955) has published a description of a photoelectric guider of similar design.

in fact, it might be necessary to wait for a year or more before a sufficiently long period of good seeing occurred.

It should be pointed out that although there is an increase in the *number* of good-quality photographs obtained in a given period, there is no improvement in the *quality* of the very best photographs. This is because the quality of the best photographs is limited, not by the seeing, but by the theoretical resolution of the 5-inch objective.

To date, the seeing monitor has been used only as an aid in high-resolution direct photography. However, it seems likely that it would prove equally beneficial in other types of observations, e.g., in spectroscopic and magnetic observations of sunspots, where exposure times of several seconds may be required. In such cases it might be necessary to 'build up' the exposure over a number of periods of good seeing, since periods of good seeing last on the average for only a second or so—at any rate at the site of the Sydney photoheliograph. This could be achieved by modifying the shutter triggering system and providing an auxiliary integrating circuit to control the total exposure time.

Finally, it may be remarked that the seeing-monitor technique is still in its infancy, and that parallel with the development of 'extra-terrestrial' observing techniques, there will no doubt be a continual improvement in ground-level methods. The seeing monitor we have described merely allows the telescope to select the moments of best seeing. The possibility of *compensating* for the seeing by correcting phase distortions in the incoming wave by means of so-called 'deformable optical elements' has been briefly discussed by Linnik (1957) and Babcock (1958). DeWitt, Hardie, and Seyfert (1957) have constructed a seeing compensator for planetary photography which corrects for image motion; it does not, however, correct for degradation which, unfortunately, often accompanies image motion.

The application of such advanced electronic techniques in the years to come will no doubt result in further improvement in the performance of ground-level telescopes.

2.3.3 SOLAR PHOTOGRAPHY FROM A MANNED BALLOON

We have seen in the previous sections that even when local telescope heating is eliminated by special devices, there remains the problem of refractive index inhomogeneities in the lower atmosphere due to convection currents resulting from ground heating. In an attempt to eliminate poor definition due to this source, Blackwell, Dewhirst, and Dollfus (1957a, b, 1959) constructed a balloon-borne solar telescope and obtained photographs of the granulation from heights of 20,000 feet.[10]

A refracting telescope was used, the objective being a visual doublet of 29 cm aperture working at $f/10$; the primary image was magnified by an enlarging lens to give a final image diameter of 46 cm (first flight) and 14 cm (second flight).

---

[10] Earlier experiments by Blackwell and Dewhirst from an aircraft at 22,000 feet were unsuccessful owing to mechanical vibration and disturbed air near the door of the aircraft.

With the latter image diameter, the exposure time was 0·2 milliseconds. The exposures were made with a flying slit shutter of original design. Owing to thermal changes in the telescope and changes in the refractive index of the air with height, it was not possible to predict the position of best focus with an accuracy better than plus or minus 2 mm: accordingly, during the flights the enlarging lens was repeatedly shifted by small amounts throughout this range in order that a certain proportion of the photographs would be in good focus. The telescope was supported by an altazimuth mounting underneath the basket carrying the two observers. One of the observers controlled the pointing of the telescope at the Sun; under reasonably calm conditions it was found possible to keep the telescope directed towards the Sun for a few minutes at a time with an error not exceeding 20–30′ of arc. The other observer controlled the telescope mechanisms, e.g., shutter, camera wind-on, focusing, etc. The techniques of astronomical observations from manned balloons have been discussed by Dollfus (1959) in some detail.

Photographs of sunspots were not obtained. The granulation photographs, although much better than those obtained at random from the ground, are inferior to the best ground-level photographs (see, for example, Plate 3.1*b*). It appears, therefore, that in spite of the great height from which the photographs were obtained, some residual seeing disturbances were still present. The residual seeing can probably be attributed to the fact that (*a*) convection currents are still present at 20,000 feet—well within the troposphere—although no doubt the associated temperature inhomogeneities are much smaller than those near ground level, and (*b*) the atmospheric density is sufficiently high at this level that heating of the telescope (and, perhaps, the balloon) can still give rise to harmful currents of hot air.

2.3.4 SOLAR PHOTOGRAPHY FROM AN UNMANNED STRATOSPHERIC BALLOON

Following the pioneering work of Blackwell, Dewhirst, and Dollfus, a team of individuals and organizations under the direction of M. Schwarzschild built a 12-inch balloon-borne solar telescope for automatic operation in the stratosphere. Several successful flights at heights in the vicinity of 80,000 feet were carried out. At these heights, the telescope is above 96 per cent of the Earth's atmosphere. Some of the granulation photographs obtained (cf. Plate 3.1*a*) appear to be completely devoid of seeing effects. These photographs are of higher quality than any that have yet been obtained from the ground; for comparison, Plate 3.1 (*b*) shows a granulation photograph obtained with the 5-inch ground-level photoheliograph described in Section 2.3.1.

The telescope (Rogerson, 1958; Schwarzschild, 1959; Schwarzschild and Schwarzschild, 1959) was basically a Newtonian reflector. The main mirror had an aperture of 12 inches and a focal length of 8 feet; to minimize thermal distortion, it was made of fused quartz (cf. Section 2.4.4) 3 inches thick. A layer of heat insulating material ('Styrofoam') surrounded the back and sides of the

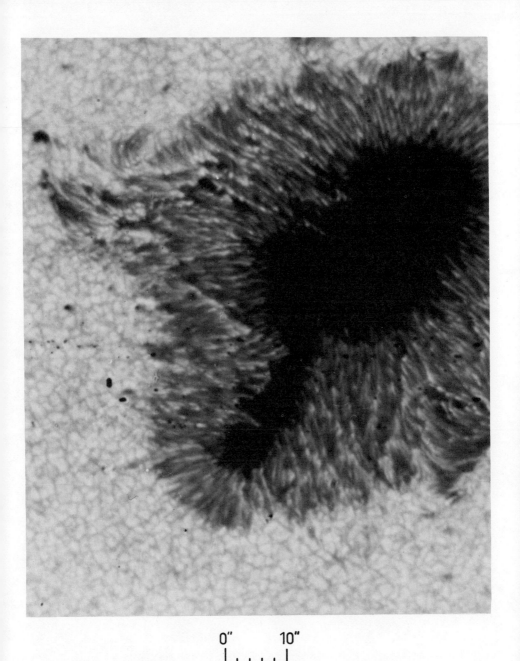

0"   10"

PLATE 2.3. Sunspot photographed with a 12-inch balloon-borne stratospheric telescope (Danielson, 1961). (*Photo by courtesy of Project Stratoscope of Princeton University, sponsored by ONR, NSF, and NASA.*)

mirror to a depth of 4 inches. Thus all heat losses and gains occurred exclusively at the aluminized front surface of the mirror. Under these conditions the isothermal surfaces within the mirror remain very nearly parallel to the front surface. Such a temperature distribution produces a slight change of focal length but very little optical aberration: the resolving power would therefore remain unaffected.[11] At a height of 80,000 feet, the mirror—being continuously exposed to full sunlight—would attain a temperature 50°C greater than that of the ambient air. Although the air density at this height is only 4 per cent of that at sea level, the hot mirror was a potential source of damaging convection currents (cf. Section 2.4.4). However, computations by Rogerson showed that any such disturbances were just within a tolerable limit.

The secondary mirror, also of fused quartz, was carried on an arm which, in synchronism with the shutter and the camera wind-on, rotated once per second. This mirror spent only 3 per cent of its time in the hot primary image, so its presence did not give rise to harmful currents. The axis of rotation was inclined 45° to the main optical axis (cf. Plate 2.2), so no image motion occurred as a consequence of rotation of the mirror. A magnifying lens produced a final image on the gate of a 35 mm camera holding 1000 feet of film; the effective diameter of the solar image was 57 cm.

In order to stabilize the distance between the primary and secondary mirrors as much as possible, the telescope tube was made of Invar. Nevertheless, it was not possible to predict the position of best focus with sufficient accuracy from ground experiments; accordingly, provision was made for automatically moving the enlarging lens from exposure to exposure throughout the expected range. One exposure out of twenty could be expected to be in good focus.

The telescope was mounted in a gimbal structure, which in turn was carried by a large 'Skyhook' balloon of a type used in cosmic-ray research. Guiding motions in altitude and azimuth were provided by electric motors driving through magnetic clutches. The necessary guiding signals were provided by several pairs of photodiode 'eyes'. Because of the fast exposure times possible in direct solar photography (in this case about one millisecond) some swinging of the telescope is permissible. The rate of swing during the actual flights did not exceed 200″ of arc per second of time.

All operations after launching were entirely automatic; a timer was set to release the telescope 6 hours after launching and it was brought safely to earth by parachute. In flight the balloon was tracked both optically and by means of a radio beacon; the distance between the launching site and the place of landing varied from 80 to 230 miles. The useful camera time for each flight was 140 minutes.

Of the 8000-odd photographs obtained during the first flight, 400 could be expected to be in good focus. As it happened, however, only a handful were found to have the definition of the photograph illustrated in Plate 3.1 (a). The

[11] The expected temperature difference of 6°C between the front and back surfaces would lead to a change in focal length of about 0·3 mm.

poor definition of the other photographs was attributed to the transmission of mechanical vibrations in the various motors to the optical elements. A somewhat larger number of good frames were obtained during the second flight, after the mounting of the enlarging lens had been stiffened.

During the first two flights no sequences of photographs of the same region of the Sun were obtained; nor did the flights yield photographs of sunspots. However, in 1959 four additional flights were made with substantially improved equipment (Danielson, 1961). In the first place a television link was added. This presented to the observers on the ground the same picture as that simultaneously being photographed by the telescope, whose focus and orientation could be adjusted by means of radio command signals. Secondly, in an attempt to eliminate poor definition caused by mechanical vibrations, certain modifications were made to the telescope itself; in particular, the rotating quartz secondary mirror was replaced by a fixed mirror made of 'Stellite'. With this improved equipment a number of very high-quality photographs of sunspots were obtained, an example of which is shown in Plate 2.3. However, the duration of the longest time sequence of a single sunspot was only 40 minutes and, owing to faults in the guiding system, the high-definition photographs were not uniformly distributed in time. During the last flight of Project Stratoscope (September 24, 1959), a few high-definition photographs of sunspot umbrae were obtained, but at the time of writing (March, 1963) no details of these photographs have been published.

The relatively small number of sunspots photographed and the shortness of the sequences obtained limited the amount of new information yielded by Project Stratoscope. In fact, as Danielson (1961) has himself concluded, it would be necessary to substantially improve the guiding system in order to utilize the full capabilities of the telescope.

2.4 Factors Affecting the Performance of Ground-level Solar Telescopes; Possible Improvements in Design

2.4.1 INTRODUCTION

It is too early to predict with any certainty the future role of balloon and satellite techniques in observing the fine detail on the Sun. It must be pointed out that present-day balloons and satellites do not really constitute effective vehicles for carrying the complicated and bulky auxiliary equipment, such as spectrographs and magnetographs, often associated with ground-based telescopes. Even in the case of direct photography, a balloon-borne telescope must be equipped with provision for accurate guiding and focusing on objects of interest by ground observers (e.g., by means of television and radio links) if it is to compete effectively with ground-level telescopes in all but a small range of programmes. As is often the case in astronomy, the new technique, rather than superseding the older methods, will probably play a complementary role. In this connection it may be remarked that, to date, only direct photographic

TABLE 2.1

High-Resolution Observations of Sunspots and the Solar Photosphere
Obtained with Ground-level Telescopes of Various Sizes[1]

REFERENCE	OBSERVA-TORY	APER-TURE (CM)	TYPE OF TELE-SCOPE	NATURE OF OBSERVATIONS
Janssen (1896)	Meudon	13·5	Refractor	Photospheric granulation; sunspots.
Hansky (1905)	Pulkovo	—	Refractor	Diameters of photospheric granules.
Hansky (1906)	Pulkovo	—	Refractor	Photospheric granulation.
Hansky (1908)	Pulkovo	—	Refractor	Photospheric granulation; sunspots.
Chevalier (1908)...........	Zô-Sè (China)	30	Refractor	Photospheric granulation.
Chevalier (1914)...........	Zô-Sè (China)	36	Refractor	Photospheric granulation.
Chevalier (1919)...........	Zô-Sè (China)	—	Refractor	Wilson effect in sunspots.
Strebel (1932)	Munich	20	Reflector	Photospheric granulation; sunspots.
Strebel (1933)	Munich	20	Reflector	Polygonal nature of photospheric granules.
ten Bruggencate (1938)	Potsdam	10	Tower	Photospheric granulation.
Keenan (1938)	Yerkes	19	Refractor	Diameters of photospheric granules.
Keenan (1939)	Yerkes	19	Refractor	Photometry of photospheric granules.
Pettit *et al.* (1939)..........	Mt Wilson	—	Tower	Enumeration of photospheric granules.
ten Bruggencate (1940)	Potsdam	12	Tower	Lifetime of facular granules.
Waldmeier (1940)	Zürich	—	—	Size, lifetime, and contrast of facular granules.
ten Bruggencate (1942)	Potsdam	12	Tower	Photometry of facular granules.
von Klüber and Müller (1948)	Potsdam	10	Tower	Photospheric granulation.
Richardson and Schwarz-schild (1950)	Mt Wilson	30	Tower	Doppler velocities of photospheric granules.
Macris (1953)	Pic-du-Midi	23	Refractor	Lifetimes of photospheric granules and of bright regions in sunspot penumbrae.

[1] Observations are included in this table only when there is clear evidence in the published paper that a resolution of 1–2″ of arc was actually achieved; purely visual observations are excluded.

TABLE 2.1 (*contd.*)

REFERENCE	OBSERVA- TORY	APER- TURE (CM)	TYPE OF TELE- SCOPE	NATURE OF OBSERVATIONS
McMath, Mohler, and Pierce (1955)............	McMath– Hulbert	—	Tower	Doppler velocities of photospheric granules.
McMath, Mohler, Pierce, and Goldberg (1956)	McMath– Hulbert	—	Tower	Doppler velocities of photospheric granules.
Bray and Loughhead (1957)..	Sydney	13	Refractor	Bright regions at sunspot umbra–penumbra boundary.
Leighton (1957)	Mt Wilson	30	Tower	Lifetime of photospheric granules.
Rösch (1957)...............	Pic-du-Midi	23	Refractor	Diameters of photospheric granules; sunspot umbra granulation.
Bray and Loughhead (1958a).	Sydney	13	Refractor	Lifetime of sunspot penumbra filaments.
Bray and Loughhead (1958b).	Sydney	13	Refractor	Lifetime and evolution of photospheric granules.
Loughhead and Bray (1958)..	Sydney	13	Refractor	Wilson effect in sunspots.
Bray and Loughhead (1959)..	Sydney	13	Refractor	Lifetime and cell size of sunspot umbra granules; umbral 'loops'.
Dunn (1959)	Sacramento Peak	41	Refractor	Limb darkening at the extreme limb.
Loughhead and Bray (1959)..	Sydney	13	Refractor	Faculae bordering sunspot pores near limb.
Rösch (1959b)	Pic-du-Midi	23	Refractor	Cell size and photometry of photospheric granules.
Rösch and Hugon (1959)	Pic-du-Midi	23	Refractor	Lifetime of photospheric granules.
Rösch and Hugon (1959)	Pic-du-Midi	38	Refractor	Evolution of photospheric granules.
Severny (1959)	Crimea	—	Tower	Fine structure of sunspot magnetic fields.
Kiepenheuer (1960)	Anacapri	11	Refractor	Sunspots.
Loughhead and Bray (1960a).	Sydney	13	Refractor	Lifetime, cell size, and evolution of sunspot umbra granules.
Loughhead and Bray (1960b).	Sydney	13	Refractor	Granulation near the extreme limb.

TABLE 2.1 (*contd.*)

REFERENCE	OBSERVA-TORY	APER-TURE (CM)	TYPE OF TELE-SCOPE	NATURE OF OBSERVATIONS
Miller (1960)	Manila	10	Refractor	Granule counts in neighbourhood of sunspots.
Steshenko (1960)	Crimea	—	Tower	Fine structure of photospheric magnetic fields.
Bray and Loughhead (1961). .	Sydney	13	Refractor	Lifetime of facular granules.
Loughhead and Bray (1961). .	Sydney	13	Refractor	Birth of sunspot pores.
Macris and Banos (1961)	Athens	30	Refractor	Cell size of photospheric granules.
Suzuki (1961)	Kyoto	10	Refractor	Fine structure of sunspots.
Macris and Prokakis (1962) . .	Athens	40	Refractor	Granule diameter measurements in neighbourhood of sunspots.

observations have benefited from the construction of telescopes free from internal seeing and controlled by seeing monitors. Yet the marriage of the high *spatial* resolution of which such a telescope is capable to the high *spectroscopic* resolution of a modern blazed grating is certainly possible in principle. Such a combination would undoubtedly lead to a great increase in our knowledge of the Sun in general and sunspots in particular; some of the problems involved are discussed in this section.

In discussing the design features of such a high-resolution telescope–spectrograph combination we can profit by the experience gained by various solar observers in obtaining high-resolution observations of sunspots and the solar photosphere. As a useful preliminary, Table 2.1 gives a summary of successful observations of this type—made at ground level—in which an effective resolution of 1–2″ of arc was unmistakably achieved. The table gives the source reference, the type, location, and aperture of the telescope, and the nature of the work carried out. Purely visual observations are excluded.

Several interesting conclusions emerge from Table 2.1. In the first place, it is clear that on the whole more high-resolution work has been accomplished with refractors than with tower-type reflectors. Secondly, in general this resolution has usually been achieved with very moderate apertures. Finally, comparatively little work other than direct photography has been successfully accomplished at this level of resolution. (Most of the successful work of a spectroscopic nature which *has* been accomplished has attained a resolution nearer 2″ than 1″ of arc.)

The reason for the relative lack of success in achieving high resolution in observations other than direct photography is of course partly the longer exposure times required. However, it is also partly a consequence of a telescope design unsuitable for high definition, telescopes used for this type of work being invariably of the reflecting tower-type. Some of the disadvantages of this type of solar telescope are discussed in Section 2.4.4 below. It is clear that no refinements in the spectrograph alone—e.g., the provision of a vacuum tank to eliminate 'spectrograph seeing'—can compensate for lack of definition in the solar image on the spectrograph slit.

We now turn to a consideration of some of the design features of solar telescopes (for use in a variety of observing programmes) which have a bearing on the problem of achieving high resolution.

2.4.2 LOCATION OF OBSERVATORY SITE

In the past, choice of the location of many astronomical observatories, including solar observatories, has unfortunately often been governed by mere convenience or historical accident. Nowadays, however, careful consideration is usually given to the choice of a site.

It seems on the basis of the rather limited amount of information available that better solar seeing prevails over flat terrain. There is no evidence that would lead one to favour any particular type of terrain, for example grassy, or alternatively, wooded. It may well be that certain types of terrain, for example, flat grassy plains, particularly favour the intermittency effect in the temperature traces and in the solar seeing, discussed in Section 2.2.5, whereas at other locations the intermittency is not so marked, the seeing being bad or mediocre all the time.

No data has been published concerning the quality of solar seeing over water. However, atmospheric temperature fluctuations over water are known to be substantially less than those over land,[12] so some consideration might be given to establishing a solar telescope on a very small (perhaps artificial) island at the centre of a large lake or reservoir.

A solar telescope should certainly be located far from any built-up area, since a building provides a permanent source of convection currents damaging to good definition.

The authors have consistently obtained high-resolution observations at a site only 200 feet above sea level. Other things being equal, however, the higher above sea level the better. On the other hand, it should be pointed out that very disturbed air is often associated with extensive mountain ranges containing many peaks, so in some cases the advantage accruing from a lower atmospheric density may not be realized in practice.[13]

Freedom from scattered light is of importance in some types of solar observa-

---

[12] This conclusion is based on measurements carried out by the C.S.I.R.O. Division of Meteorological Physics, Melbourne, Victoria, Australia. We are indebted to Dr E. L. Deacon for communicating the results of this work.
[13] The dependence of the quality of the solar seeing and its variation during the day on the local topography at five European sites has been discussed by Kiepenheuer (1962).

tions; however, in most cases a site selected with regard to the considerations discussed above will automatically satisfy this condition.

Finally, it may be remarked that in the years to come seeing monitors equipped with chart recorders (Section 2.2.4) will probably play a valuable part in selecting new sites for solar observatories.

2.4.3 HEIGHT OF TELESCOPE ABOVE GROUND LEVEL

On simple physical considerations one could predict that the amplitude of the temperature fluctuations in the boundary layer of the Earth's atmosphere—which, as we have seen in Section 2.2.5, are the cause of bad solar seeing—would decrease with height, although prediction of the *rate* of decrease would not be easy. Measurements at the site of the Kitt Peak National Observatory, Arizona, U.S.A. (McMath and Pierce, 1960), at heights ranging from $1\frac{1}{2}$ to 55 feet, indicate that at this site the rate of decrease is fairly rapid, the short period fluctuations decreasing from about $3°C$ near the ground to only $0.4°C$ at a height of 55 feet. However, measurements made at Edithvale, Victoria, Australia, at heights ranging from 5 feet to 105 feet (Priestley, 1959: Fig. 20), show a much less pronounced decrease. The rate of decrease no doubt depends not only on the site but also on meteorological factors—for example, on the surface wind velocity.

In view of this uncertainty, therefore, it is not possible to quote a single figure for the optimum height of a solar telescope which would be universally applicable. Since the cost of a solar installation can be expected to increase sharply with height, at any given site there is an 'economic height' above which any further improvement in the seeing becomes too small to justify the additional expense. Measurements of temperature fluctuations alone are really insufficient to determine the variation of the seeing with height. Measurements should also be made with a seeing monitor, particular attention being paid to the duration and frequency of occurrence of the periods of *good* seeing, and the quality of the seeing during these moments. Less interest is attached to the periods of bad or mediocre seeing. The measurements should be carried out at various heights from a tower, whose design should be such that its presence does not give rise to thermal disturbances.

2.4.4 REFLECTOR OR REFRACTOR?

The traditional form of the solar telescope is the vertical solar tower. In this type of telescope a coelostat mounted on the top of a tower sends the beam downwards into a fixed telescope (which usually, but not always, has all-reflecting optics) mounted vertically. A large, stationary image of the Sun is produced at or near ground level. The 60-foot tower at Mt Wilson was the first telescope of this type to go into operation; it was built in 1907 according to the design of G. E. Hale and C. G. Abbot. Similar solar tower telescopes are now widely distributed throughout the world. In more recent years certain variants of the tower type, such as the Cassegrain coudé at Oxford (Plaskett,

1955), have appeared. Good accounts of solar telescopes of the tower type, together with their associated equipment, are to be found in the following references: Abetti (1929); Dimitroff and Baker (1945); McMath (1953); King (1955); McMath and Mohler (1962).

The popularity of the reflecting, tower-type solar telescope stems from three distinct advantages which it possesses:

(1) it provides an easily accessible solar image at a fixed position at ground level, enabling one to employ spectrographs and other auxiliary equipment of a size and complexity which would preclude their attachment to a conventional equatorial telescope;

(2) it is strictly achromatic, enabling one to obtain observations in spectral regions ranging from the infra-red to the ultra-violet;

(3) it provides a large, bright solar image at a relatively moderate cost.

Nevertheless, the tower-type telescope possesses a number of features which as far as providing a *high-definition* solar image is concerned somewhat handicap it in comparison with a refractor of the photoheliograph type:

(1) a minimum number of three and a usual number of four or five mirrors are required. Unless the optical surfaces are unusually good, and each mirror is mounted substantially free from strain, the large number of optical surfaces involved can cause some impairment of the definition;

(2) heating of the mirrors can also have a deleterious effect on the definition. The reflectivity of freshly evaporated aluminium is approximately 90 per cent in the visible region of the spectrum, so that 10 per cent of the radiation incident on each mirror is absorbed. Energy is therefore absorbed by a 40 cm mirror at a rate of about 12 watts, assuming a radiation flux at the Earth's surface of 1 kilowatt per square metre. The energy absorption could easily be doubled if the reflection coating were a few months old. In the absence of an effective mirror cooling system, two undesirable effects can result: (*a*) the mirrors can become distorted, causing deterioration of the image (a mère change of focus could be tolerated: cf. Section 2.3.4), and (*b*) the heated mirrors can give rise to damaging convection currents in the path of the light beam. Thermal distortion of the mirrors can be minimized by employing a material having a low coefficient of expansion, such as fused silica (cf. Plaskett, 1939, 1955).[14] Mirror cooling systems have been devised, but they are not without their own complications. With regard to the convection currents, experience shows that it is not sufficient merely to displace the hot air in the neighbourhood of the mirror surfaces by means of fans;

(3) the long optical path which is usually a feature of the tower-type instrument is difficult to maintain at a uniform temperature: we recall from Sections 2.2.2 and 2.2.5 that temperature differences of only a few degrees can have a

[14] McMath and Pierce (1960) have suggested using the new low-expansion ceramic material 'Pyroceram' for the mirrors of the large solar telescope at Kitt Peak. The properties of Pyroceram have been described in some detail by Meinel (1960a) with reference to its suitability for stellar telescope mirrors.

serious effect on the definition of the image. Such temperature differences are difficult to avoid in an optical path which may be as long as 50 or 100 feet or more;

(4) the scattered light intensity is often rather high in the tower-type telescope, particularly when the mirror coatings are no longer fresh. This makes observations of low contrast fine detail rather difficult, and in many types of measurements uncertain and tedious corrections have to be made. In this type of telescope the high value of the scattered light intensity is partly due to the fact that light from *all* parts of the solar image is scattered from all of the mirrors. In the case of a photoheliograph-type telescope, on the other hand, a small diaphragm can be placed at the prime focus in front of the enlarging lens, so that light enters the rest of the telescope only from the small region of the image being studied: only the main objective can contribute scattered light to a given point of the final image (see below).

A refractor of the photoheliograph type, as described in Section 2.3.1, suffers none of the disadvantages (1)–(4) listed above. This type of telescope, in addition, possesses the following advantages for high-resolution work:

(1) the number of large optical components is smaller than is usually the case for a tower-type instrument. Moreover, it is well known that it is far easier to make and mount moderately-sized lenses which perform close to the theoretical limit than mirrors;

(2) heating of the optical components is no problem: the small amount of radiation absorbed by an objective lens exposed to the Sun has a negligible effect on its performance. Heating of the other parts of the telescope can be avoided by the use of special cooling devices (cf. Section 2.3.1). The provision of a cooling system is much easier for a refractor than for a tower-type reflector because the telescope and its mounting present a much smaller area to the oncoming beam;

(3) the use of an enlarging lens provides a relatively large solar image without the long optical path usually associated with tower-type telescopes. This is a decided advantage, since there is a better chance of achieving a uniform temperature throughout the telescope;

(4) by providing a diaphragm at the prime focus to allow light only from the small region of the image under study to enter the enlarging lens, scattered light can be almost entirely eliminated (see the description (Section 3.6.2) of the technique used in photographing the granulation in sunspot umbrae). Apart from scattered light from the Earth's atmosphere, the only remaining scattered light is that originating at the main objective. This is small, particularly if a cemented doublet is used and the optical surfaces are kept clean—which is much easier for a refractor than for a reflector.[15]

[15] To estimate the scattered light in the 5-inch photoheliograph described in Section 2.3.1, the authors obtained a series of graded exposures of the limb, using a prime-focus occulting disk. Only when an exposure of 16 times normal was reached could a faint trace of scattered light be detected beyond the limb.

The disadvantages of a refractor are mainly in connection with spectroscopic observations of the Sun:

(1) a refractor is not strictly achromatic over a wide spectral region;

(2) an equatorial refractor on a tower 50 feet high, say, would not provide a stationary solar image at ground level: therefore, in contrast to the case of the tower-type reflector, it is not possible to use a fixed spectrograph or other auxiliary equipment, designed without consideration of bulk;

(3) as far as cost is concerned, the refractor cannot compete with the tower-type telescope for apertures greater than, say, 20 inches. Therefore, if a large, bright image is required and high definition is of secondary importance, a reflector is to be preferred. For apertures of less than 15 inches, however, the cost of a refractor should be competitive. (The choice of objective diameter for high-resolution observations is discussed in Section 2.4.6 below.)

To summarize, it seems that while the refractor is well suited to providing a high-definition solar image, the tower-type reflector is better suited to obtaining spectroscopic observations (particularly those requiring high dispersion) with an image of only moderate definition. Nevertheless, it cannot be disputed that future advances in the study of the hydrodynamic and magnetic structure of sunspots and the photosphere will be critically dependent on the possibility of *systematically* obtaining spectral observations having a spatial resolution of the order of 1″ of arc. How is this to be achieved?

One suggestion would be to *accommodate the spectrograph and a refracting telescope of the photoheliograph type on the same equatorial mounting.* A 20- or 25-foot 'equatorial spar' (i.e., about twice the length of the mounting illustrated in Plate 2.1) could accommodate a 40-foot auto-collimating grating spectrograph in addition to a 12-inch photoheliograph, provided one were prepared to bend the light beam in the spectrograph as well as in the telescope—a scheme which does not seem to introduce any essential difficulty. Since the spectrograph would lie along the spar in a direction parallel to the optical axis of the telescope, the additional area facing the Sun's rays would be quite small; this area could be shielded in the same way as the telescope itself (cf. Section 2.3.1).

Let us now consider what limitations the non-achromaticity of such a telescope would place upon the type of observations which could be undertaken. It would clearly be impossible to make simultaneous observations over a wide spectral region. However, it is certainly possible to design an objective in such a way that it images within the Rayleigh tolerance over a spectral region of several thousand Ångstroms, it being understood that in making observations at any particular wavelength refocusing is permissible and appropriate filters are employed.[16] In the case of magnetic or velocity measurements of sunspots, for example, this is really all that is required, since one is usually interested in a

[16] Mr E. Braithwaite, of the Optical Design Department, Sir Howard Grubb Parsons and Co., Ltd., has calculated a design for an $f/15$ air-spaced doublet, of 12 inches aperture, which will give images having a residual spherical aberration less than the Rayleigh tolerance over a range of 4000 to 7000Å, when used under these conditions. We are indebted to Mr G. M. Sisson for communicating this information.

single Fraunhofer line or a group of lines spanning only a small spectral region. For simultaneous observations over a wide spectral region, however, the more versatile tower-type telescope–spectrograph combination is to be preferred.

2.4.5 SHIELDING OF SOLAR INSTALLATIONS FROM WIND; PREVENTION OF HEATING EFFECTS

One of the problems facing the designer of a solar telescope is the problem of shielding the telescope and its mounting or tower from wind. Wind shake can be ignored in the case of direct photography, where the exposure times are only of the order of a millisecond. In the case of spectroscopic observations, on the other hand, exposure times of several seconds may be necessary and movement caused by wind must be avoided if the definition is not to suffer

Considering firstly solar telescopes of the tower type, two examples of the solution of the problem are provided by the 150-foot tower of the Mt Wilson Observatory (King, 1955: see p. 338) and the 50- and 75-foot tower telescopes of the McMath–Hulbert Observatory (McMath, 1953). The 150-foot tower is a skeleton structure, each of the steel members being enclosed within the corresponding member of a second skeleton tower mounted on independent foundations. The coelostat is mounted on the inner tower and the dome protecting it is supported by the outer tower. In the McMath–Hulbert telescopes, on the other hand, a cylindrical construction has been adopted for both the inner and outer structures; as before, the coelostat is carried by the inner and the dome by the outer structure. The firm of Zeiss favour the use of a heavy wooden structure for the inner (coelostat) tower; this is found to be satisfactory in damping out any residual vibrations.

Although the above methods of construction have proved successful in eliminating wind shake, they have one unsatisfactory feature: the dome at the top of the tower, becoming hot under the action of the Sun's rays, can give rise to damaging convection currents in the path of the light beam falling on the coelostat mirrors; and, in general, the presence of the dome prevents the attainment of thermal equilibrium between the telescope and the outside air—so necessary for good definition. A cylindrical outer tower of the McMath–Hulbert type also behaves as a source of convection currents. The open skeleton construction of the Mt Wilson tower is preferable in this connection, although even a structure of this type must give rise to some undesirable currents.[17] The problem of protecting a solar telescope from wind without simultaneously introducing surfaces which become heated is not an easy one to solve.

Let us now consider how the problems of wind and thermal protection could be tackled in the case of an equatorial solar telescope of the photoheliograph type. We assume that we are concerned with, say, a 12-inch telescope, and that

[17] The double cylinder mode of construction has also been adopted for the very large solar telescope planned for the Kitt Peak Observatory (McMath and Pierce, 1960). However, in this telescope the outer cylinder and other parts exposed to sunlight are to be cooled by the forced flow of a suitable refrigerant.

both the telescope and a large grating spectrograph are carried on the same 20- or 25-foot equatorial spar, as suggested in Section 2.4.4. Let us further assume that seeing tests indicate an 'economic height' at the chosen site of about 50 feet. A suitable mode of construction for the supporting tower would be a double skeleton stucture similar to the 150-foot Mt Wilson tower. The inner tower would support the mounting for the equatorial spar. The outer tower, instead of carrying a dome to house the telescope and its mounting, would support a horizontal *apron* or *platform* situated just below the spar. The radius of the apron would be so chosen that during the normal hours of solar observation the 50-foot tower was kept continuously in shade: thus no heating of the legs or cross-members of the tower could occur. To eliminate convection currents from the apron itself, air would be sucked through numerous perforations in its upper surface. The telescope, spectrograph, and the spar carrying them, would be protected from the direct rays of the Sun by suction shields similar to those described in Section 2.3.1.

The apron would be made robust enough to serve as a platform for use by the observers in the initial setting up of the telescope and spectrograph, or other auxiliary equipment, for a day's observations.

By avoiding the use of a dome, thermal equilibrium is maintained between the telescope and the surrounding air—a very desirable feature. However, some provision must be made for protection of the telescope itself from wind. One possible solution would be to install a number of panels hinged to the circumference of the platform: these would be raised either singly or several at a time according to the strength and direction of the wind. Like the apron and the telescope shields, they would be cooled by air suction. In this way the telescope would be protected from wind but would nevertheless be effectively mounted in the open air.

Finally, there is the problem of housing the spar and the instruments mounted on it when the telescope is not in use. This problem raises certain difficulties not normally encountered in the design of domes; however, these difficulties are purely of an engineering nature, and a discussion of them would be out of place here.

2.4.6 CHOICE OF APERTURE, IMAGE DIAMETER, AND EMULSION

What is the optimum objective diameter for a solar telescope designed for high-resolution observations at ground level, including observations of a spectroscopic nature? This is a very complex question to which, owing to insufficient data, it is not yet possible to give a final and unambiguous answer. In this section, therefore, we shall confine ourselves to discussing some of the factors involved rather than making definite recommendations.

Neglecting seeing, let us first consider what aperture is required to give sufficient *theoretical* resolution to study the known fine structures found on the Sun. The majority of the photospheric granules have diameters in the range 1–2″ of arc, although there are some less than 0″.5 (cf. Section 3.2.2). The sunspot umbra granules have a somewhat smaller average separation than the photo-spheric granules (Section 3.6.4); their diameters are probably also less. The width of the sunspot penumbra filaments is unknown, but very probably it is less than 0"5 (Section 3.5.4). A solar installation capable of systematically producing an effective resolution of, say, 0"4 of arc, would in fact enable observations to be made of nearly all the fine structures on the Sun known at the present time. The figure 0"4 corresponds to the theoretical resolution of a telescope of aperture 12 inches (30 cm).

In theory, an increase in the diameter of the objective should result in an improvement in the definition.[18] In practice, however, owing to the effect of atmospheric seeing, this is no longer true once a certain diameter is reached; above this diameter the definition actually *decreases* with increasing aperture. As an illustration, we quote the experience of Keenan (1953): 'Tests at the Yerkes Observatory, extending over many months [of solar photography], showed that when the aperture of the 40-inch refractor was reduced by a dia-phragm to 10 inches, the theoretical resolving power of about 0"5 was always below the limit set by atmospheric smearing. At that location the improvement in the image with decreasing aperture continued down to an opening of 6 or 8 inches, even under the best conditions of seeing, presumably because the gain in making the lens area as small as possible in comparison to the size of the convective elements in the lower atmosphere more than counter-balanced the loss in optical resolving power. It is probable that even at the most favourable terrestrial locations a telescopic aperture of 12 or 15 inches (30–38 cm) gives optimum results on the Sun.'

As Keenan remarks, one reason for preferring a relatively small aperture is to make the lens area small in comparison to the size of the atmospheric disturbances responsible for bad seeing. Another—perhaps equally important—reason is to keep the heat flux through the telescope as small as possible. Keenan's telescope was not provided with cooling devices, and it follows that his results may not be generally applicable: it may well be that for a telescope provided with adequate cooling devices the optimum aperture exceeds the value of 6–8 inches found by Keenan.

In the absence of sufficient data concerning high-resolution observations made with larger apertures, it is not possible to give a reliable figure for the optimum aperture of a solar telescope. However, we can agree with Keenan that an aperture in the range 10–15 inches (25–38 cm) would probably be a good choice at the present time. When used for direct photography, such an aperture would probably yield a smaller proportion of photographs showing the full theoretical resolution than a 5-inch aperture, but the best photographs would probably be superior. Until 1" of arc resolution can be systematically obtained for all types of solar observations (including spectroscopic) it would seem unwise to try to attain a resolution limit very much smaller than this figure;

[18] The theoretical resolving limit $\simeq \lambda/D$ radians, where λ is the effective wavelength and D is the objective diameter, λ and D being in the same units.

therefore, at the present time it seems that quite moderate apertures are to be preferred.

Nevertheless, when we consider the question of obtaining *spectroscopic* observations having a spatial resolution of 1″ of arc, it must be emphasized that there are two conflicting factors: on the one hand, a relatively modest aperture seems to have a better chance of producing a high-definition solar image; on the other hand, a larger aperture produces a *brighter* image, thereby reducing the exposure times—and the effect of exposure time on the spatial resolution actually achieved may be crucial.[19] In view of the absence of data, it is not certain that an aperture of 10–15 inches would be a good choice for spectroscopic work as well as for direct photography. However, it does seem likely that the best chance of success lies in making the aperture (and the spectroscopic resolving power) *as small as possible* consistent with obtaining observations of the required spatial (and spectral) resolution.

We now turn to the question of image size. In the case of a tower-type telescope the image size is determined by the focal length of the image-forming mirror or lens; in the case of a refractor of the photoheliograph type the image size can be chosen at will by an appropriate choice of magnifying lens. (In both cases, of course, the effective focal ratio of the telescope should equal that of any spectrograph which may be employed.) The choice of image diameter is intimately related to the resolving power of the photographic emulsions available. As an example, we may cite the case of the photoheliograph described in Section 2.3.1: here, the image diameter is 20 cm, so that 1″ of arc (which corresponds to the theoretical resolution of the 5-inch objective) is equivalent to about 0·1 mm on the negative. The film used is Recordak Micro-File, a fine-grained emulsion having a resolution of 170 lines/mm. It is clear that even with this quite modest image size the effective resolution is very far from being emulsion-limited. With a telescope of twice the aperture it might be desirable to either double the image size or alternatively, to use an emulsion with an even finer grain. There are a number of emulsions available suitable both for direct photography and spectroscopic work which have a resolution of several hundred (or even a thousand) lines per millimetre, so in general it is possible to use quite small image sizes without loss of resolution. There is everything to be gained in using the *smallest* image size consistent with avoidance of limitation of the resolution by the emulsion, since the exposure times are then as small as possible.

However, in view of the number of opposing factors involved, the actual emulsion to employ for a given type of observation with a given instrument can, of course, be determined only by experiment. If a very fine-grained (and therefore slow) emulsion is used, then the greater exposure times may involve some loss of resolution due to seeing effects; if a coarse-grained emulsion is selected, then the exposure times are lower, but the resolution may be limited by the emulsion rather than by the seeing. Finally, we may remark that there is a

[19] In the years to come, the use of electronic image converters may play a part in reducing exposure times. To date, however, image converters have not been employed in solar observation.

constant (and successful!) effort on the part of the manufacturers of photographic materials to improve the grain and speed characteristics of their products. Like other astronomers, the solar observer must always be ready to take advantage of any new materials, as well as new techniques, which will enable him to improve the quality of his observations.

REFERENCES

ABETTI, G. [1929] 'Solar physics', *Handbuch der Astrophysik*, ed. G. EBERHARD, A. KOHL-SCHÜTTER, and H. LUDENDORFF, vol. 4, p. 57. Berlin, Springer.

BABCOCK, H. W. [1958] 'Deformable optical elements with feedback', *J. Opt. Soc. Amer.* **48,** 500.

BLACKWELL, D. E., DEWHIRST, D. W., and DOLLFUS, A. [1957a] 'Photography of solar granulation from a manned balloon', *Observatory* **77,** 20.

BLACKWELL, D. E., DEWHIRST, D. W., and DOLLFUS, A. [1957b] 'Solar granulation and its observation from a free balloon', *Nature* **180,** 211.

BLACKWELL, D. E., DEWHIRST, D. W., and DOLLFUS, A. [1959] 'The observation of solar granulation from a manned balloon. I. Observational data and measurement of contrast', *Mon. Not. R.A.S.* **119,** 98.

BRAY, R. J., and LOUGHHEAD, R. E. [1957] 'Bright regions in sunspots', *Observatory* **77,** 201.

BRAY, R. J., and LOUGHHEAD, R. E. [1958a] 'The lifetime of sunspot penumbra filaments', *Aust. J. Phys.* **11,** 185.

BRAY, R. J., and LOUGHHEAD, R. E. [1958b] 'Observations of changes in the photospheric granules', *Aust. J. Phys.* **11,** 507.

BRAY, R. J., and LOUGHHEAD, R. E. [1959] 'High resolution observations of the granular structure of sunspot umbrae', *Aust. J. Phys.* **12,** 320.

BRAY, R. J., and LOUGHHEAD, R. E. [1961] 'Facular granule lifetimes determined with a seeing-monitored photoheliograph', *Aust. J. Phys.* **14,** 14.

BRAY, R. J., LOUGHHEAD, R. E., and NORTON, D. G. [1959] 'A "seeing monitor" to aid solar observation', *Observatory* **79,** 63.

BREKKE, K. [1959] 'A high-speed rotating disk shutter for photographing the solar disk', *Inst. Theor. Astrophys. Oslo*, Rep. No. 5.

BRUGGENCATE, P. TEN [1938] 'Beitrag zur Technik von Granulationsaufnahmen', *Z. Astrophys.* **16,** 374.

BRUGGENCATE, P. TEN [1940] 'Über die Natur der Fackeln auf der Sonnenscheibe. I. Fackelgranulen und ihre mittlere Lebensdauer', *Z. Astrophys.* **19,** 59.

BRUGGENCATE, P. TEN [1942] 'Über die Natur der Fackeln auf der Sonnenscheibe. II. Photometrie von Fackelgebieten', *Z. Astrophys.* **21,** 162.

CHANDRASEKHAR, S. [1952] 'A statistical basis for the theory of stellar scintillation', *Mon. Not. R.A.S.* **112,** 475.

CHEVALIER, S. [1908] 'Contribution to the study of the photosphere', *Astrophys. J.* **27,** 12.

CHEVALIER, S. [1914] 'Étude photographique de la photosphère solaire', *Ann. Obs. Zô-Sè* **8,** C1.

CHEVALIER, S. [1919] 'Recherches sur les taches solaires. 2. Effet Wilson', *Ann. Obs. Zô-Sè* **11,** B10.

DANIELSON, R. E. [1961] 'The structure of sunspot penumbras. I. Observations', *Astrophys. J.* **134,** 275.

DANJON, A., and COUDER, A. [1935] *Lunettes et Télescopes.* (Éditions de la Revue d'Optique Théorique et Instrumentale: Paris).

DEWITT, J. H., HARDIE, R. H., and SEYFERT, C. K. [1957] 'A seeing compensator employing television techniques', *Sky and Tel.* **17,** 8.

DIMITROFF, G. Z., and BAKER, J. G. [1945] *Telescopes and Accessories*. Harvard, Blakiston.

DOLLFUS, A. [1959] 'Observations astronomiques en ballon libre', *L'Astronomie* **73,** 345; **73,** 411; **73,** 467.

DUBOV, E. E. [1955] 'An automatic guider for the coronograph of the Crimean Astrophysical Observatory', *Izv. Crim. Astrophys. Obs.* **13,** 155.

DUNN, R. B. [1959] 'Solar limb darkening near λ6563 from 0·9 to 1·00 R', *Astrophys. J.* **130,** 972.

ELLISON, M. A. [1956] 'The effects of scintillation on telescopic images', *Astronomical Optics*, ed. Z. KOPAL, p. 293. Amsterdam, North-Holland Publishing Company.

ELLISON, M. A., and SEDDON, H. [1952] 'Some experiments on the scintillation of stars and planets', *Mon. Not. R.A.S.* **112,** 73.

ELSÄSSER, H. [1960] 'Die Szintillation der Sterne', *Naturwiss.* **47,** 6.

ELSÄSSER, H., and SIEDENTOPF, H. [1959] 'Zur Theorie der astronomischen Szintillation. I.' *Z. Astrophys.* **48,** 213.

FÜRTH, R. [1956] 'Statistical analysis of scintillations of stars', *Astronomical Optics*, ed. Z. KOPAL, p. 300. Amsterdam, North-Holland Publishing Company.

GAVIOLA, E. [1948] 'On seeing, fine structure of stellar images, and inversion layer spectra', *Astron. J.* **54,** 155.

GIFFORD, F., JOHNSON, H., and WILSON, A. [1955] 'Heights of scintillation layers in the Earth's atmosphere', *Astron. J.* **60,** 161.

HANSKY, A. [1905] 'Photographies de la granulation solaire faites à Poulkovo', *Pulkovo Mitt.* **1,** 81.

HANSKY, A. [1906] 'Photographies de la granulation solaire', *Bull. Soc. Astr. France* **20,** 178.

HANSKY, A. [1908] 'Mouvement des granules sur la surface du Soleil', *Pulkovo Mitt.* **3,** 1.

HOSFELD, R. [1954] 'Comparison of stellar scintillation with image motion', *J. Opt. Soc. Amer.* **44,** 284.

JANSSEN, J. [1896] 'Mémoire sur la photographie solaire', *Ann. Obs. Meudon* **1,** 91.

KEENAN, P. C. [1938] 'Dimensions of the solar granules', *Astrophys. J.* **88,** 360.

KEENAN, P. C. [1939] 'Photometry of the solar granules', *Astrophys. J.* **89,** 604.

KEENAN, P. C. [1953] 'Photography of the Sun's disk in integrated light', *The Sun*, ed. G. KUIPER, p. 597. Univ. Chicago Press.

KELLER, G. [1953] 'Astronomical "seeing" and its relation to atmospheric turbulence', *Astron. J.* **58,** 113.

KIEPENHEUER, K. O. [1960] 'Über die Struktur der gestörten Sonnenatmosphäre', *Z. Astrophys.* **49,** 73.

KIEPENHEUER, K. O. [1962] 'Experiences and experiments with solar seeing', *Symposium on Solar Seeing*, p. 49. Rome, National Research Council of Italy.

KING, H. C. [1955] *The History of the Telescope*. London, Griffin.

KLÜBER, H. VON, and MÜLLER, H. [1948] 'Bemerkungen zur Technik direkter Sonnenaufnahmen mit langbrennweitigen Instrumenten', *Z. Astrophys.* **24,** 207.

LAND, G. [1954] 'Anomalies in atmospheric refraction', *Astron. J.* **59,** 19.

LEIGHTON, R. B. [1957] 'Some observations of solar granulation', *Pub. Astr. Soc. Pac.* **69,** 497.

LINNIK, V. P. [1957] 'The possibility, in principle, of reducing atmospheric influence on the star image', *Optika i Spektroskopiya* **3,** 401.

LOUGHHEAD, R. E., and BRAY, R. J. [1958] 'The Wilson effect in sunspots', *Aust. J. Phys.* **11,** 177.

LOUGHHEAD, R. E., and BRAY, R. J. [1959] 'Observations of faculae bordering small sunspots near the limb', *Aust. J. Phys.* **12,** 97.

LOUGHHEAD, R. E., and BRAY, R. J. [1960a] 'The lifetime and cell size of the granulation in sunspot umbrae', *Aust. J. Phys.* **13,** 139.

LOUGHHEAD, R. E., and BRAY, R. J. [1960b] 'Granulation near the extreme solar limb', *Aust. J. Phys.* **13,** 738.

LOUGHHEAD, R. E., and BRAY, R. J. [1961] 'Phenomena accompanying the birth of sunspot pores', *Aust. J. Phys.* **14**, 347.

LOUGHHEAD, R. E., and BURGESS, V. R. [1958] 'High resolution cinematography of the solar photosphere', *Aust. J. Phys.* **11**, 35.

MACRIS, C. [1953] 'Recherches sur la granulation photosphérique', *Ann. Astrophys.* **16**, 19.

MACRIS, C. J., and BANOS, G. J. [1961] 'Mean distance between photospheric granules and its change with the solar activity', *Mem. Nat. Obs. Athens: Series* 1, No. 8.

MACRIS, C. J., and PROKAKIS, T. J. [1962] 'Sur une différence des dimensions des granules photosphériques au voisinage et loin de la pénombre des taches solaires', *C.R. Acad. Sci.* **255**, 1862.

MAYER, U. [1960] 'Beobachtungen der Richtungsszintillation', *Z. Astrophys.* **49**, 161.

MCMATH, R. R. [1953] 'Tower telescopes and accessories', *The Sun*, ed. G. KUIPER, p. 605. Univ. Chicago Press.

MCMATH, R. R., and MOHLER, O. C. [1962] 'Solar instruments', *Handbuch der Physik*, ed. S. FLÜGGE, vol. 54, p. 1. Berlin, Springer.

MCMATH, R. R., MOHLER, O. C., and PIERCE, A. K. [1955] 'Doppler shifts in solar granules', *Astrophys. J.* **122**, 565.

MCMATH, R. R., MOHLER, O. C., PIERCE, A. K., and GOLDBERG, L. [1956] 'Preliminary results with a vacuum solar spectrograph', *Astrophys. J.* **124**, 1.

MCMATH, R. R., and PIERCE, A. K. [1960] 'The large solar telescope at Kitt Peak', *Sky and Tel.* **20**, 64; **20**, 132.

MEINEL, A. B. [1960a] 'Design of reflecting telescopes', *Telescopes*, ed. G. P. KUIPER and B. M. MIDDLEHURST, p. 25. Univ. Chicago Press.

MEINEL, A. B. [1960b] 'Astronomical seeing and observatory site selection', *Telescopes*, ed. G. P. KUIPER and B. M. MIDDLEHURST, p. 154. Univ. Chicago Press.

MIKESELL, A. H., HOAG, A. A., and HALL, J. S. [1951] 'The scintillation of starlight', *J. Opt. Soc. Amer.* **41**, 689.

MILLER, R. A. [1960] 'Observations on photospheric brightness surrounding sunspots', *J. Brit. Astr. Ass.* **70**, 146.

PETTIT, E. (*et al.*) [1939] Annual Report Mt Wilson Obs. 1938–9, p. 12.

PLANK, V. G. [1959] 'Convection and refractive index inhomogeneities', *J. Atm. Terr. Phys.* **15**, 228.

PLASKETT, H. H. [1939] 'The Oxford solar telescope and Hartmann tests of its performance', *Mon. Not. R.A.S.* **99**, 219.

PLASKETT, H. H. [1955] 'The Oxford 35 m solar telescope', *Mon. Not. R.A.S.* **115**, 542.

PRIESTLEY, C. H. B. [1959] *Turbulent Transfer in the Lower Atmosphere*. Univ. Chicago Press.

RICHARDSON, R. S., and SCHWARZSCHILD, M. [1950] 'On the turbulent velocities of solar granules', *Astrophys. J.* **111**, 351.

ROGERSON, J. B. [1958] 'Project Stratoscope', *Sky and Tel.* **17**, 112.

RÖSCH, J. [1955] 'Fluctuations de température dans la tube d'une lunette astronomique', *J. Phys. et Rad.* **16**, 54S.

RÖSCH, J. [1956] 'Étude de l'agitation des images télescopiques par la méthode de Hartmann', *Astronomical Optics*, ed. Z. KOPAL, p. 310. Amsterdam, North-Holland Publishing Company.

RÖSCH, J. [1957] 'Photographies de la photosphère et des taches solaires', *L'Astronomie* **71**, 129.

RÖSCH, J. [1959a] 'Observations sur la photosphère solaire. I. Technique des observations photographiques (objectif de 23 cm)', *Ann. Astrophys.* **22**, 571.

RÖSCH, J. [1959b] 'Observations sur la photosphère solaire. II. Numération et photométrie photographique des granules dans le domaine spectral 5900–6000Å', *Ann. Astrophys.* **22**, 584.

RÖSCH, J., and HUGON, M. [1959] 'Sur l'évolution dans le temps de la granulation photosphérique', *C.R. Acad. Sci.* **249**, 625.

SALANAVE, L. G. [1957] 'Observing at Junipero Serra Peak', *Sky and Tel.* **16**, 320.

SCHWARZSCHILD, M. [1959] 'Photographs of the solar granulation taken from the stratosphere', *Astrophys. J.* **130**, 345.

SCHWARZSCHILD, M., and SCHWARZSCHILD, B. [1959] 'Balloon astronomy', *Sci. Amer.* **200**, No. 5, p. 52.

SCORER, R. S. [1958] *Natural Aerodynamics.* London, Pergamon.

SEVERNY, A. B. [1959] 'The fine structure of the magnetic field and the depolarization of radiation in sunspots', *Astron. J. U.S.S.R.* **36**, 208 (*Sov. Astron. AJ* **3**, 214).

SIEDENTOPF, H. [1939] 'Über Luftunruhe und Szintillation', *Die Sterne* **19**, 145.

SIEDENTOPF, H., and WISSHAK, F. [1948] 'Die Szintillation der Strahlung terrestrischer Lichtquellen und ihr Gang mit der Tageszeit', *Optik* **3**, 430.

SISSON, G. M. [1960] 'On the design of large telescopes', *Vistas in Astronomy*, ed. A. BEER, vol. 3, p. 92. London, Pergamon.

SMITH, A. G., SAUNDERS, M. J., and VATSIA, M. L. [1957] 'Some effects of turbulence on photographic resolution', *J. Opt. Soc. Amer.* **47**, 755.

STEAVENSON, W. H. [1955] 'Air disturbance in reflectors', *Vistas in Astronomy*, ed. A. BEER, vol. 1, p. 473. London, Pergamon.

STESHENKO, N. V. [1960] 'On the determination of magnetic fields of solar granulation', *Izv. Crim. Astrophys. Obs.* **22**, 49.

STREBEL, H. [1932] 'Sonnenphotographische Dokumente', *Z. Astrophys.* **5**, 36.

STREBEL, H. [1933] 'Beitrag zum Problem der Sonnengranulation', *Z. Astrophys.* **6**, 313.

SUZUKI, Y. [1961] 'The fine structures of sunspots', *Bull. Kyoto Gakugei Univ.: Series B*, No. 19.

WALDMEIER, M. [1940] 'Die Feinstruktur der Sonnenoberfläche', *Helv. Phys. Act.* **13**, 13.

WOODWARD, B. [1960] 'Penetrative convection in the sub-cloud régime', *Cumulus Dynamics*, ed. C. A. ANDERSON, p. 28. London, Pergamon.

CHAPTER 3

The Morphology of Individual Sunspots

3.1 Introduction

The introduction of modern high-resolution observing methods, described in the previous chapter, has led in recent years to a substantial increase in our knowledge of the fine structure of sunspots. As we shall see, most of the fine detail has a scale of 1–2″ of arc, although some features appear to be less than 0″.5 in width. Compared to the photospheric granulation, which has a lifetime of about 10 minutes, the sunspot fine detail is relatively long-lived; in fact, individual features sometimes persist for many hours. Consequently, the dynamical properties of the sunspot fine detail can be effectively studied only with the aid of *sequences* of high-quality photographs lasting for periods of several hours, each photograph showing an effective resolution of at least 1″. When such observations are obtained they reveal a wealth of information, some of it new and quite unexpected, which promises to inaugurate a new era in sunspot research.

In this chapter we shall attempt to give a systematic account of present knowledge of the fine detail in sunspots. In doing so we shall largely ignore the fact that spots occur in groups, and shall treat each spot as an individual entity in its own right. (A discussion of the properties of spot groups is reserved for Chapter 6.) Furthermore, we shall confine ourselves to describing the information obtained by high-resolution direct observations, leaving the results of spectroscopic and magnetic observations of spots and spot groups to be dealt with in later chapters.

One of the most important achievements of the modern observing techniques has been the successful photography of the fine detail in sunspot umbrae and, in particular, of the umbral granulation. The wealth of fine detail in spots is much greater than has hitherto been generally realized, although the origin of many of the features remains quite unexplained. For example, spot penumbrae often contain, in addition to the normal filamentary structure, bright features of various sizes, shapes, and brightness, whose relationship to other fine detail is sometimes quite remarkable. Not all the fine detail described in this chapter necessarily occurs in any single sunspot, and consequently a comprehensive description of the fine structure of sunspots must be based on observations of a large number of different spots. Much of the observational material presented in this chapter was obtained by the authors with the Sydney 5-inch photoheliograph, described in Section 2.3.1. During the period May 1957 to June 1960, some 50 different spot groups of various sizes and at different stages of development were photographed with this instrument, over 100,000 photographs being obtained in all. This period included the very intense solar maximum of 1957-8, and in fact during more than 50 per cent of the period the level of solar activity (measured by the Zürich relative sunspot number) remained continuously above the peak of the 1947-8 maximum—itself the highest recorded since reliable observations began. With the exception of Plate 3.1 (a), all the illustrations in this chapter have been selected from these sunspot films; they provide a clear picture of the type and variety of fine detail revealed by sunspot observations made with an effective resolution of 1″ of arc.

A knowledge of the properties of the photospheric granulation is an essential pre-requisite to an understanding of sunspots. The presence of granulation in the umbrae of sunspots shows that the basic *convective* processes responsible for the photospheric granulation also operate in sunspots, although until recently it was believed that the sunspot magnetic field would suppress the convection. Modern observations indicate that an understanding of the nature of the interaction between solar convection currents and magnetic fields may help to elucidate the actual process of spot formation. Accordingly the chapter begins with a description of the basic properties of the photospheric granulation (Section 3.2), our knowledge of which has increased greatly in recent years as a result of improved observations made both with ground-level photoheliographs and balloon-borne telescopes.

The appearance of the photosphere around spots is then discussed (Section 3.3); apart from a slight increase in brightness around spots at the shorter wavelengths, the character of the granulation generally remains unaltered right up to the very boundaries of the spots. However, disturbances in the granulation accompanying the birth and development of sunspot pores have occasionally been observed. The disturbances take the form of a number of dark lanes lying between pores of opposite magnetic polarities; they are interpreted as evidence of rising loops of magnetic flux (Section 3.3.4).

Sunspot pores are small, long-lived, dark regions in the granulation pattern showing no penumbral structure. Their properties are discussed in Section 3.4. The process of birth and development of a pore is described in Section 3.4.4; it is believed that all sunspots begin their lives as pores. The vast majority of pores never develop beyond this stage, but sometimes pores in an isolated cluster increase in size and develop penumbrae, thus resulting in the appearance of a well-developed spot group.

The structure of sunspot penumbrae is described in Section 3.5. Pores exceeding about 5″ of arc in diameter display a strong tendency to develop at least some rudimentary penumbral structure. Observations of small spots showing rudimentary penumbrae are discussed in Section 3.5.2, where the process of formation and dissolution of such a penumbra is also described. The detailed structure of well-developed penumbrae is then considered, and the remarkable variation in the nature of the filamentary detail from spot to spot is pointed out. The properties of the penumbral filaments are discussed in Section 3.5.4, while in Section 3.5.6 an account is given of the many diverse bright features which occur in spot penumbrae. The question of the origin of these bright features and of the role they may play in the evolution of sunspots raises a new field of inquiry which, as yet, has scarcely been touched.

Section 3.6 deals with the structure of sunspot umbrae. The technique of umbral photography is first described and then an account is given of the typical detail observed in sunspot umbrae. It is found that an *umbral granulation* is invariably present. Apart from their smaller brightness, the umbral granules resemble the ordinary photospheric granules in appearance, but have substantially longer lifetimes and are more closely packed. This difference in properties is probably due to the influence of the spot magnetic field. Sunspot light-bridges are next discussed (Section 3.7). These display a great diversity in shape, size, and brightness. Observations show that the presence of a light-bridge or bright streamer across the umbra of a spot often corresponds to the existence of a 'saddle' region in the umbral intensity isophotes. Consequently they are an integral part of the umbral structure and sometimes persist throughout most of the lifetime of a spot. On the other hand, the appearance of light-bridges in the umbra of a spot during the later stages of its evolution is frequently a sign of impending division or final dissolution.

In Section 3.8 we turn to the question of the changing appearance of spots as they approach the limb. Observations are presented of the Wilson effect in regular spots (Section 3.8.2), and attention is directed to the peculiar diffuseness of the umbra–penumbra boundary sometimes observed on the side away from the limb (Section 3.8.3). Neither an elevation nor a depression is observed when a spot reaches the limb.

Finally, in Section 3.9, we summarize the available quantitative data on the lifetimes of the various sunspot and photospheric fine structures.

3.2 The Photospheric Granulation

3.2.1 INTRODUCTION

The early high-resolution photographs of the solar photosphere obtained by Janssen, Hansky, and Chevalier (cf. Chapter 1) showed a well-defined pattern of bright granules with diameters lying mostly in the range 1–2″ of arc. The similarity of this structure to laboratory convection patterns led to the view that the granules represent convection cells of the Bénard type. However, although some modern observers (for example, Keenan, 1938; Macris, 1953; Rösch, 1955, 1956) also found that the photosphere has a granular structure with the same narrow distribution of sizes, others reached contrary conclusions. In particular, on the basis of photometric measurements of photographs obtained with the 60-foot solar tower at Mt Wilson diaphragmed to an aperture of 4 inches, some workers concluded that the solar surface actually presents the appearance of random brightness fluctuations (for references, see Loughhead and Bray, 1959). These brightness fluctuations were identified with so-called 'turbulent eddies' and the view was for a time widely held that the photosphere is in a state of aerodynamic turbulence. In fact, Richardson and Schwarzschild (1950) even went to the extent of using Kolmogoroff's law to derive a complete 'spectrum of turbulence' for the photosphere. These authors found the most energetic 'turbulent elements' to have a diameter of about 150 km—this figure being raised to 300 km by Schwarzschild in 1955—and predicted the existence of small, very bright granules with which the most energetic 'turbulent elements' were to be identified.

However, these ideas were later discredited by results obtained by Leighton (1957) and by the present authors (Bray and Loughhead, 1958b) from *sequences* of high-quality photographs of the photospheric granulation. Both sets of observations showed the granules as bright features of various shapes separated by lanes of darker material. The majority of the granules have diameters in the range 1–2″ of arc. Leighton, who made his observations in the early morning using the full 12-inch aperture of the 60-foot solar tower at Mt Wilson, found the apparent width of the dark lanes to be comparable with the resolving limit of his telescope, 0″4. Both Leighton and the present authors found that the basic appearance of individual granules remained unaltered from photograph to photograph. These observations indicated that the granules were to be identified with convection cells and not with the 'eddies' of any large-scale turbulence.

The observations made by the authors were of sufficient quality to enable the development of individual granules (classified according to brightness, size, and shape) to be followed from photograph to photograph over an average period of nearly 7 minutes. Most of the granules selected for study showed a remarkable absence of systematic change over this period. This stability strongly supported Leighton's hypothesis that the motions within the granules themselves are basically laminar rather than turbulent. Moreover, the Australian observations, while not capable of showing granules as small as the very bright elements predicted by Schwarzschild, revealed no correlation between the brightness and size of individual granules: in fact, it was found that bright granules are just as frequently larger as smaller than average.

Meanwhile, Blackwell, Dewhirst, and Dollfus (1959) and Schwarzschild (1959) had obtained a number of high-quality photographs of the granulation from balloon-borne, 12-inch telescopes at heights of 20,000 and 80,000 feet respectively (cf. Sections 2.3.3 and 2.3.4). Some of the photographs obtained by Schwarzschild are of unsurpassed definition. Although there are a few features whose diameter does not exceed 0″5 of arc, the majority of the granules have diameters in the range 1–2″. The granules are separated by narrow lanes of darker material, whose apparent width is often comparable with the limit of resolution of the telescope (0″4). These photographs show that the granules have irregular, often polygonal, outlines. There is no predominance of small, very bright granules. The balloon photographs therefore in general strikingly confirmed the results of the modern ground-level work, although no *sequences* of photographs of the granulation were obtained in these flights.

It is evident that modern observations completely discredit any suggestion of the existence of large-scale turbulence in the photosphere, and indicate that the cellular granulation pattern actually observed has a convective origin (cf. Loughhead and Bray, 1959). Further evidence in favour of the convective interpretation will emerge in the course of the description of the properties of the granulation given below.

3.2.2 GRANULATION PATTERN: DIVERSITY OF GRANULES IN SHAPE, SIZE, AND BRIGHTNESS

The appearance of the photospheric granulation in the central region of the solar disk is illustrated by Plate 3.1 (*a*), which is an enlargement from a stratospheric photograph obtained by Schwarzschild. It shows that the granulation consists of a *cellular* pattern of bright granules on a darker background. The majority of the granules have diameters in the range 1–2″ of arc and are separated by narrow dark lanes, whose apparent width often does not exceed a few tenths of a second of arc. In places, however, there are relatively large areas of dark intergranular material, which seem to result from the temporary absence of one or more granules. These occasional dark regions are characteristic features of the granulation pattern, and should not be confused with *pores*, which are small sunspots with no penumbra. As we shall see in Section 3.4, pores are darker and very much longer-lived than the dark elements of the granulation pattern. It is interesting to compare the appearance of the granulation on Plate 3.1 (*a*) with that on Plate 3.1 (*b*), which is a ground-level photograph of somewhat lower resolution taken by the authors with the Sydney 5-inch photoheliograph. One effect of the better definition of Plate 3.1 (*a*) is to make the dark lanes between the granules appear narrower than on Plate 3.1 (*b*)—thus accentuating the cellular appearance of the granulation pattern.

It is evident from Plate 3.1 (*a*) that most of the granules have very *irregular shapes*: elongated and even polygonal outlines are very common. The fact that many granules show polygonal outlines was first pointed out by Strebel (1933), whose best white-light photographs (e.g., Strebel, 1932a: cf. Plate 1) show areas where the resolution is comparable to that of modern stratospheric photographs. A number of later observers (Macris, 1953; Bray and Loughhead, 1958b; Blackwell, Dewhirst, and Dollfus, 1959) also found a significant proportion of the granules to be non-circular. Elongated granules show no preferred direction (Bray and Loughhead, 1958b; Rösch, 1959).

Although Plate 3.1 (*a*) shows a few granules of diameter less than 0″.5 of arc, the majority of the granules on this photograph have dimensions in the range 1–2″. Moreover, large granules are seen to be individual structures and not complexes of smaller granules packed closely together. It is thus evident that earlier observations made with an effective resolution of 1″ (or better) did in fact reveal the true nature of the photospheric granulation, though not with the clarity of stratospheric photographs like Plate 3.1 (*a*). Good granulation photographs often show isolated regions where small granules seem to predominate (Blackwell, Dewhirst, and Dollfus, 1959; Edmonds, 1960); these appear to be chance associations which gradually disappear as new granules form and others decay.

Individual granules show a considerable diversity in *brightness*, just as they do in size and shape. The brightness differences among the granules are shown better in Plate 3.1 (*b*) than in 3.1 (*a*): among granules of comparable size it is easy to find some which are distinctly brighter than average and others distinctly fainter than average.

The distribution of granules according to *brightness*, *size*, and *shape* has been studied by the authors (Bray and Loughhead, 1958b), using a sequence of good-quality photographs of the same region taken with the Sydney 5-inch photoheliograph. As a consequence of limited resolving power, it is possible to describe the granules only in a rather crude and qualitative way; therefore the brightness categories were restricted to 'bright', 'medium', and 'faint', and the size categories to 'large', 'average', and 'small'. Average-sized granules were classified as either 'circular' or 'elongated', whereas large granules were permitted an additional category, namely, 'irregular'. No shape classification was made for small granules. The 140 granules selected for study were classified according to their appearance on a very good-quality photograph near the middle of the sequence, using as supporting evidence three other good photographs occurring within 35 seconds.

TABLE 3.1

Classification of Photospheric Granules

BRIGHTNESS	SIZE			
	Large	Average	Small	Total
Bright	12	23	2	37
Medium	23	50	19	92
Faint	1	6	4	11
Total	36	79	25	140

Table 3.1 gives the results of the classification of the granules. Considering large and average-sized granules together, Table 3.1 indicates that bright granules are about one-half as numerous as those of medium brightness and five times as numerous as faint ones. Particularly noteworthy is the almost complete absence of large, faint granules. With regard to shape, 63 granules were found to be circular[1], 36 elongated, and 16 irregular. In the case of small granules, bright granules are very rare and less numerous than faint ones. While the results for small granules may be affected by limited resolution, it is significant

[1] With greater resolution many of these granules would probably appear as more or less regular polygons (cf. Plate 3.1 *a*).

that stratospheric photographs taken with a telescope of twice the resolving power also show that small, bright granules are rather rare (cf. Plate 3.1 *a*). (Changes in the brightness, size, and shape of granules during their lifetimes are discussed in Section 3.2.7.)

3.2.3   CELL SIZE DISTRIBUTION

It is sometimes necessary to have a reliable, quantitative measure of the scale of the granulation pattern, e.g., in comparing the scale of the photospheric granulation with that of the granulation in sunspot umbrae. Unfortunately, direct measurements of the diameters of the individual granules are of little use

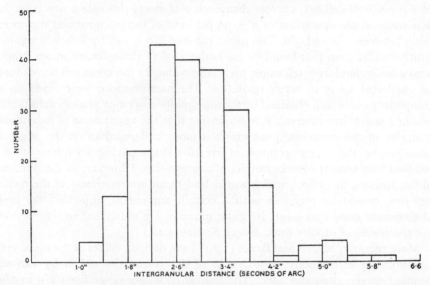

Fig. 3.1. Cell size of the photospheric granulation pattern. The histogram gives the distribution of the distances between the centres of adjacent granules (corrected for foreshortening), measured for a group of 92 granules. It is evident that the granulation has a well-defined and rather narrow distribution of sizes: 70 per cent of the values lie between 2″2 and 3″8 of arc. The mean value is 2″9.

for this purpose as they provide only a rough guide to the true dimensions of the granules. In fact, the measurement of granule size is a complex photometric problem involving not only the photographic contrast, but also the instrumental profile of the telescope and, in the case of ground-level observations, of the atmosphere. However, measurements of the *mean cell size* of the pattern, defined as the average distance between the centres of adjacent granules, are independent of these effects, provided individual granules are actually resolved. Consequently, the mean cell size has been used by the authors and by Rösch as a convenient, quantitative parameter for characterizing the scale of the pattern.

Figure 3.1 (Bray and Loughhead, 1959) gives the distribution of the distances between the centres of adjacent granules, derived from measurements of a group of 92 granules on a good-quality photograph taken with the Sydney 5-inch photoheliograph. It is evident that the granulation pattern has a well-defined and rather narrow distribution of sizes: in fact, 70 per cent of the values lie between $2''2$ and $3''8$ of arc. The mean value is $2''9$. The long tail extending out to $6''2$ reflects the presence in the pattern of the occasional dark elements described in Section 3.2.2. The distribution is truncated at $1''0$, the effective resolving limit of the telescope.

The cell size distribution obtained by Rösch (1959) is qualitatively similar to that obtained by the authors. It is more symmetrical, but still shows a narrow tail extending out to about $5''0$ of arc. The granulation pattern is again found to have a well-defined, narrow distribution of sizes: the values are strongly concentrated about a mean of $2''0$, 95 per cent of the intergranular distances lying between $1''0$ and $3''0$. The mean value of $2''0$ found by Rösch is significantly smaller than that found by the authors ($2''9$). However, although Rösch used a somewhat larger telescope, his smaller value for the mean cell size cannot be attributed solely to better resolution. The measurements were made on a composite photograph obtained by the superposition of many photographs taken within a short time interval. Rösch remarks that the appearance of individual granules on the composite photograph is more uniform than on the original photographs, the biggest granules of irregular shape having a tendency to be resolved into several round granules of average size. However, in the opinion of the authors, this effect, which would lead to an underestimate of the mean cell size, should be regarded with a certain amount of suspicion: the best observations show that large, irregular granules are individual structures and not complexes of smaller granules (cf. Section 3.2.2).

More recently, Macris and Banos (1961) have derived values of the mean cell size from 14 high-quality photographs of the granulation taken by various people between 1880 and 1960. The individual values range from $2''0$ to $2''9$ of arc, the mean being $2''5$. In the absence of a determination based upon the stratospheric photographs obtained by Schwarzschild and his collaborators, the figure $2''5$ probably represents the best estimate of the mean cell size at present available.

3.2.4 GRANULAR CONTRAST

Many estimates have been made of the magnitude of brightness variations in the photosphere. The technique frequently adopted is to make a microphotometer tracing along a line across a photograph of the granulation and measure either the r.m.s. brightness variation of points along the line, or the average peak-to-peak intensity difference between the points of maximum and minimum brightness. However, neither of these quantities actually represents the contrast of the granulation: a unidimensional tracing—despite the relative ease with which it can be obtained—does not reproduce the actual observed brightness of the individual granules except in cases where, by chance, the scanning spot passes through the centre of a granule. It is therefore hardly surprising that, although the granulation has the appearance of a well-defined cellular pattern (Section 3.2.2), the intensity variations derived in this way fit a Gaussian curve (Schwarzschild, 1959: cf. Fig. 5).

A more realistic approach has been adopted by Blackwell, Dewhirst, and Dollfus (1959)[2] and by Rösch (1959). These authors made isophotometric contour maps of the granulation, from which they derived mean values for the observed contrast between granular and intergranular areas. These values then have to be corrected for the finite resolving power of the telescope. The most thorough discussion of the problem of correcting the measurements for instrumental distortion is that given by Blackwell *et al.*, who measured the contrast transmission function of their whole telescope—objective, magnifying lens, and photographic emulsion. In spite of the apparently satisfactory results of conventional visual tests made on the 12-inch objective, the performance of the whole telescope was found to be markedly inferior to that of an aberration-free system: the contrast transmission function at a spatial frequency corresponding to $1''.0$ of arc was only about one-third of its value for an ideal telescope. In the case of the granular contrast the magnitude of the correction for instrumental distortion depends mainly on the width of the *dark lanes* in the granulation pattern, which often does not exceed $0''.5$. The value of the contrast transmission function at the spatial frequency corresponding to $0''.5$ was found to be only about $0·05$. This means that the observed contrast of a unidimensional, sinusoidal intensity distribution with this spatial frequency would have to be multiplied by a factor of 20 to obtain the true value. Hence it is clear that even a telescope 12 inches in aperture, while quite capable of *resolving* the granules, is far too small to permit reliable measurement of the granular contrast.

It should also be noted that the conventional contrast transmission function refers to unidimensional brightness distributions, whereas the granulation pattern constitutes a two-dimensional brightness distribution. Consequently it is not possible to correct for instrumental distortion simply by multiplying an observed value of the granular contrast by the reciprocal of the value of the contrast transmission function at some (arbitrarily) selected spatial frequency.

Except in the case of the stratospheric photographs (from which only r.m.s. brightness variation estimates have been derived), the important—but largely unknown—effects of seeing must also be considered. This is true even in the case of the balloon photographs obtained by Blackwell and his co-workers; the quality of these photographs is inferior to the best ground-level photographs obtained with smaller instruments (cf. Section 2.3.3), showing that seeing effects were in fact still present.

At the moment it is not possible to quote a reliable estimate of the granular contrast.

3.2.5   GRANULATION NEAR THE EXTREME SOLAR LIMB

The visibility of the photospheric granulation decreases towards the limb until,

[2] This paper gives a review of the results of earlier work.

finally, no trace of the granulation pattern can be detected. A determination of the distance from the limb at which it finally disappears provides an estimate of the height of the top of the convection zone. Rösch (1957) found that on good photographs the granulation remains visible to within less than 10″ of arc from the limb, and sometimes to less than 5″. This result, which was later confirmed by the present authors (Loughhead and Bray, 1960b), indicates that the convection currents extend at least up to an optical depth of 0·1 (de Jager, 1959: cf. p. 83).[3]

3.2.6 SPECTROSCOPIC OBSERVATIONS: GRANULE VELOCITIES AND MAGNETIC FIELDS

As a consequence of the difficulty of obtaining the necessary high spatial resolution in spectroscopic observations (cf. Section 2.4.1), the direct measurement of the radial velocities of the granules is no easy matter. Moreover, it is complicated by the presence of large-scale photospheric velocity fields (Hart, 1956; Plaskett, 1959; Stepanov, 1961; Leighton, Noyes, and Simon, 1962), which tend to mask the small-scale velocity fluctuations. Nevertheless, Plaskett (1954) and Stuart and Rush (1954) established a significant correlation between brightness and line-of-sight velocity in the photosphere, the bright areas moving upwards and the dark areas downwards. The results obtained by Stuart and Rush are based on an analysis by Richardson and Schwarzschild (1950) of a fine spectrum obtained with the 150-foot solar tower at Mt Wilson. This spectrum shows appreciably better spatial resolution than the spectra analysed by Plaskett; in fact, Stuart and Rush deduced a value for the cell size of the granulation pattern of 3″ of arc, in fairly good agreement with the value derived from high-resolution direct photographs (cf. Section 3.2.3). For this reason the correlation between brightness and line-of-sight velocity found by Stuart and Rush is stronger than that found by Plaskett. After the elimination of the effects of large-scale velocity fields, values somewhat less than 0·2 km/sec were obtained in both investigations for the observed radial velocities of the granules.[4] It should be noted that these values were derived from the measurement of lines originating in the upper photosphere and therefore do not necessarily correspond with the granule velocities at the lower levels of the photosphere shown on direct photographs of the central region of the disk. The extension of the granules into the upper photosphere is confirmed by direct limb photographs, which show that individual photospheric granules remain visible to within 5″ of the limb (cf. Section 3.2.5).

The relationship between brightness and line-of-sight velocity in the photo-

[3] Edmonds (1962a) does not agree with this estimate. On the basis of an examination of stratospheric photographs he has concluded that the granulation disappears at 15–10″ of arc from the limb, thus revising his earlier estimate of 33–21″ (Edmonds, 1960).
[4] In practice, spectroscopic measurements of this type yield only *lower limits* to the true velocities of individual granules except in cases where, by chance, the spectrograph slit passes through the centre of a granule. The situation is exactly analogous to that encountered in trying to derive a value for the granular contrast from a unidimensional microphotometer tracing (cf. Section 3.2.4).

sphere has also been investigated by a number of subsequent workers (Servajean, 1961; Bernière, Michard, and Rigal, 1962; Edmonds, 1962b; Evans and Michard, 1962a, b; Evans, Main, Michard, and Servajean, 1962; Leighton, Noyes, and Simon, 1962) who find, on the whole, a *weaker* correlation than that found by Stuart and Rush. One important new result has been the discovery of vertical *oscillatory* motions in the upper levels of the photosphere. These were first detected by Leighton and have since also been observed by Evans and Michard (1962b) and Howard (1962). The observed periods lie in the range 4–5 minutes, while the oscillations themselves may persist for upwards of three periods (Leighton *et al.*, 1962). However, it is not yet clear to what extent, if any, these oscillations represent the actual motions of the individual granules. Indeed, it is doubtful whether the spatial resolution of the observations concerned was adequate to resolve the granulation: see, for example, Evans and Michard (1962a: Fig. 1). The spatial resolution of this spectrogram would seem to be decidedly inferior to that of the Mt Wilson spectrogram upon which Stuart and Rush's results are based. A proper understanding of the time variation of granule velocities must await the advent of spectroscopic observations with a spatial resolution comparable to that of high-quality direct photographs.

The possible existence of magnetic fields in the individual granules has been investigated by Steshenko (1960), who found an upper limit of 50 gauss for any possible magnetic field within individual granules. The subsequent studies of Howard (1962) and Semel (1962) have yielded even lower upper limits to the possible field.

3.2.7   LIFETIME AND EVOLUTION OF THE GRANULES

To obtain an accurate estimate of the *lifetime* of the photospheric granules it is necessary to make a careful analysis of a sequence of good-quality photographs of the same region extending over a period of at least 10 minutes and preferably longer, say 20 minutes. The first adequate determination was made by Macris (1953), using a 22-minute sequence of the granulation obtained by Lyot in 1943. Macris found a value of 7–8 minutes for the most probable lifetime, though individual values as high as 15–16 minutes were recorded. Another determination was later made by the present authors (Bray and Loughhead, 1958b), whose results are in broad agreement with those obtained by Macris. However, owing to the relatively short duration of the authors' sequence (10 minutes), the starting and ending times of many of the granules fell outside the period of observation, suggesting that the most probable lifetime is somewhat greater than the value of 7–8 minutes actually found. In agreement with this conclusion, Rösch and Hugon (1959) reported that many granules last for about 10 minutes, but did not publish a detailed analysis of their observations. More recently, Bahng and Schwarzschild (1961) derived a mean lifetime of 8·6 minutes from a correlation analysis of two sequences of stratospheric photographs. Taken together these results show that the lifetime of the photospheric granules is of the order of 10 minutes. This value is considerably greater than that indicated by earlier estimates.

A more difficult observational problem (but one of great interest for its bearing on the dynamics of the granulation) is the study of the *evolution* of the granules. The greater difficulty lies in the fact that the description of the granules demands photographs much better than those required for mere identification. The first systematic attempt to detect changes in the brightness, size, and shape of the individual granules during their observed lifetimes was made by the authors (Bray and Loughhead, 1958b) using the 10-minute sequence of granulation photographs mentioned above. As explained in Section 3.2.2, 140 granules were first classified in regard to brightness, size, and shape according to their appearance on a very good 'master' photograph near the middle of the sequence (cf. Table 3.1). These granules were then described according to their appearance on each of a number of other photographs occurring before and after the master, thus enabling the development of the individual granules to be followed from photograph to photograph over an average period of nearly 7 minutes. The results are summarized in Table 3.2.

TABLE 3.2

Changes in Photospheric Granules

TYPE OF CHANGE		NO. OF GRANULES
No change		71
Brightness {	increase	12
	decrease	12
	increase and decrease[1]	4
Size ..... {	increase	16
	decrease	7
Change of shape		17

[1] These granules showed both an increase and a decrease in brightness during the period of observation.

Table 3.2 indicates that in general the granules display remarkable stability: of the 125 granules for which sufficient data were obtained, 57 per cent showed no detectable change in brightness, size, or shape over an average period of nearly 7 minutes, an additional 14 per cent showing only minor changes of shape. Moreover, while there is some tendency among granules showing change for size increases to predominate over decreases, brightness increases and decreases occur with equal frequency. There is no correlation between the two types of change, nor was any tendency found for brightness or size variations to occur during any particular part of the life cycle. Several examples of the types of change sometimes observed in individual granules have been given by Rösch and Hugon (1959) who fail, however, to reach any definite conclusion about the general mode of evolution of the granules.

The stability of the granulation pattern over a period of 10 minutes is illustrated in Plate 3.2, which shows a sequence of photographs obtained with the

Sydney 5-inch photoheliograph (Bray and Loughhead, 1958b). Although there are apparent differences from one photograph to another due to seeing, several granules can be followed without difficulty over almost the entire sequence. A similar illustration has recently been published by Rösch and Hugon (see December 1960 issue of *l'Astronomie*: front cover); it shows a sequence of 18 photographs covering a period of 17 minutes, which was obtained with the new 15-inch photoheliograph at the Pic-du-Midi Observatory (cf. Section 2.3.1). The resolution achieved is better than that of Plate 3.2.

Changes in the photospheric granules are particularly difficult to detect during their periods of formation and decay, when the granules cannot easily be identified as such. For this reason the observations made by the authors provide little information about the modes of formation and dissolution. Only 26 cases of well-defined births or deaths were recorded among the granules whose lifetimes were determined; from these the impression was gained that in general a granule develops from a vague patch of diffuse bright material, which originates in a hitherto dark area. These diffuse patches are very difficult to distinguish from granules smeared by poor seeing. The dissolution of a granule appears to occur by the reverse process, although occasionally a granule loses its identity by coalescing with another granule. Similar modes of formation and dissolution are shown by the granules within sunspot umbrae (cf. Section 3.6.4).

3.2.8   CONVECTIVE ORIGIN OF THE GRANULATION

The cellular nature of the granulation pattern, the narrowness of its cell size distribution, and the observed stability of the individual granules during their lifetimes all strongly suggest the existence of ordered *convection currents* resulting from thermal instability in the sub-photospheric layers. The observed correlation between brightness and line-of-sight velocity (Section 3.2.6) provides strong additional support for this view, although the manner in which the granule velocities may vary with time is not yet properly understood.

Since solar convection occurs under conditions vastly different from those encountered in the laboratory, it seems pointless to try and identify the granulation with any particular form of liquid convection familiar from laboratory experiments. Not only do the temperature and density in the photosphere increase rapidly with depth, but in addition radiative transfer tends to smooth out temperature fluctuations over any horizontal plane (cf. Spiegel, 1957). Moreover, above the layer of optical depth unity the dynamical conditions may be greatly modified by the onset of radiative cooling (cf. Plaskett, 1955; Oster, 1957). At the moment, although there are theories of the onset of convection, a theory of established convection under solar conditions is entirely lacking.[5] However, it is our belief that further studies of the evolution of the granules (including their modes of formation and decay) may well help to guide the development of such a theory.

[5] A comprehensive review of existing theoretical work on solar convection has been given by Pecker (1959).

3.3 The Photosphere around Sunspots

3.3.1 INTRODUCTION

As we shall see in later chapters a sunspot is a region, pervaded by a strong magnetic field, where the physical conditions are very different from those in the surrounding photosphere. What effect does the presence of spots (and their associated magnetic fields) have on the granulation around them? Observations show that, except in the case of new and developing spots, the presence of a spot generally has no effect on the surrounding granulation, whose properties remain unaltered right up to the boundary of the spot, irrespective of its size or complexity. The only significant change sometimes observed is a slight increase in intensity, particularly at the shorter wavelengths; this gives rise to the appearance of *bright regions* around spots, which are often detected in the violet or ultra-violet but are rarely seen in the green-orange region of the spectrum now usually used in sunspot photography. The properties of these bright regions are discussed in Section 3.3.3.

On the other hand, disturbances in the granulation pattern have occasionally been observed in the neighbourhood of new and developing sunspot pores. In one case, the disturbance lasted for about 3 hours and took the form of a number of dark lanes lying between two groups of pores whose location with respect to the rest of the spot group suggests that they were of opposite magnetic polarities. This phenomenon has been interpreted as evidence of the existence of a rising loop of magnetic flux accompanying the birth of the new pores (Section 3.3.4).

3.3.2 APPEARANCE OF THE GRANULATION AROUND SUNSPOTS

The *normal appearance* of the granulation in the immediate neighbourhood of sunspots is illustrated by Plate 3.3. This is a photograph taken with the Sydney 5-inch photoheliograph in green light centred on 5400Å, which shows a group of small spots in the central region of the solar disk. It is evident from Plate 3.3 that the granulation remains unaltered right up to the boundaries of the spots. These boundaries are remarkably sharp; in fact, the edge of the pore in the lower, left-hand corner of the photograph actually appears notched by the surrounding granules. There is no indication of any abnormal brightness in the photosphere close to the spots. The conclusion that the appearance of the granulation in general remains unaltered right up to the boundaries of sunspots is also convincingly demonstrated by the stratospheric photographs of spots obtained by Schwarzschild and his co-workers (cf. Danielson, 1961a). In addition, Bahng and Schwarzschild (1961) find that granules in regions close to spots have the same lifetime as granules in regions remote from spots.

The question of whether there is any real variation in the *size* or *packing* of the granules close to spots has been investigated quantitatively by a number of workers. Macris (1953) found a mean diameter of $1''29 \pm 0''27$ of arc for the granules close to a spot and $1''56 \pm 0''34$ for granules further away; however, the difference between these values is only of the order of their mean errors and

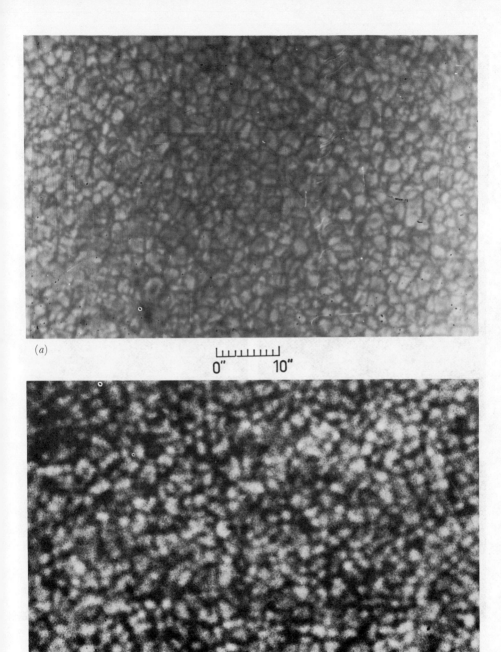

(a)

0"   10"

(b)

PLATE 3.1. The photospheric granulation.

(a) Photograph obtained by Schwarzschild with a 12-inch balloon-borne telescope. The granulation consists of a cellular pattern of bright granules, mostly 1–2″ of arc in diameter, on a darker background. There is a considerable diversity in the brightness, size, and shape of individual granules. (*By courtesy of Project Stratoscope of Princeton University, sponsored by ONR, NSF, and NASA.*)

(b) Photograph obtained with the Sydney 5-inch photoheliograph. The cellular appearance of the granulation is again well shown, despite the lower resolution.

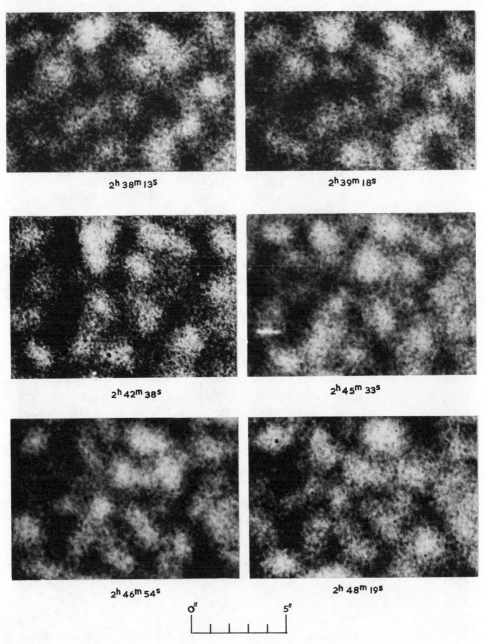

2ʰ 38ᵐ 13ˢ

2ʰ 39ᵐ 18ˢ

2ʰ 42ᵐ 38ˢ

2ʰ 45ᵐ 33ˢ

2ʰ 46ᵐ 54ˢ

2ʰ 48ᵐ 19ˢ

0″   5″

PLATE 3.2. Stability of the photospheric granulation pattern. Despite apparent changes due to seeing, some granules can be followed without difficulty over the entire sequence of photographs, a period of 10 minutes.

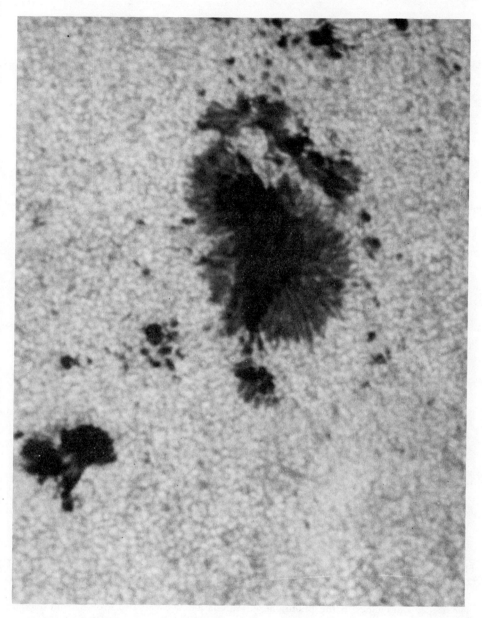

0"   10"
|⌊⌊⌊⌊⌊⌊⌊⌊⌋

PLATE 3.3. Appearance of the granulation in the neighbourhood of a small sunspot group (September 23, 1957). The granulation remains unaltered right up to the boundaries of the spots. The main spot has a serrated outline, lanes of dark material projecting a short distance into the surrounding photosphere. The edge of the large pore in the lower, left-hand corner is notched by adjacent granules.

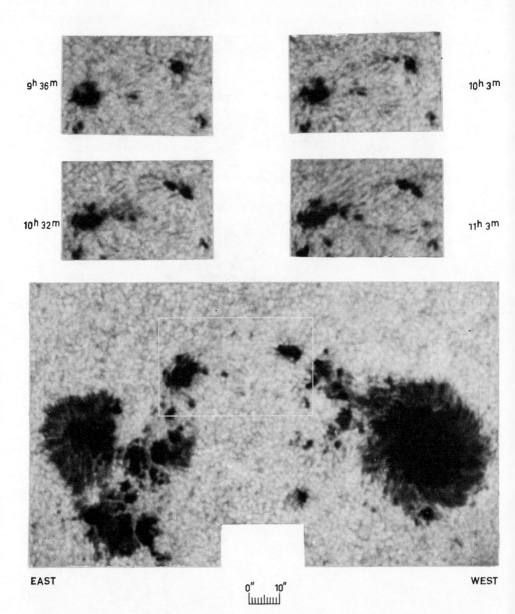

9h 36m

10h 3m

10h 32m

11h 3m

EAST

0"   10"

WEST

PLATE 3.4. Dark lanes in the photospheric granulation between two groups of developing sunspot pores (February 4, 1958). The lanes are prominent between 10^h 3^m and 11^h 3^m but have disappeared by 12^h 42^m. Note the curvature of the lanes at 10^h 32^m. The location of the two groups of pores with respect to the rest of the spot group (cf. 12^h 42^m) suggests that they have opposite magnetic polarities. The arrows at 10^h 3^m point to a new pore which grows during the course of the sequence, ultimately coalescing with a pre-existing pore just to the right of it; this event is shown in more detail in Plate 3.6.

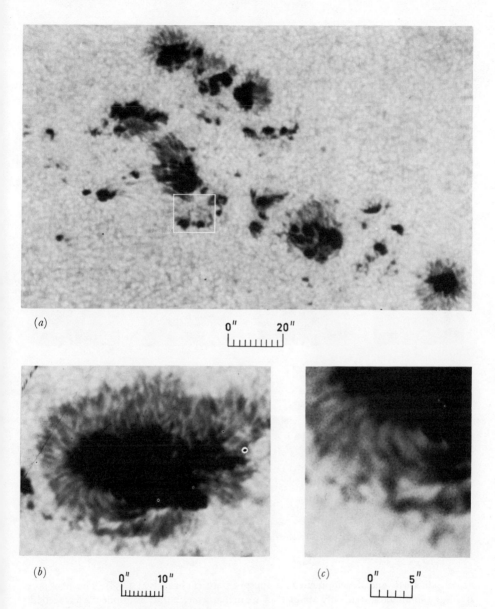

PLATE 3.5. (*a*) Group of small spots and attendant pores (June 2, 1957). Most of the umbral areas larger than 5″ of arc show some evidence of penumbral structure. The white rectangle encloses a developing penumbral region.

(*b*) Alignment of penumbral filaments (September 13, 1957). Some of the filaments on the lower, left-hand side curve around towards a region of umbral material intruding into the penumbra.

(*c*) Enlargement of portion of (*b*), on twice the scale.

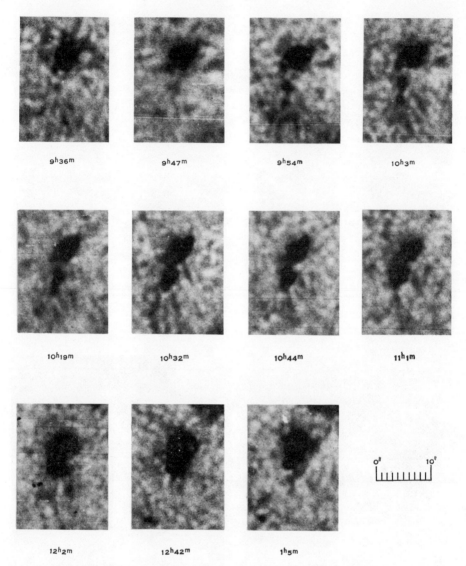

9ʰ36ᵐ 9ʰ47ᵐ 9ʰ54ᵐ 10ʰ3ᵐ

10ʰ19ᵐ 10ʰ32ᵐ 10ʰ44ᵐ 11ʰ1ᵐ

12ʰ2ᵐ 12ʰ42ᵐ 1ʰ5ᵐ

0″ 10″

PLATE 3.6. Birth and development of a sunspot pore (February 4, 1958). The new pore starts as a small dark area, only about 1″ of arc in diameter, in the photosphere a few seconds below a pre-existing pore (9ʰ 36ᵐ). For a detailed description of its subsequent growth and development, see text (Section 3.4.4).

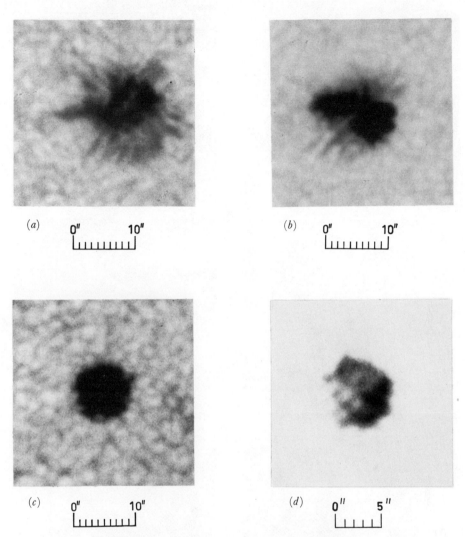

PLATE 3.7. (*a*) Small spot with rudimentary penumbra (January 16, 1960). The few penumbral filaments present are faint and poorly defined. Note the peculiar spike of dark material on the left-hand side.

(*b*) Small spot with rudimentary penumbra (January 24, 1958). The degree of penumbral development is less than in (*a*), no clearly defined filaments being present.

(*c*) Large sunspot pore (August 27, 1958). Note the sharpness of the boundary and the normal appearance of the photospheric granules just outside.

(*d*) The same pore some 42 minutes later, shown with enhanced exposure and greater magnification. Several umbral granules are clearly visible, although those in the upper part are somewhat smeared by mediocre seeing. On the original negative a faint umbral granule is visible in the lower, right-hand part, which is darker than the rest.

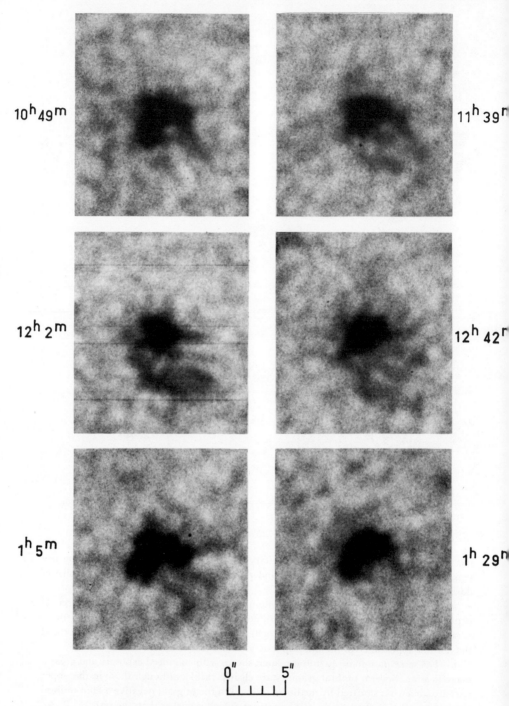

10ʰ 49ᵐ

11ʰ 39ᵐ

12ʰ 2ᵐ

12ʰ 42ᵐ

1ʰ 5ᵐ

1ʰ 29ᵐ

0″   5″

PLATE 3.8. Formation and dissolution of a rudimentary penumbra (February 4, 1958): During the period covered by the sequence, a rudimentary penumbra developed around the small spot and then quickly dissolved. At 10ʰ 49ᵐ the only penumbral material visible is the dark spike penetrating into the photosphere near the lower, right-hand side of the spot. For a detailed description of the subsequent growth and dissolution of the penumbra, see text (Section 3.5.2). The total lifetime of the transitory penumbra was less than three hours.

therefore does not appear to be statistically significant. Rösch (1959) and Miller (1960b) counted the granules in regions near to and away from spots but failed to find any significant difference in the number per unit area. More recently, Macris and Prokakis (1962) claim to have found a significant difference in the mean granule diameter between regions near to and remote from two complex spot groups. However, similar measurements made in the case of two small, isolated spots revealed no significant difference. Schröter (1962) claims a definite decrease in the mean granule diameter in the immediate vicinity of the penumbral border.

In our films of some 50 different spot groups there is only one case where a difference has been noticed in the appearance of the granulation around a spot. The spot concerned is illustrated in Plate 3.10 (c); in this case the larger (∼2″ of arc) granules of the normal pattern seem to be absent in the region near the periphery of the spot on the upper, left-hand side. A very faint bright ring is visible around the spot on the original negative, but fails to appear in Plate 3.10 (c). However, this spot is in many ways exceptional: not only is its penumbral structure unusually diffuse (see Section 3.5.3) but also the granulation near the top of the spot contains an abnormal amount of dark material, which seems to be aligned more or less concentrically with the spot.

A very similar disturbance has been observed by Danielson (1961a: cf. Figs. 4 and 7) in the granulation near a small, fairly regular sunspot which formed part of a rapidly changing, complex group. The disturbance is again characterized by lanes of dark intergranular material aligned more or less concentrically with the spot. The lanes are more prominent in Danielson's Fig. 4 than in his Fig. 7, taken some 57 minutes later, thus suggesting that the phenomenon is of relatively short duration. In addition, the larger granules of the normal pattern appear to be absent around the part of the spot bordering the disturbed region. In the case of both Danielson's observation and our own, the appearance of the dark lanes is rather similar to that of the disturbed pattern occasionally observed in the granulation near new and developing sunspot pores (cf. Section 3.3.4).

3.3.3   BRIGHT REGIONS AROUND SUNSPOTS

Routine solar photographs taken in violet or blue-violet light show that, at these wavelengths, sunspots are commonly surrounded by regions of somewhat enhanced brightness (Waldmeier, 1939; Das and Ramanathan, 1953). According to Waldmeier a spot group is normally surrounded by a *bright zone* which, when the development of the group has reached the stage at which there remains only a single spot with possibly some minor attendants, takes the form of a *bright ring* around the spot (Waldmeier, 1939: cf. Fig. 1). The boundaries of the rings are found to be rather diffuse and ill-defined.[6]

A detailed study of the properties of the bright rings around isolated, regular

[6] Tests made with small obscuring disks show that the presence of the bright rings is not due to the Eberhard effect (see, for example, Miller, 1960b).

spots has been made by Waldmeier (1939). He measured the ratio of the diameter of the penumbra (P) to the diameter of the bright ring (R) for 41 spots and found a mean value of 0·72; the value of the ratio P/R increases with P. In addition, Waldmeier made microphotometer tracings across 4 large spots near the centre of the disk and found that the intensity of the bright ring is greatest immediately outside the penumbra and decreases slowly outwards. At the effective wavelength of the observations (4000Å) the intensity at the brightest parts of the rings was 3 per cent greater than that of the surrounding photosphere. This agrees with the more recent results of Das and Ramanathan (1953), who made detailed spectroscopic measurements of the intensity distribution across a large sunspot at the centre of the disk. They found that in the continuum at λ3842·5 the bright ring around the spot was 2 to 3 per cent brighter than the surrounding photosphere. The value of the ratio P/R for this spot was 0·76. According to Maltby (1960) the bright rings are unusually intense around spots showing large *Evershed velocities* (cf. Section 4.4.2). Maltby found that the rings surrounding three such spots were 3–5 per cent brighter than the photosphere at 4000Å.

According to Waldmeier the visibility of the bright rings decreases towards the limb: he estimates that the contrast between a ring and the photosphere at a heliocentric angle of $\theta = 60°$ is only one-half its value at the centre of the disk. Nearer the limb the photospheric faculae, which usually extend over a wide area around spot groups, become increasingly prominent and make the presence of a bright ring more difficult to detect.

The bright zones around sunspots are much less prominent at higher wavelengths (Waldmeier, 1939), and are rarely seen on solar photographs taken in the green-orange region of the spectrum. An examination of our films of some 50 spot groups photographed in green light (5400Å) has revealed the presence of bright rings around only 5 spots. These are all small or medium-sized spots which are either isolated or accompanied by only a few minor attendants. Bright rings are faintly perceptible around these spots on the original films but fail to appear on reproductions. The widths of the rings appear to be somewhat less than the widths of the spot penumbrae. In the case of one of these spots, which was observed during almost its entire passage across the disk, the bright ring was seen both when the spot was near the centre of the disk and also when it was fairly near the limb ($\theta = 64°$).

The apparent absence of bright rings around the other spots in our records has been confirmed quantitatively in one case: photometric measurements of a large spot showed no evidence of the presence of a bright ring. However, the problem is complicated by the existence of large-scale brightness variations in the photosphere. With a circular scanning aperture some 2″ of arc in diameter, the r.m.s. brightness fluctuation due to these large-scale variations was found to be ± 2 per cent of the mean photospheric brightness. Consequently, the presence of a bright ring can easily be masked unless its average intensity exceeds, say, 4 per cent of the background—the estimated limit of detection.

A similar phenomenon has also been observed in the chromosphere. Hale and Ellerman (1906) detected the presence of narrow bright rings partially or completely encircling certain sunspots on spectroheliograms taken at the centre of the $H\delta$ line. They found that the form of the rings does not change materially over a period of several hours. Royds (1925) observed a similar effect in the $H\alpha$ line: by examining spectroheliograms taken with a slit $0\cdot5$Å wide set on one wing of the line he found that a bright ring, often complete and unbroken, occurs around nearly all spots of medium size. The width of the ring is approximately equal to the radius of the spot. Royds observed the same effect on both sides of the $H\alpha$ line, generally without any essential differences. He also observed bright rings around spots on spectroheliograms taken at the centre of $H\alpha$ but found that these are never so conspicuous as those seen in the wings of the line. Das and Ramanathan (1953), in the investigation referred to above, measured the intensity of the bright ring around a large spot at the centre of the disk in the absorption lines $\lambda3820\cdot5$ (Fe I), $\lambda3937$ (K_{1R} of Ca II), and $\lambda3971$ (H_{1R} of Ca II); they concluded that, in all cases, the ring was about 10 per cent brighter than the background. Moreover, they found that the brightness shows a remarkable increase in the K_2 portion of the K line, where the ring may be nearly 50 per cent brighter than the background.

No satisfactory explanation of the presence of the bright rings in the photosphere around spots at the shorter wavelengths has yet been given. The decrease in contrast towards the limb observed by Waldmeier would seem to imply that the enhanced emission originates mainly in the *deeper* layers of the photosphere (cf. Waldmeier, 1939). This raises the question of to what extent, if any, the photospheric and chromospheric phenomena are related. The whole problem merits further attention on both the theoretical and observational sides.

However, it is possible to draw one conclusion from the observations of some importance to the theory of sunspots: the increase in emission observed in the bright rings around spots is inadequate to explain the so-called 'flux deficit' (Sweet, 1955), i.e. the difference between the radiant energy output of a spot and that of an equal area of undisturbed photosphere. It seems unlikely that the flux deficit of a spot is made good by a slight increase in intensity, too small to be detected, over an area around the spot much larger than that implied by Waldmeier's value for the width of the bright ring. The flux deficit is of great importance to the question of the energy balance of sunspots, a topic which is discussed in Chapter 8.

3.3.4 DISTURBANCES IN   THE   GRANULATION   BETWEEN   GROWING SUNSPOT PORES

In contradiction to the generally unaltered appearance of the granulation in the neighbourhood of quiescent sunspots, a definite *disturbance* has been observed by the authors (Loughhead and Bray, 1961) in the granulation between new and developing sunspot pores. The region of interest is indicated by the white rectangle in the lowest photograph of Plate 3.4; it lies between the leader and follower components of a typical bipolar sunspot group, which was photographed with the Sydney 5-inch photoheliograph for a period of six hours on February 4, 1958. During the period covered by the sequence of photographs shown in Plate 3.4, a pore is born in the upper, right-hand side of the region and subsequently coalesces with a pre-existing pore; this event is described in detail in Section 3.4.4. At the same time, similar events occur near the left-hand side of the region. The location of the region strongly suggests that the left- and right-hand pores are areas of opposite magnetic polarities (magnetic data are not available).

At 9^h 36^m (Plate 3.4) the granulation between the pores appears to be more or less normal. However, a number of dark lanes are visible at 10^h 3^m, when the birth of the new pore on the right-hand side has already begun, at the point indicated by the arrows. The lanes appear to be somewhat darker than the ordinary intergranular material. Between the lanes individual photospheric granules can still be distinguished, but they now lie along lines parallel to the lanes. The direction of the lanes is roughly parallel to a line joining the two groups of pores. They are even more evident on the third photograph (10^h 32^m); on this photograph they exhibit a decided curvature. They are most prominent at 11^h 3^m, by which time the two pores on the right-hand side have coalesced. By 12^h 42^m the period of activity of the pores has come to an end; at the same time the disturbance in the granulation has died away, the pattern reverting to normal. No further disturbance was observed between 12^h 42^m and the close of observations at 3^h 28^m. The total time interval between the appearance and disappearance of the dark lanes is about 3 hours.

The possibility that the disturbance illustrated in Plate 3.4 might be due to the effect of seeing has been excluded by an examination of many other good-quality photographs taken during the same period. The persistence of the dark lanes from photograph to photograph leaves no doubt that the phenomenon is real.

In the opinion of the authors, the disturbance in the granulation accompanying the birth and growth of the pores is to be interpreted as evidence of the existence of *a rising loop of magnetic flux*. At the ends of the loop, where the field is strong and predominantly vertical, the photosphere is radically altered by the field, as is shown by the appearance of sunspot pores at these points. Between the pores, on the other hand, the field is predominantly horizontal, and the observations described in this section show that the field merely tends to align the granules in a preferred direction without affecting their brightness, although the material between the granules becomes somewhat darker than normal. The alignment lasts only while the magnetic loop is presumably pushing through the photospheric layers—a period of about 3 hours in the present case. Although the mode of penetration of the field is not yet fully understood, the observations do suggest that the photospheric convection currents may play a more important role than has hitherto been supposed.

Miller (1960a) has published a photograph showing a number of curved dark lanes lying between regions of spot debris whose location with respect to the associated spot group is very similar to that of the pores illustrated in Plate 3.4. According to magnetic observations made at Mt Wilson these areas of spot debris were probably of opposite polarities. The resolution of Miller's photograph is insufficient to show the granulation, and other supporting photographs are not available. Nevertheless, the similarity of the dark lanes photographed by Miller to those illustrated in Plate 3.4 (in particular the photograph at $10^h 32^m$) leaves no doubt that the phenomenon described by Miller is identical to that described above. Miller suggests that the dark lanes follow lines of magnetic force.

Although the phenomenon has so far been observed on only two occasions (once by Miller and once by the authors), it seems very likely that it is a general accompaniment of the birth of sunspot pores.

3.4 Sunspot Pores

3.4.1   INTRODUCTION

Good-quality photographs often reveal the presence in the photospheric granulation of small regions, only a few seconds of arc in diameter, which are much darker than the surrounding photosphere and are referred to as *pores*. It has long been known that pores are simply small sunspots with no penumbral structure. Pores are the simplest type of sunspot and, in fact, it seems that the vast majority of spots do not develop beyond this stage.

It is important to realize that pores are quite distinct from the ordinary dark elements of the granulation pattern, described in Section 3.2.2, from which they are easily distinguished by their lower intensities and much longer lifetimes. The indiscriminate use of the term 'pore' by some authors to describe both phenomena is unjustified and misleading. In this book the word 'pore' will be applied to any spot, irrespective of size, which has no penumbral structure.

3.4.2   OCCURRENCE

Large, well-developed spot groups usually contain a large number of attendant pores, the occurrence of 30 or more within a single group being quite common. In most cases the pores have a tendency to follow rather than to precede the leader spot and are usually scattered over a wide area between the leader and follower spots. Sometimes the penumbra of the leader spot in a bipolar group shows a bulge on the rear side, many of the pores in the group being clustered around the bulge or strung out behind it like a 'tail' (cf. Secchi, 1875: p. 91). This effect is clearly shown by the leader spot of the bipolar group illustrated in the lowest photograph ($12^h 42^m$) of Plate 3.4.

Another illustration of the occurrence of pores in spot groups is given by Plate 3.5 (*a*). This shows a spot group consisting of a number of small spots with many attendant pores. Apart from the one on the extreme right, the spots have only rudimentary penumbral structure: in each case the penumbra is present only around a limited part of the umbra of the spot. (Note the curved lanes of dark material in the region marked by the white rectangle. These form part of a developing penumbral region which eventually linked the pores at the bottom and to the right of the marked rectangle to the neighbouring spot above.)[7] However, the occurrence of pores is not restricted to the vicinity of spot groups. They are sometimes found, either alone or in small clusters, in regions of the photosphere far removed from spot groups. Owing to the small size of most pores, high-resolution photographs are required to detect their presence. However, since such photographs usually cover only a limited region, little information is available about the distribution of pores over the solar disk. In particular, it is not known whether the occurrence of pores is restricted to the sunspot zones. The fragmentary data available at present suggest that pores occur fairly frequently in regions outside spot groups at times of sunspot maximum. It would be of considerable interest to determine whether pores are still present at times of sunspot minimum.

3.4.3 PROPERTIES OF PORES: SIZE, BRIGHTNESS, AND LIFETIME

The majority of pores have *diameters* in the range 2–5″ of arc, while some are no larger than photospheric granules, having diameters of only 1–2″. Although pores with diameters of 10″ or more are occasionally observed, pores with diameters exceeding 6″ or 7″ are rather uncommon. Above this size pores show an increasing tendency to develop into small spots having at least some rudimentary penumbral structure. A number of pores showing different degrees of development are shown in Plate 3.5 (*a*); it will be noticed that most of the umbral areas greater than 5″ in diameter show some evidence of penumbral structure.

The *brightness* of pores is less than that of the dark elements of the surrounding photospheric granulation (cf. Plate 3.5a), but is greater than that of the umbrae of larger sunspots. Photometric measurements of a few small pores accidentally recorded on a stratospheric photograph of the granulation have been made by Bahng (1958), who finds their central brightness to be about 65 per cent of the average photospheric background. However, this figure is uncorrected for instrumental profile, so that the true intensity must be somewhat lower (cf. Section 3.2.4).

Pores have very much longer *lifetimes* than the photospheric granules and often remain substantially unchanged for many hours. Even the smallest pores persist for at least several hours and, in fact, in most cases a single day's observations are not adequate to determine their lifetimes.

3.4.4 PROCESS OF BIRTH AND DEVELOPMENT

There is no doubt that both pores and well-developed sunspots owe their origin

---

[7] The properties of rudimentary sunspot penumbrae and their mode of formation are discussed in Section 3.5.2.

to the penetration of internal solar magnetic fields into the photospheric layers (cf. Chapter 8). Observations of the birth and initial development of a pore can be expected to throw some light on the mode of penetration of the field since, during this period, the field is presumably in the act of pushing through the visible layers. However, such observations are very difficult to obtain and the only published account of the birth of a pore is that of the authors (Loughhead and Bray, 1961).

The event in question took place near the right-hand side of the region indicated by the white rectangle in the lowest photograph of Plate 3.4; this region lies between the leader and follower components of a typical bipolar sunspot group. During the period covered by the sequence of photographs shown in Plate 3.4, a pore is born in the upper, right-hand side of the region; this grows in size and ultimately coalesces with a pre-existing pore. During the same period disturbances occurred in the nearby granulation: these are described in Section 3.3.4. The birth and development of the new pore are illustrated in more detail in Plate 3.6. (Note: the photographs in Plate 3.6 are orientated at right angles with respect to those in Plate 3.4.)

The first sign of the advent of the new pore is the appearance of a small dark region, only about $1''$ of arc in diameter, in the photosphere a few seconds below the existing pore. At first ($9^h 36^m$) this dark region is not distinguished in any way from normal intergranular material. However, it soon becomes darker and larger and, by $9^h 54^m$, a second such region has appeared, about $3''$ below the first. At $9^h 54^m$ and $10^h 3^m$ the two small dark regions are seen to be joined together by a narrow lane of dark material. By $10^h 19^m$, the two regions have completely merged to form a typical sunspot pore of irregular shape, some $5''$ long and $2''$ wide at its narrowest point. The whole process of the formation of the new pore from a region of hitherto normal photosphere is completed in a period of some 45 minutes.

The new pore now enters into the second phase of its development. Initially ($10^h 3^m$) it is separated from the upper, pre-existing pore by a narrow strip of photospheric granulation. However, the granules in this region gradually disappear until, finally, only a single chain of granules separates the two pores ($10^h 32^m$). This chain is soon broken ($10^h 44^m$) and the pores merge into a single pore with a narrow waist; the amalgamation of the two pores takes place in about 25 minutes. Finally ($11^h 1^m - 1^h 5^m$), the granules in the neighbourhood of the waist are gradually engulfed and the composite pore loses its dumbbell appearance.[8]

Two conclusions emerge from the observations described above. Firstly, significant changes in the shape and area of pores during their birth and development take place within periods of half an hour or so. A similar time-scale

[8] The dark material visible in the top, right-hand corners of the last two photographs of Plate 3.6 is portion of an elongated region of spot debris trailing the leader spot of the group (cf. Plate 3.4: $12^h 42^m$). During the day the distance between this debris and the pore illustrated in Plate 3.6 steadily decreased.

was derived by Secchi (1875: cf. p. 56) from visual observations of changes occurring during the *dissolution* of a pore. This time-scale is short compared with the total lifetime of a pore which, as we have seen in Section 3.4.3, is at least several hours and often much longer.

Secondly, the process of birth and development of a pore appears to take place by a mechanism involving *individual* photospheric granules. The first indication that a new pore is being formed is the appearance of a small dark region, no larger than a single photospheric granule. The growth of the new pore, once formed, then seems to occur through the gradual disappearance of individual granules around its periphery. This is particularly evident during the amalgamation of the two pores; as shown in Plate 3.6, the granules in the narrow region separating the pores gradually dissolve until only a single chain of granules remains. These in turn gradually disappear (cf. $10^h\ 19^m$ — $11^h\ 1^m$).

It seems as if the action of the rising magnetic field is to inhibit the individual photospheric convection columns one at a time, eventually producing a region much darker and therefore cooler than the surrounding photosphere, namely a pore. However, as we shall see in Section 3.6.3, convection within the pore itself is not entirely suppressed, *umbral* granules invariably being present. The observed sharpness of the boundaries of sunspot pores (see, for example, Plate 3.7c) shows that the field can affect one granule but leave a neighbouring granule unaltered (cf. Section 3.6.3).

3.4.5 DEVELOPMENT OF PORES INTO LARGER SUNSPOTS

It has been known since the time of the early visual observers that all sunspots begin their lives as pores. The vast majority of pores never develop further, but sometimes pores in an isolated cluster increase in size and develop into larger spots (cf. Chevalier, 1916: p. B18). The transformation from a pore to a large sunspot can apparently occur with great rapidity: Secchi (1875: p. 62) has recorded the appearance of a large spot some $76''$ of arc in diameter in a region where only three pores had been visible the preceding day.

Observations of the development of a spot during the first day or so of its life are of critical importance, for it is during this initial period that the magnetic flux of a new spot is increasing most rapidly (cf. Section 5.4.7). Unfortunately, observations of the development of a spot during the first 24 hours or so of its life cannot be made from a single observatory, unless it is situated in the polar regions. A successful programme would require the cooperation of a number of observatories well distributed in longitude. A proposal for international cooperation in studying long-period changes in sunspots was, in fact, put forward by Waldmeier in 1956, prior to the start of the International Geophysical Year. Ideally, such a programme should be carried out with a number of identical instruments, perhaps similar to the 5-inch photoheliograph described in Section 2.3.1, each photographing the same region of the solar disk.

3.5 The Structure of Sunspot Penumbrae

3.5.1 INTRODUCTION

As stated in Section 3.4.5, it is believed that all sunspots begin their lives as pores, which by definition have no penumbral structure. Although some pores attain diameters of 10″ of arc or more, most pores with diameters exceeding 5″ show a strong tendency to develop into small sunspots possessing at least some penumbral structure. The process of transition from a pore to a large spot with a well-developed penumbra is exemplified by the occurrence of small spots with rudimentary or transitory penumbrae. In this section we shall first describe observations of small spots showing such rudimentary or transitory penumbrae (Section 3.5.2) and will then discuss in some detail the filamentary structure of well-developed penumbrae (Section 3.5.3).

Although there is always a sharp distinction between the penumbra and the umbra of a spot, penumbrae often contain projections from the umbra or even isolated areas of umbral material. The structure is usually further complicated by the presence of bright features of various sizes and shapes. Hitherto little attention has been paid to these bright features, which often seem to be just as integral a part of the penumbral structure as the penumbral filaments themselves. Despite their diversity in brightness, size, and shape, the bright regions seem to fall naturally into three distinct classes, which are described in Section 3.5.6. The question of the origin of the bright features and of the role that they play in the evolution of sunspots raises a new field of inquiry which, as yet, has scarcely been touched.

3.5.2 RUDIMENTARY AND TRANSITORY PENUMBRAE

The normal basic structure of sunspot penumbrae consists of a system of narrow bright filaments on a darker background; although the filaments are bright compared to the penumbral background, they are, of course, darker than the surrounding photosphere. Even in well-developed penumbrae there is a considerable diversity in the clarity of the filamentary detail from spot to spot, as we shall see in Section 3.5.3. In the case of rudimentary penumbrae the dark background is dominant, the bright filaments being few or, in some cases, entirely absent. This is illustrated in Plate 3.7. The first photograph (Plate 3.7a) shows the leader spot of a tiny bipolar group; the rudimentary penumbra is most developed around the lower border of the umbra. Its dominant features are lanes of dark material, which at places penetrate quite deeply into the surrounding granulation. The few bright filaments present are faint and, for the most part, poorly defined. Note the peculiar dark spike penetrating into the photosphere on the left-hand side of the spot, having a brightness intermediate between that of the penumbra and umbra.

An example of an even more rudimentary penumbra is shown in Plate 3.7 (b). This is a photograph of an isolated small spot which has a faint fringe of penumbral material along limited portions of its umbral boundary. No bright filaments are visible and, in fact, individual photospheric granules can still be distinguished in the penumbral fringe at the bottom of the spot. The appearance of the small spots in Plates 3.7 (*a*) and (*b*) may be contrasted with that of the large pore in Plate 3.7 (*c*), which shows no trace of any penumbral structure whatsoever.

The actual process of the formation and dissolution of a rudimentary sunspot penumbra is illustrated in Plate 3.8. This shows a sequence of photographs of a small spot which was located some distance behind the leader spot of the bipolar group illustrated in the lower photograph of Plate 3.4. (Note: the photographs in Plate 3.8 are orientated at right angles to those in Plate 3.4.) During the period covered by the sequence, a rudimentary penumbra developed around the spot and then quickly dissolved. At the time of the first photograph ($10^h 49^m$) the only sign of penumbral material is a dark spike penetrating into the photosphere near the lower, right-hand side of the spot. Between $10^h 49^m$ and $11^h 39^m$ the *intergranular* material in the region below and to the left of the spike becomes darker, although granules are still present and appear more or less unchanged. At $12^h 2^m$ the penumbral region is both larger and darker and contains, in addition to two apparently normal granules, two bright filaments.

Between $12^h 2^m$ and $12^h 42^m$ the area of penumbral material around the spot becomes larger, but the two filaments disappear. Some granules can be seen within the penumbral region at $12^h 42^m$. The penumbra attains its greatest development around $12^h 42^m$ and thereafter dissolves quite rapidly. By $1^h 5^m$ only three or four curved lanes of dark material are present and these have disappeared by $1^h 29^m$, when the only trace of penumbral material remaining is a small dark spike directed towards the lower, right-hand corner of the photograph. The total lifetime of the transitory penumbra is less than three hours, during which time the umbra of the spot also undergoes many changes in shape and size.

These observations show very clearly that the fundamental process in the formation of a spot penumbra is the darkening of the *intergranular* material in the immediate neighbourhood of the umbra. For a while granules are still present in the region and appear more or less unchanged. Bright filaments appear only at a later stage of the development. Conversely, in the process of dissolution the bright filaments disappear first and then the whole penumbral region gradually reverts to undisturbed photosphere.

A similar phenomenon sometimes observed is the development of penumbral material between an existing spot and a nearby pore (or pores). In such cases the development again takes place by the formation of dark lanes in the intervening granulation. An illustration of a developing penumbra of this type is given in Plate 3.5 (*a*): the curved lanes of dark material in the region enclosed by the white rectangle form part of a growing penumbral area which ultimately linked the pores at the bottom and to the right of the marked rectangle to the neighbouring spot above. Sometimes, on the other hand, dark lanes linking a pore or group of pores to the penumbra of a well-developed spot are observed to disappear.

The whole process of the development of a penumbral region is reminiscent of the disturbance in the granulation in the neighbourhood of growing sunspot pores described in Section 3.3.4. This suggests that the formation of a penumbra is due to the penetration of part of the umbral magnetic field into the surrounding photosphere, resulting first in a darkening of the intergranular material and then in the replacement of the granules by thin bright filaments directed more or less radially outwards from the spot.

3.5.3   THE STRUCTURE OF WELL-DEVELOPED PENUMBRAE

The complexity of the structure of a well-developed penumbra depends partly on the size of the spot and partly on the stage of its evolution. The varying degree of complexity is well illustrated by Plate 3.9: the upper photograph shows an isolated elliptical spot with a fairly regular penumbral structure, while the lower photograph shows a very large spot with an extremely complex penumbra. In general, the most regular penumbral structure is observed in leader spots which have reached the stationary stage of their development and are accompanied by only a few (or no) minor attendants.

The filamentary detail illustrated in Plate 3.9 (a) is typical of regular penumbrae: the structure consists basically of a pattern of narrow bright filaments on a darker background. The more regular the spot, the more the filaments tend to run radially outwards from the umbra to the photosphere (cf. Secchi, 1875: p. 92). However, even in a regular spot like the one illustrated in Plate 3.9 (a) the filaments frequently branch and coalesce. By contrast, the filamentary detail of the spot shown in Plate 3.9 (b) is complicated and confused, except over limited regions near the top and bottom of the spot. There is no penumbra along portion of the upper, left-hand part of the umbra and, at other places, the filaments seem to run in more or less arbitrary directions. However, the inner ends of the filaments are always more or less at right angles to the border of the adjacent part of the umbra. Particularly prominent are the three long filaments located a little to the right of the centre of the photograph.

Even in fairly regular spots the clarity of the filamentary detail can vary to a marked extent not only from one spot to another but even around the periphery of a single spot. Both these effects are illustrated in Plate 3.10. The first photograph (Plate 3.10a) shows a large spot which is distinguished by the unusual clarity of its filamentary detail; even so, the filaments are appreciably better defined in the lower, right-hand quadrant than elsewhere in the penumbra. Plate 3.10 (b) is an enlargement of this region on twice the scale of Plate 3.10 (a). By contrast, the filamentary detail of the spot shown in Plate 3.10 (c) is diffuse and ill-defined. In fact, an examination of an enlargement of the lower, right-hand part of this spot (Plate 3.10d) shows that the penumbral structure has a blobby rather than a filamentary appearance. This is confirmed by other good-quality photographs of the spot, so that the lack of filamentary detail certainly cannot be attributed to mediocre seeing. The bearing of these observations on the interpretation of the Evershed effect is discussed in Section 4.4.3.

As we shall see in Section 3.5.4, the boundary between the umbra and penumbra of a spot is sharp and well-defined. Nevertheless, it is quite common for thin lanes of umbral material to penetrate deeply into the penumbra of a spot. In some cases these umbral projections even extend through to the photosphere outside the spot (cf. Plate 3.13*d*). In addition, small isolated areas of umbral material are very frequently found in spot penumbrae. Sometimes the presence of umbral material within a penumbra appears to influence the alignment of the neighbouring penumbral filaments: an example of this is shown in Plate 3.5 (*b*), where some of the filaments in the lower, left-hand part of the penumbra are seen to curve around towards an umbral projection. This region of the spot is shown on twice the scale of Plate 3.5 (*b*) in Plate 3.5 (*c*). The curvature of the filaments may be related to the magnetic field configuration in their neighbourhood.

In addition to penumbral filaments and dark umbral material, most spot penumbrae contain bright features of various shapes, whose brightness sometimes exceeds that of the neighbouring photosphere. Bright regions may occur anywhere in the penumbra and are often found at the border of the umbra, of an umbral projection, or of an isolated umbral area. Several bright features are, for example, present in the penumbra of the spot shown in Plate 3.9 (*a*), such as the diffuse blob at the top and the prominent bright streaks on the left-hand side. A detailed description of the bright features occurring in the penumbrae of sunspots is given in Section 3.5.6.

The boundary between the penumbra of a spot and the surrounding photosphere is exceedingly sharp. Nevertheless, the penumbra often has a serrated outline since the lanes of dark material between the filaments tend to project some distance into the photosphere; in some cases these have been observed to project for a distance equal to about one-third of the average width of the penumbra. A good illustration of the typical appearance of the outer boundary of a penumbra is afforded by the spot shown in Plate 3.3. In addition, note how this spot is linked to several nearby pores on the right-hand side by dark penumbral extensions; a better example of a pore being linked to a spot by dark penumbral material is shown in Plates 3.10 (*a*) and (*b*).

3.5.4 PROPERTIES OF THE PENUMBRAL FILAMENTS; DETAIL AT THE UMBRA–PENUMBRA BOUNDARY

Individual filaments are very narrow, often appearing to be less than 1″ of arc in *width*. In fact, the scale of the filamentary detail in sunspot penumbrae is even finer than that of the photospheric granulation (1–2″), so that it is correspondingly more difficult to photograph successfully. The present authors (Bray and Loughhead, 1958a) found that the apparent width of the filaments lies between 0″.5 and 1″.0, while Danielson (1961a) found that their apparent width on good-quality stratospheric photographs of sunspots is frequently only about 0".4—a figure comparable with the diffraction limit of his 12-inch telescope. The extreme narrowness of the penumbral filaments is well illustrated in Plate 2.3, Chapter 2, which is a reproduction of a stratospheric photograph published by Danielson. The *lengths* of individual filaments vary according to the size of the spot and the complexity of its penumbral structure: a representative value for larger spots is about 10" or 7500 km (cf. Plate 3.10b).

So far, the only systematic attempt to determine the *lifetimes* of the penumbral filaments is that made by the present authors (Bray and Loughhead, 1958a). Measurements were made on a number of unusually distinct filaments in the penumbra of the large spot shown in Plate 3.10 (a); an enlargement of the region of the penumbra selected for study, on twice the scale of Plate 3.10 (a), is given in Plate 3.10 (b). From films of this spot taken with the Sydney 5-inch photoheliograph on June 7, 1957, it was possible to select a sequence of 27 good-quality photographs fairly uniformly distributed over a period of 4 hours 38 minutes. The lifetimes of six particularly distinct filaments were estimated by carefully intercomparing maps of the filaments corresponding to different photographs in the sequence. The values obtained were $1^h 32^m$, $3^h 57^m$, $1^h 32^m$, 6^m, 47^m, and 24^m. These results clearly indicate that some filaments remain identifiable for periods of the order of hours. They are thus much longer-lived than the photospheric granules, whose lifetime is of the order of 10 minutes (cf. Section 3.2.7).

The filamentary structure of sunspot penumbrae has an important bearing on the interpretation of the *Evershed effect*. For example, the authors (Bray and Loughhead, 1958a) have suggested that the Evershed effect consists of a laminar flow of matter outwards from the umbra along the filaments, the latter probably being shallow structures of depth comparable to their width. It may be significant that, with a mean Evershed velocity of 1 km/sec, the time taken by matter to flow along the entire length of a filament is comparable with the observed lifetimes. The interpretation of the Evershed effect is discussed in Section 4.4.3.

No systematic study of the *evolution* of individual penumbral filaments has yet been made. Because of their narrow width and long lifetime, a sequence of photographs with a resolution of 0".5 of arc or better—extending over a period of at least several hours—would be required. For the moment, the possibility of studying changes in individual filaments must await the development of improved observational techniques.

The intensity of the penumbra usually decreases steadily inwards from the photosphere to the border of the umbra, where the value may be about 70 per cent of the photospheric intensity (see Section 4.2.7). Despite this, the filaments are found to end abruptly, so that the umbra–penumbra border is sharply delineated. However, in order to photograph the detail along the boundary it is usually desirable to use a somewhat greater exposure than normal. The typical appearance of the umbra–penumbra boundary on such an enhanced exposure is illustrated by the second photograph in Plate 3.16. The first photograph (Plate 3.16a) shows a large spot photographed with the Sydney

5-inch photoheliograph on September 19, 1958. The second photograph (Plate 3.16b), taken with twice the exposure of the first, shows the sharply-defined ends of the penumbral filaments on the right-hand side of the spot (the detail in a light-bridge visible on the original negative of Plate 3.16 (a) is also well shown).[9] Several bright umbral granules (cf. Section 3.6.3) are visible adjacent to the ends of some of the filaments, but detached from them. Examination of other films of sunspot umbrae taken with the Sydney photoheliograph has shown that these granules adjacent to the inner ends of the penumbral filaments are characteristic features of the umbra–penumbra boundary. They are much brighter than those further within the umbra; their existence was first pointed out by Secchi (1875: cf. p. 84).

It is important to realize that the apparent umbra–penumbra boundary of a spot on a *normal* photograph does not necessarily correspond to the physical boundary described above. This is due to the fact that photographic contrast, coupled with the steady decrease in intensity inwards from the photosphere, tends to make the area of the umbra on a normal photograph appear larger than it really is (cf. Chevalier, 1916: p. B6). This effect is well illustrated in Plate 3.16. The lower photographs were taken through a diaphragm whose mean position with respect to the spot is shown by the white circle in Plate 3.16 (a). It will be noticed that although the diaphragm includes the inner edge of the penumbra on Plates 3.16 (b) and (c), on Plate 3.16 (a) it appears to lie wholly within the umbra. This means that, unless proper precautions are taken, routine measurements of the umbral areas of sunspots derived from normal photographs may have little or no physical significance.

3.5.5 RELATIVE SIZES OF PENUMBRA AND UMBRA

The question of the relative contributions of the umbra and the penumbra to the total area of a spot has been discussed by a number of workers (Nicholson, 1933; Waldmeier, 1939; Jensen, Nordø, and Ringnes, 1955; Tandberg-Hanssen, 1956; Edwards, 1957; Jensen and Ringnes, 1957a, b). Most of this work has been based on statistical analyses of the data for the total and umbral areas of individual spots given in the *Greenwich Photoheliographic Results*. Waldmeier, however, made direct measurements of the penumbral and umbral diameters of a number of regular spots. Unfortunately, the results obtained by the various authors are rather inconclusive and, to some extent, contradictory. However, two results deserve mention: (a) the average value of the ratio of the area of the umbra to the total area of the spot is about 0·17, although the individual values vary widely from spot to spot; and (b) the average value of this ratio around sunspot maximum is somewhat smaller than the average value around sunspot minimum.

It seems doubtful whether any further statistical work along these lines would be warranted: as we have explained in the preceding section, routine

[9] This photograph was taken through a prime-focus diaphragm in order to reduce the effect of scattered light. The technique of umbral photography is described in Section 3.6.2.

measurements of the apparent umbral area of a spot have little meaning unless proper precautions are taken to ensure that the true umbra–penumbra boundary appears on the photograph.

3.5.6 BRIGHT REGIONS IN SUNSPOT PENUMBRAE

The penumbral structure of most spots is complicated by the presence not only of umbral projections and isolated regions of umbral material (cf. Section 3.5.3) but also of *bright features* of various kinds. Despite their diversity in brightness, size, and shape, it seems possible to divide the bright features into at least three distinct morphological types:

(1) diffuse bright blobs which occur wholly within the penumbra and show no obvious relationship with the umbra proper or other umbral material;

(2) bright streaks and loops which frequently, but not always, border the umbra or other umbral material within the penumbra; and,

(3) small bright regions of the type previously described by the authors (Bray and Loughhead, 1957), which invariably occur close to the border of the umbra or other umbral material.

Although the physical factors underlying this classification are not understood, we shall use it as a convenient means of cataloguing the available observational material.

Bright Features of Type 1. The diffuse blob-like regions are the most common of all bright features in spot penumbrae. They are found in practically all spots in greater or lesser numbers. The small diffuse blobs dotted throughout the penumbrae of the two spots illustrated in Plate 3.10 are typical examples; most of them are rather faint but a few are as bright as the surrounding photosphere. The diameters of most regions of this type are found to lie in the range 1–3″ of arc, but occasionally larger ones do occur, such as the large blob of bright material in the upper, right-hand part of the penumbra of the large spot illustrated in Plate 3.17 (*a*). The lifetimes of 20 of these features, located in the penumbra of a fairly large spot photographed in the central region of the disk, were measured by Macris (1953); the individual values ranged from 14^m to $1^h\ 3^m$, the mean lifetime being nearly 30 minutes.[10] Evidently, the bright regions, like the penumbral filaments (cf. Section 3.5.4), are longer-lived than the photospheric granules. The contrast of the bright regions increases towards the limb; this is well illustrated by Plate 7.1 and the Frontispiece, where some of these regions are seen to be as bright as the nearby facular granules.

Bright Features of Type 2. Bright streaks or loops are frequently found in spot penumbrae. In most cases they occur in association with umbral material in the form of a bright border to part of the umbra proper, to an umbral projection, or to an isolated region of umbral material within the penumbra. However, streaks and loops do occur which show no obvious connection with umbral material. It must be emphasized that although we describe the features as

---

[10] This result has been incorrectly quoted by de Jager (1959: cf. p. 151), Danielson (1961b), and Schröter (1962) as the lifetime of the penumbral *filaments*.

'streaks' or 'loops', it frequently happens that on the best photographs they are resolved into chains of individual bright blobs.

A typical example of a bright streak is shown in Plate 3.11 (a); it runs out from the boundary of the umbra across the penumbra and is bordered along half its length by an umbral projection. This streak shows a typical segmented appearance. It was observed to persist for at least one hour. Plate 3.11 (b) shows another example, in this case a loop of bright material bordering an isolated umbral area located in the penumbra of a spot. Despite changes of size and shape, this bright border was observed to persist for at least several hours. On some frames it is resolved into separate bright blobs the size of ordinary photospheric granules.

An extraordinary example of a long bright streak showing no obvious connection with umbral material is visible in the upper, left-hand region of the spot illustrated in Plate 3.17 (a) (Bray and Loughhead, 1959: cf. footnote on p. 323). The region around the streak is reproduced on a somewhat greater scale in Plate 3.11 (c), from which it can be seen that the streak has a sharp boundary on the side facing the umbra and a diffuse boundary on the other side. Other photographs of the same spot taken on the same day indicate that this feature persisted for at least three hours. An attempt to relate the streak to any overlying chromospheric feature was unsuccessful since the resolution of the available Hα photographs was inadequate for the purpose of comparison.

Bright loops are also observed in spots near the limb. One example is shown in Plate 7.1, where a bright loop can be seen in the upper part of the penumbra of the main spot. This loop is as bright as some of the faculae outside the spot and is evidently composed of a chain of small bright blobs.

Plate 3.12 illustrates another phenomenon occasionally observed, namely the presence of a bright streak bordering part of the umbra of a spot (cf. Strebel, 1932b; Ananthakrishnan, 1951; Das and Ramanathan, 1953). On both photographs a bright streak is visible along the lower central border of the umbra. At 10^h 21^m the bright streak is resolved into a number of individual bright segments and appears to be connected at either end to light-bridges across the spot.[11] Although the shape of the umbra changes in the intervening period, the bright border is still present at 3^h 1^m. A photograph obtained by Rösch showing an even more prominent bright border to part of the umbra of a spot has been published by de Jager (1959: cf. Fig. 35b). Faint bright streaks are also sometimes seen bordering the umbrae of spots near the limb; in such cases they always seem to be located on the *limb* side of the umbra.

The existence of bright regions along the umbral borders of spots was reported by Secchi (1875: cf. pp. 82–3) who, however, erroneously concluded that an inner 'bright ring' around the border of the umbra was a characteristic feature of sunspot penumbrae. As we have stated above, bright streaks occur only rather

---

[11] Other cases of apparent connections between light-bridges and bright material bordering the umbrae of spots have been recorded, but the physical significance, if any, of this relationship is not understood.

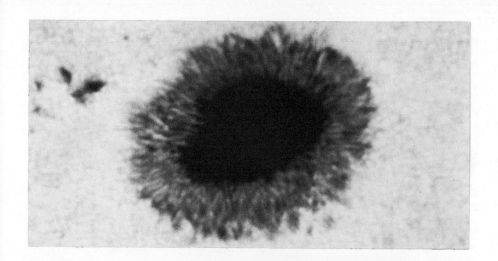

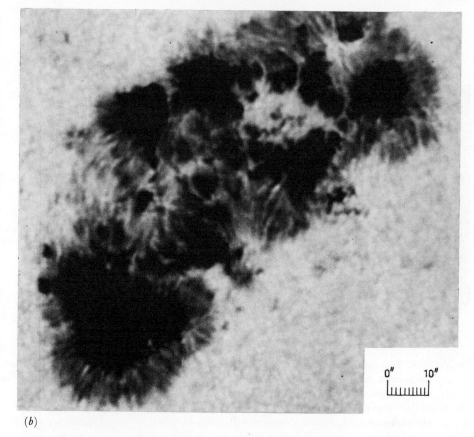

(b)

PLATE 3.9. The varying degree of complexity of sunspot penumbrae.

(a) Isolated elliptical spot (October 4, 1957) showing a fairly regular penumbral structure, which consists basically of a pattern of narrow bright filaments on a darker background. The filaments run more or less radially outwards from the umbra to the photosphere, but frequently bifurcate and coalesce.

(b) Large spot (December 2, 1957) showing a very irregular penumbral structure. There is no penumbra along portion of the upper, left-hand part of the umbra and, at other places, the filaments seem to run in more or less arbitrary directions.

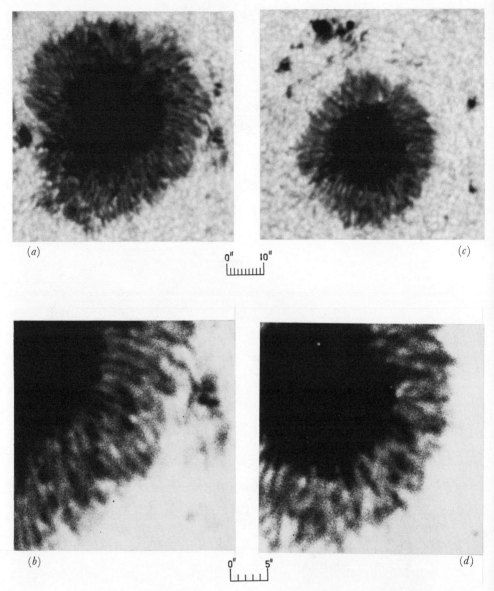

PLATE 3.10. The varying degree of clarity of the filamentary structure in sunspot penumbrae.

(*a*) Large spot showing unusually clear filamentary detail (June 7, 1957); even so, the filaments are appreciably better defined in the lower, right-hand quadrant than elsewhere. Note the dark penumbral material linking the spot to the pore on the right-hand side.

(*c*) Medium-sized spot showing diffuse and ill-defined filamentary detail (February 3, 1959). The penumbral structure has, in fact, a blobby rather than a filamentary appearance.

(*b*) and (*d*) Enlargements of portions of (*a*) and (*c*) respectively.

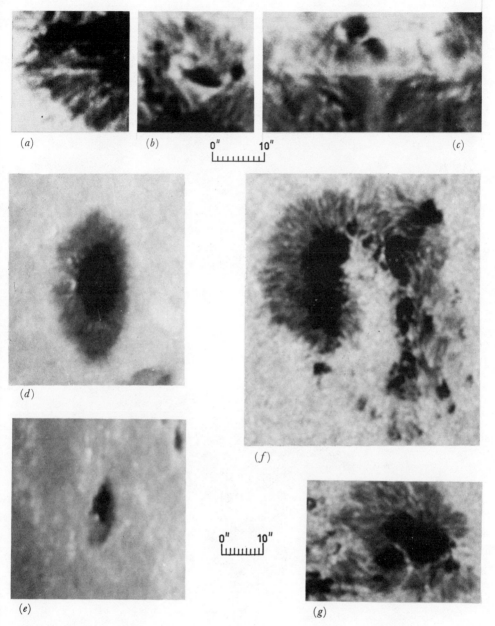

PLATE 3.11. (a)–(c) Bright regions of Type 2 (Section 3.5.6). The long bright streak in (c) is sharp on the side facing the umbra, diffuse on the other.

(d)–(e) Bright regions of Type 3 bordering sunspot umbrae near the limb (May 20, 1957 and June 7, 1960). In (d) the more prominent one is brighter than the surrounding photosphere, while in (e) the brightness is comparable to that of nearby facular granules.

(f) Bright streamers crossing a spot umbra (May 31; 1957); these link the penumbra at the top of the spot to a large invading tongue of bright material, almost indistinguishable from ordinary photosphere. Note the segmented appearance of some of the streamers.

(g) Bright streamer crossing the lower, left-hand portion of a spot umbra (June 6, 1957); it consists of four bright blobs.

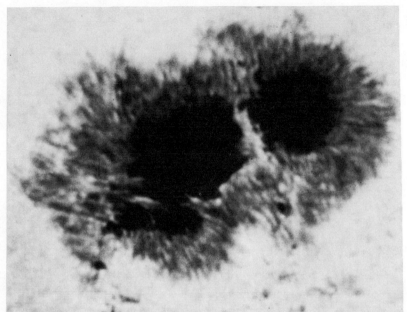

10ʰ21ᵐ

0″   10″

3ʰ1ᵐ

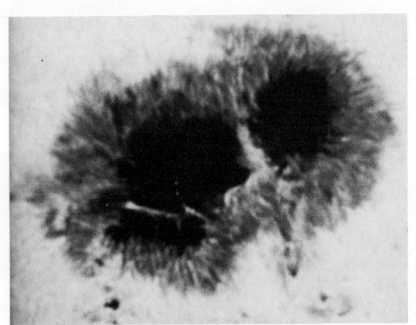

PLATE 3.12. Bright streak bordering the lower, central part of a spot umbra (January 14, 1958). At 10ʰ 21ᵐ it is resolved into a number of individual bright segments. It is connected at both ends to light-bridges and is still present at 3ʰ 1ᵐ. The main light-bridge shows little change.

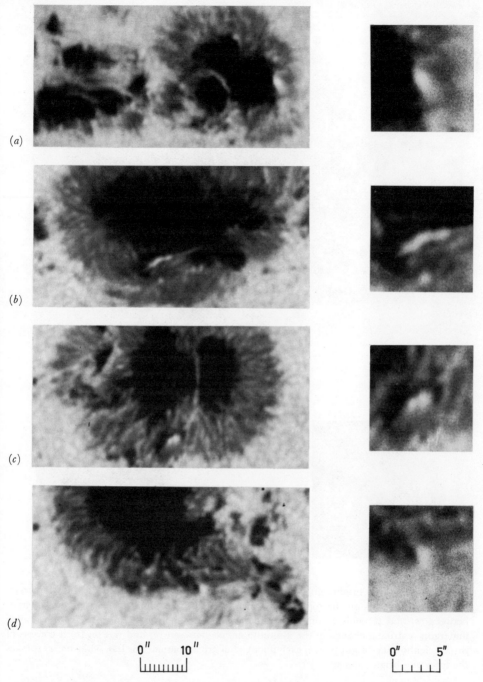

PLATE 3.13. Bright regions of Type 3 (Section 3.5.6): June 6 and September 13, 1957; February 26 and 4, 1958. In (a) and (b) the bright regions border the umbra proper, whereas in (c) and (d) they border umbral material located in the penumbrae; each region is brighter than the surrounding photosphere. Enlargements are shown to the right.

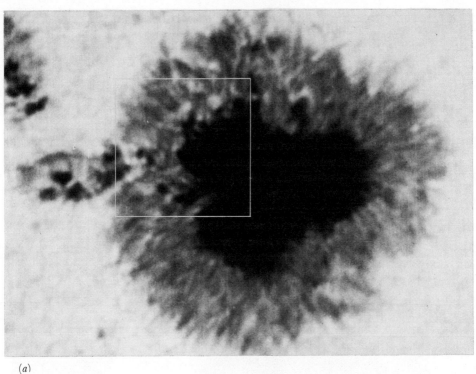

(a)

0″   10″

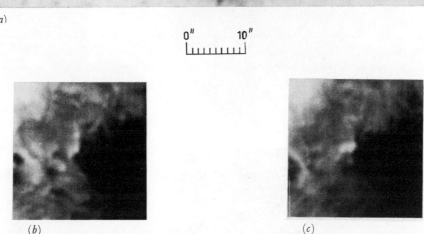

(b)   (c)

PLATE 3.14. Sudden brightening of a Type 3 bright region (Section 3.5.6): May 9, 1957. The feature develops on the top, left-hand rim of the umbra during the 75-minute interval between (a) and (b), only a faint diffuse illumination being initially present. Its appearance undergoes a striking change in the 4-minute period between (b) and (c): on (b) it consists of three neighbouring bright points, each about 1″ of arc in diameter or less, while on (c) it takes the form of a bright elongated jet.

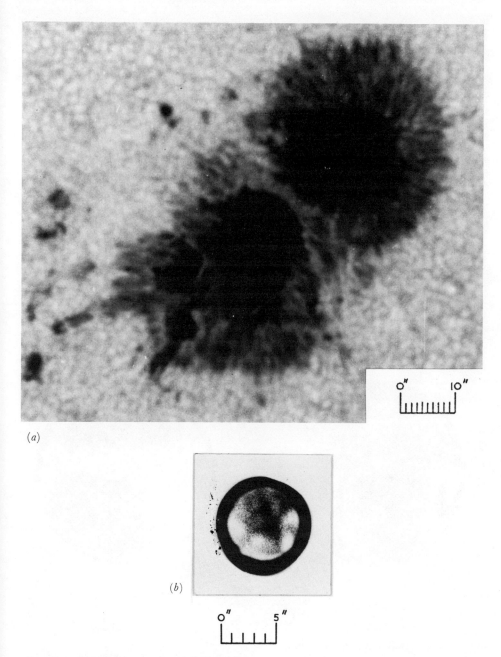

PLATE 3.15. Detail in a sunspot umbra.

(a) Sunspot photographed at 1ʰ 36ᵐ on September 1, 1958. Note the granular appearance of the light-bridge across the middle of the spot.

(b) Fourfold overexposure taken at 11ʰ 44ᵐ through a small diaphragm centred on the upper portion of the umbra. Several umbral granules are distinctly shown, including three faint ones at the core of the umbra. The scale of (b) is twice that of (a).

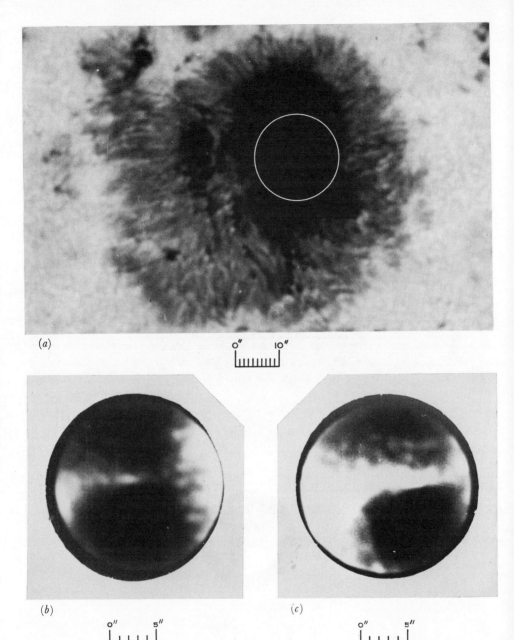

(a)

0″ ⎣⎨⎨⎨⎨⎨⎨⎨⎦ 10″

(b)

0″ ⎣⎨⎨⎨⎨⎦ 5″

(c)

0″ ⎣⎨⎨⎨⎨⎦ 5″

PLATE 3.16. Detail in a sunspot umbra.

(a) Sunspot photographed at 12ʰ 25ᵐ on September 19, 1958. On the original negative a faint light-bridge can be seen across the centre of the umbra.

(b) Twofold overexposure taken at 10ʰ 39ᵐ through a diaphragm. Note the discrete structure of the light-bridge and the bright granules at the umbra–penumbra boundary.

(c) Threefold overexposure taken at 11ʰ 19ᵐ. The part of the umbra above the light-bridge is brighter than that below, and shows clear umbral granulation. The light-bridge is burnt out with this exposure.

Note: the approximate position of the diaphragm relative to the spot is shown by the white circle. The scale of (b) and (c) is twice that of (a).

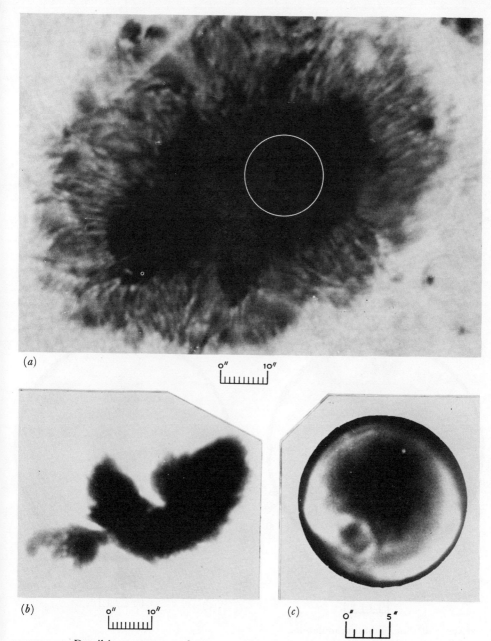

(a)

0″ |⊔⊔⊔⊔⊔⊔⊔| 10″

(b)

0″ |⊔⊔⊔⊔⊔⊔⊔| 10″

(c)

0″ |⊔⊔⊔⊔| 5″

PLATE 3.17. Detail in a sunspot umbra.

(a) Sunspot photographed at 2^h 40^m on January 12, 1959. The prominent bright streak in the penumbra (upper, left) is shown on an enlarged scale in Plate 3.11 (c).

(b) Threefold overexposure taken at 12^h 34^m without a diaphragm. With this exposure the penumbra and outer parts of the umbra are burnt out and appear white. Note that the apparent umbral outline differs from that in (a); a few umbral granules can be seen in the far left-hand portion.

(c) Sevenfold overexposure taken at 2^h 10^m through a diaphragm, whose location is shown by the white circle in (a). Note the bright 'loop' structure (lower, left) and the gradual decrease in intensity towards the core of the umbra. Granulation fails to appear owing to mediocre seeing. The scale of (c) is twice that of (a) and (b).

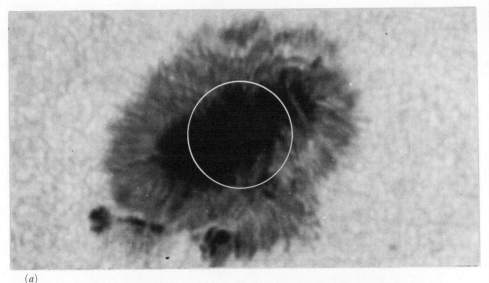

(a)

0″ 10″

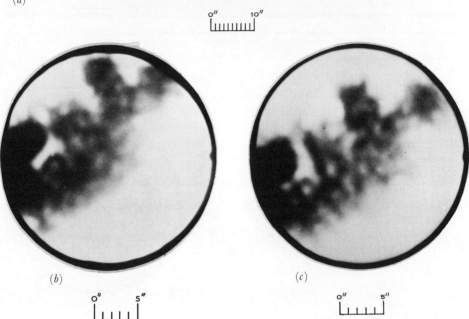

(b)   (c)

0″ 5″   0″ 5″

PLATE 3.18. The umbral granulation.

(a) Sunspot photographed at 2ʰ 13ᵐ on May 11, 1959.

(b) Threefold overexposure taken at 11ʰ 46ᵐ through a diaphragm. Granulation appears over a large part of the umbra. Note the bright 'fork' structure at the left.

(c) Fivefold overexposure taken at 1ʰ 6ᵐ. The umbral granules appear brighter with this exposure which, however, is still insufficient to reveal detail in the far left-hand portion.

Note: the approximate position of the diaphragm relative to the spot is shown by the white circle. The scale of (b) and (c) is twice that of (a).

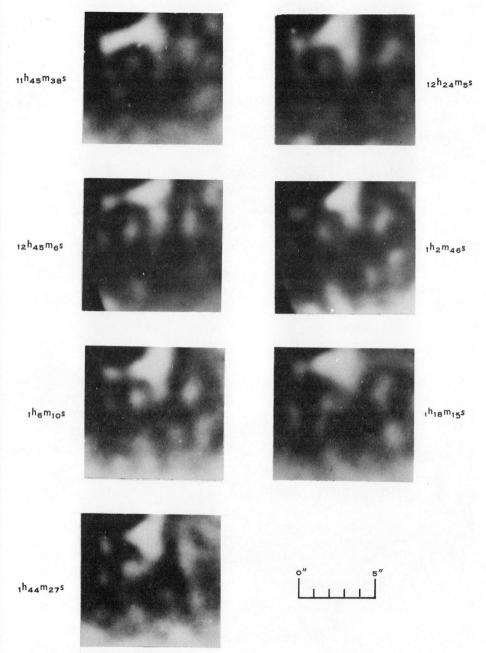

PLATE 3.19. Stability of the umbral granulation pattern. The first two photographs are three-fold overexposures, the rest are fivefold overexposures. Despite apparent changes due to seeing (e.g., the tail-like projections seen on several granules at $1^h 18^m 15^s$), some umbral granules can readily be followed over the entire sequence—a period of 2 hours. Note the granule which becomes detached from the upper prong of the 'fork' structure, and also the bright granule which ultimately attaches itself to the lower prong.

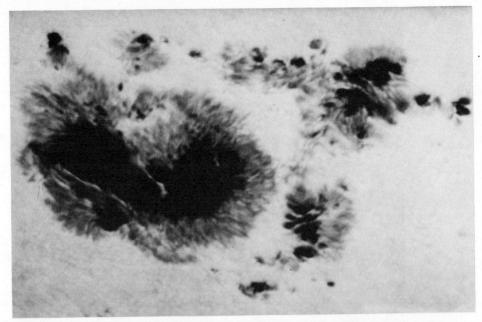

(a)

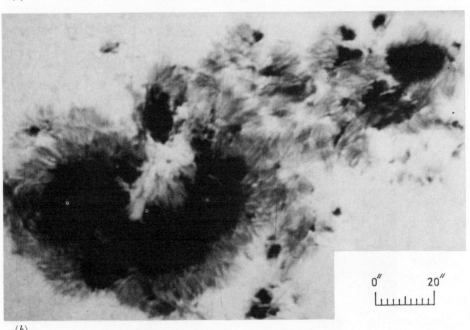

(b)

PLATE 3.20. Development of a light-bridge.

(a) Sunspot group photographed fairly close to the east limb on October 15, 1957. The umbra of the main spot is crossed by several long bright streamers which, in places, show a segmented appearance.

(b) Appearance some 52 hours later. A large tongue of bright material has now invaded the main umbra, but does not bridge it until a further 4 days have elapsed (cf. Plate 3.21).

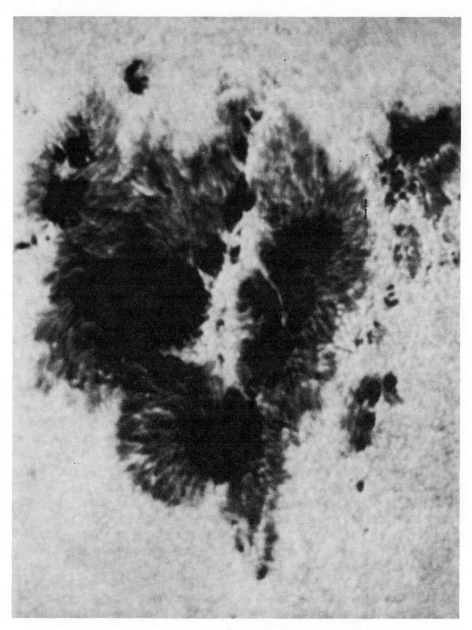

0″   10″

PLATE 3.21. Development of a light-bridge (continued). The spot of Plate 3.20 some 4 c
after Plate 3.20 (*b*). The light-bridge now extends practically right across the umbra. I
brighter than the photosphere outside the spot and shows a mottled structure similar to
photospheric granulation. Lanes of bright material link the bridge to the penumbra on
left- and right-hand sides.

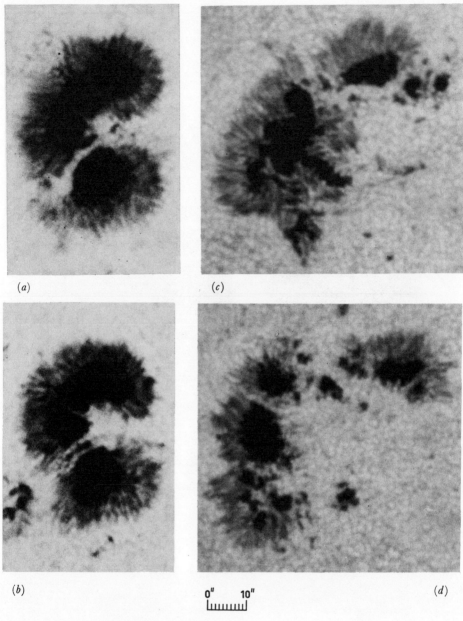

(a)   (c)

(b)   (d)

0"   10"

PLATE 3.22. Development of light-bridges.

(a) Medium-sized spot crossed by a fairly thick light-bridge (December 11, 1957). Note the incipient bridge just above the first, and above this a faint streamer running into the upper part of the umbra.

(b) Same spot some 24 hours later. The blob of bright material on the right has increased in size, and the second bridge now extends right across the umbra. Both bridges show a segmented appearance. The thin streamer is still present, although somewhat altered in shape.

(c) Medium-sized spot crossed by a complicated light-bridge (February 26, 1958). The bridge is connected to the photosphere on the right-hand side and has a granular structure.

(d) Same spot some 26 hours later. It has now divided into two parts separated by a wide strip of photospheric granulation, which has replaced the light-bridge visible in (c). Note the development of new light-bridges in the left-hand component of the original spot.

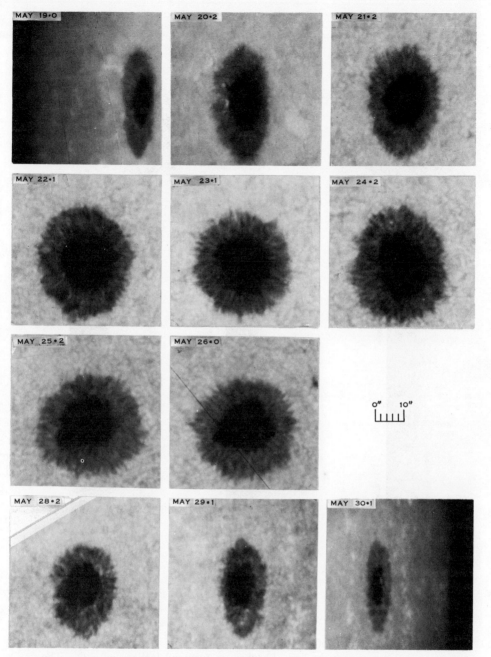

PLATE 3.23. The Wilson effect in a small regular sunspot. The sequence shows the changing appearance of the spot during its passage from the east to the west limb (May 19–30, 1957). Each photograph is orientated so that the direction of greatest foreshortening is horizontal. It is evident that, when the spot is near the limb, the apparent width of the penumbra on the side remote from the limb is less than that on the other side.

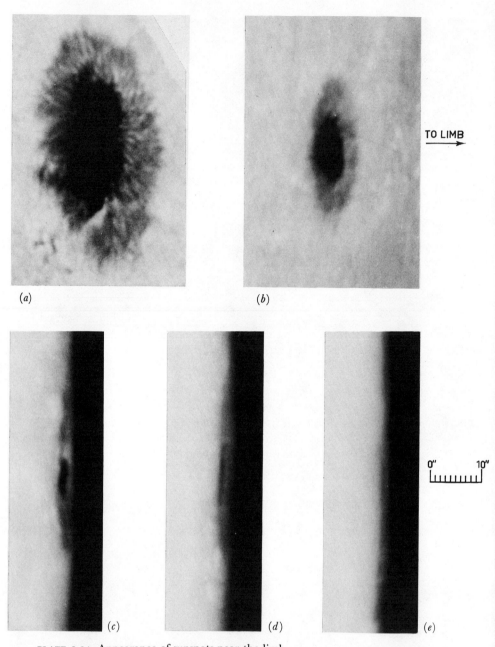

TO LIMB →

0″ 10″

(a)　　　　(b)

(c)　　　　(d)　　　　(e)

PLATE 3.24. Appearance of sunspots near the limb.

(a)–(b) Sunspots near the limb (October 15, 1957 and January 11, 1960). The umbra–penumbra boundary is diffuse on the side remote from the limb, sharp on the other. In both cases, the penumbra on the side remote from the limb is abnormally dark. Both spots show the Wilson effect.

(c)–(d) Sunspots at the extreme limb (January 15 and 6, 1960). In both cases, the distinction between umbra and penumbra is maintained despite the close proximity to the limb. Wilson effect is very marked.

(e) Same spot as (d) some 70 minutes later; it is now crossing the limb and only a slight trace of umbra remains. Note that there is no depression at the limb.

infrequently along the borders of sunpot umbrae and then only around limited arcs.

Bright Features of Type 3. In 1957 the authors drew attention to the existence of small bright regions at the borders of sunspot umbrae or umbral material, which sometimes seem to undergo big changes in periods as short as a few minutes or so (Bray and Loughhead, 1957). An examination of a large number of subsequent films has shown that these regions occur in most spots at least at some stage of their lives. They are distinguished from Type 1 regions by (a) their generally smaller size (many appear to be only 1″ of arc in diameter and may even be smaller), (b) their generally greater brightness, which often exceeds that of the neighbouring photosphere, and (c) their invariable location close to the border of the umbra or an umbral region.

Plate 3.13 illustrates some typical examples of these features. The first photograph (Plate 3.13a) shows a very bright region located on the right-hand border of the umbra of a small spot; the brightness of this region is decidedly greater than that of the surrounding photosphere. The next photograph (Plate 3.13b) illustrates a bright region on the lower border of the umbra of a somewhat larger spot. The occurrence of two neighbouring, small bright features at the border of an isolated area of umbral material in the penumbra of a spot is shown in Plate 3.13 (c). These two points are clearly brighter than the photosphere outside the spot. Finally, Plate 3.13 (d) illustrates a bright region of the same type located at the end of an umbral projection at the outer boundary of the penumbra of a spot. This region is also brighter than the surrounding photosphere.

Bright regions of Type 3 are also observed in spots near the limb. Plate 3.11 (d) shows two bright features on the border of the umbra of such a spot; the more prominent of these is considerably brighter than the surrounding photosphere at the same heliocentric angle. Another example is illustrated in Plate 3.11 (e); in this case, the brightness is comparable with that of the nearby facular granules.

Although bright features of Type 3 are usually fairly long-lived and, in fact, are sometimes observed to persist for several hours, they occasionally appear to undergo striking changes in periods as short as a few minutes or so. An example of this, previously described by the authors (Bray and Loughhead, 1957), is illustrated in Plate 3.14. On the first photograph (Plate 3.14a) the top, left-hand rim of the spot umbra is seen to be bordered merely by a faint diffuse illumination. The second photograph (Plate 3.14b) shows, bordering the umbra in the same vicinity, a bright feature which developed in the period between the two photographs (75 minutes). The appearance of this bright feature undergoes a striking change in the 4-minute period between Plates 3.14 (b) and (c): on the former it appears to consist of three neighbouring bright points, each about 1″ of arc in diameter or less, while on the latter it takes the form of a bright elongated area resembling a jet.

The invariable location of Type 3 bright regions close to the border of the umbra or other umbral material—places where there is probably a large gradient in the spot magnetic field—indicates the importance of obtaining improved observations of these features.[12]

3.6 The Structure of Sunspot Umbrae

3.6.1 INTRODUCTION

The existence of a bright granular structure in the umbrae of sunspots was first reported by Chevalier (1916: p. B10), and subsequently umbral granules were observed visually by Thiessen (1950) with the 60 cm Hamburg refractor. The first successful attempt to photograph the umbral granulation was made by Rösch (1956, 1957), who found that the umbral granules are smaller than the photospheric granules but was unable to estimate their lifetimes. Much improved photographs of the umbral granulation were subsequently obtained by the present authors as the result of a systematic programme of umbral photography carried out with the Sydney 5-inch photoheliograph (Bray and Loughhead, 1959; Loughhead and Bray, 1960a). For the first time it was clearly demonstrated that the umbral granulation forms a cellular pattern, similar in appearance (though not in properties) to the photospheric granulation; quantitative information about the lifetime and cell size was obtained. In this work special precautions were taken to ensure that the contrast of the faint umbral detail was not reduced by scattered light from the penumbra and the surrounding photosphere.

The presence of granulation in the umbrae of sunspots shows that the basic convective processes responsible for the photospheric granulation also operate in sunspots, although until recently it was believed that a sunspot magnetic field would suppress the convection. The observations described below indicate that an understanding of the nature of the interaction between solar convection currents and magnetic fields may help to elucidate the actual process of spot formation.

3.6.2 TECHNIQUE OF UMBRAL PHOTOGRAPHY

The observations of the fine detail in sunspot umbrae described in the following sections were obtained by the authors with the Sydney 5-inch photoheliograph (Bray and Loughhead, 1959; Loughhead and Bray, 1960a). The techniques employed were necessarily rather more specialized than those used in normal sunspot photography (cf. Section 2.3.1). In the first place, it is necessary to give an exposure greater than that required for the penumbra or surrounding photospheric granulation. The necessary increase in exposure depends both on

---

[12] Kiepenheuer (1960) has published good-quality photographs showing some of the small bright regions typical of the various types of bright structures commonly found in sunspot penumbrae. He identifies these features with flux tubes of a deep-seated magnetic field which have penetrated the visible layers; however, they do not appear to be related in any way to the phenomenon described in Section 3.3.4.

the size of the spot and on position in the umbra. Exposure factors ranging from two to seven were used; the latter value is normally sufficient for recording detail in the darkest region of the umbra of a very large spot. Apart from the removal of a 50 per cent neutral filter normally present in the light path of the photoheliograph, the increases in exposure were achieved by using longer exposure times.

Secondly, in the absence of any precautions, the contrast of the faint umbral detail would be reduced by scattered light from the penumbra and surrounding photosphere. To eliminate scattered light from the magnifying lens and two auxiliary prisms which divert the beam to the camera, prime-focus diaphragms were constructed, with diameters ranging from 0·06 to 0·22 mm. This range is adequate for the larger umbrae encountered on the 16 mm primary solar image of the photoheliograph. The diaphragms were made of thin aluminium foil, the holes being pierced with a fine burnished needle. Any selected diaphragm could be mounted on a small metal box containing a central aperture. This box was carried on a Hilger micrometer microscope stage, which provided fine motions in two directions at right angles to the optical axis, thus permitting the diaphragm to be positioned on any part of a spot umbra. In addition, motion of the box along the optical axis allowed the diaphragm to be brought into the focal plane; to reduce heating, air was sucked through perforations in the front surface of the box (cf. Section 2.3.1). Any remaining scattered light is due to scattering in the Earth's atmosphere and at the 5-inch objective. Atmospheric scattering is usually low at the telescope site, some 30 miles west of Sydney. In addition, the objective is a cemented doublet—a type of lens inherently free from scattered light; it was cleaned regularly during the course of this work.

The diaphragm selected for any given spot depended on the shape and size of the umbra. If this was circular, a diaphragm was usually selected which just included the inner edge of the penumbra as seen in the telescope eyepiece (the umbra appears somewhat smaller on a visual image than on normal sunspot prints such as Plates 3.15 (a), 3.16 (a), and 3.17 (a)). However, a smaller diaphragm was used when only the detail at the darkest point was required. If the spot was elliptical, a diaphragm was selected which just included the inner edge of the penumbra at the ends of the minor axis.

For a large umbra, such as that illustrated in Plate 3.17, the observing procedure was as follows: the spot was photographed without a diaphragm every 5 seconds for a period of 5 minutes, using an exposure factor of three. This sequence served to reveal the structure in the neighbourhood of the umbra–penumbra boundary and, in the subsequent examination of photographs taken through a diaphragm, helped to locate the position of the diaphragm relative to the spot. The appropriate diaphragm was then inserted and a 20-minute sequence with an exposure factor of five was obtained in order to show the structure of the outer umbra. Retaining the diaphragm, a further 20-minute sequence was obtained, using an exposure factor of seven; this sequence served to reveal the structure of the core of the umbra. Finally, a repetition of the first sequence was followed by a 15-minute sequence of normal photographs and a photometric calibration. A similar procedure was used for smaller umbrae, such as that illustrated in Plate 3.15, except that exposure factors ranging from 2 to 5 instead of from 3 to 7 were employed.

Towards the end of the observing programme, the procedure was modified in order to obtain longer sequences of overexposures suitable for determining the lifetime of the umbral granules. For this purpose the duration of each of the two sequences of overexposures taken through a diaphragm was increased to one hour.

3.6.3 THE DETAIL IN SUNSPOT UMBRAE; UMBRAL GRANULATION

During the period September 1958 to May 1959 thirteen films of seven different sunspots of various sizes were obtained with the Sydney 5-inch photoheliograph, using the technique of umbral photography described in the previous section. The typical umbral detail revealed by these observations is illustrated in Plates 3.15–3.17.

Plate 3.15 (a) is a normal photograph of a fairly large spot and the surrounding photospheric granulation. The umbra is divided into two parts by a light-bridge (cf. Section 3.7), which itself shows some granular structure. Plate 3.15 (b) shows, on twice the scale, a fourfold overexposure taken through a diaphragm centred on the upper part of the umbra. A number of umbral granules can be distinguished, including three faint ones at the dark core of the umbra. Measurements show that they are more closely packed than the photospheric granules (see Section 3.6.4).

Plate 3.16 (a) is a normal photograph of a somewhat larger spot; on the original negative a faint light-bridge can be seen across the centre of the umbra. Plates 3.16 (b) and (c) show twofold and threefold overexposures respectively, roughly centred on the light-bridge; the scale of these photographs is twice that of Plate 3.16 (a). As shown in Plate 3.16 (b), the light-bridge consists of bright segments, an elongated portion near the middle being particularly prominent. On the right-hand border of the umbra there are several bright granules, which are detached from the penumbral filaments and are much brighter than any further within the umbra. Examination of films of other spots has shown such granules to be characteristic features of the umbra–penumbra boundary (cf. Section 3.5.4).[13] In Plate 3.16 (c) the centre of the light-bridge is overexposed but its 'wings' begin to appear. Above the light-bridge a number of umbral granules are now visible. The exposure is insufficient to show any granulation in the lower portion of the umbra (which the negative shows to be darker than the upper) or in the central region of the upper part; overexposures greater than threefold are not available for this spot. It is clear that there are two points of minimum intensity separated by a 'saddle' which, on an overexposure, appears as a light-bridge. This implies that when a light-bridge is present in a spot, it is an integral

---

[13] They should not be confused with the bright regions sometimes found on the *penumbra* side of the umbra–penumbra boundary (cf. Section 3.5.6: Bright Features of Type 3).

part of the umbral structure and is not, despite its name, merely a superficial overlying feature (cf. Section 3.7.1).

Finally, Plate 3.17 (a) shows a normal photograph of the main spot of one of the largest groups which appeared during the sunspot maximum of 1957–8. On the same scale Plate 3.17 (b) shows a threefold overexposure made without a diaphragm. The outline is essentially an *isophote* of the umbra; it differs considerably from the outline of the umbra shown on the normal exposure. In addition, the far left-hand region of the umbra is now seen to be brighter than the rest; it shows several umbral granules, whereas only one particularly bright granule is visible in the main portion. Plate 3.17 (c) is a sevenfold overexposure taken through a diaphragm roughly centred on one of the points of minimum intensity; it is reproduced on twice the scale of Plates 3.17 (a) and (b). The umbral granulation is only faintly visible on the negative, probably owing to mediocre seeing, and fails to appear on the print. However, there is one striking feature, a bright *loop* which appears to consist of four elongated granules. Similar features have been seen in other umbrae, e.g., the bright 'fork' in the spot illustrated in Plate 3.18.

As illustrated in Plates 3.15–3.17, the umbra of a large spot often possesses several intensity minima, each pair being separated by a saddle which, if bright enough, appears on a normal exposure as a light-bridge. Thus, quite apart from the fine detail, the intensity contours of spot umbrae show a detailed and complex structure (cf. Chevalier, 1916: p. B10). The relationship of this structure to the umbral magnetic field distribution is discussed in Section 5.4.9. A detailed discussion of sunspot intensity contours is given in Section 4.2.7.

Granulation is also observed in the umbrae of small spots and in pores only a few seconds of arc in diameter; in such cases the granulation often appears on negatives taken with a normal exposure. An example is shown in Plate 3.7. The normal appearance of the pore and the surrounding photospheric granulation is illustrated by Plate 3.7 (c), while Plate 3.7 (d) shows umbral granules within the pore. (The scale of Plate 3.7 (d) is somewhat greater than that of Plate 3.7 (c).) Although the upper part of the pore is somewhat smeared by mediocre seeing, several umbral granules are clearly visible. On the original negative one faint umbral granule can be seen in the lower right-hand portion, which is darker than the rest; however, this granule fails to appear in the reproduction. The sharpness of the boundary between the pore and the surrounding granulation is particularly noteworthy: umbral granules within the pore exist side-by-side with ordinary photospheric granules outside. This suggests that an interaction takes place between sunspot magnetic fields and *individual* convection columns (cf. Section 3.4.4).

3.6.4 PROPERTIES OF THE UMBRAL GRANULATION: CELL SIZE, LIFETIME, AND EVOLUTION

The observations described in the previous section show that, in addition to the other fine detail in sunspot umbrae, umbral granulation is invariably present.

The umbral granules, like the photospheric granules, form a well-defined cellular pattern; this is very clearly illustrated in Plate 3.18 and Fig. 3.2.[14] The first photograph (Plate 3.18*a*) shows a fairly large spot and the surrounding photospheric granulation. Plate 3.18 (*b*) shows, on twice the scale, a threefold overexposure taken through a diaphragm at the prime focus. Granulation appears over a large part of the umbra though none is visible in the left-hand portion, which is considerably darker than the rest. Some of the granules are so close together that they can barely be distinguished. Plate 3.18 (*c*) shows a fivefold overexposure taken through the diaphragm. The umbral granules

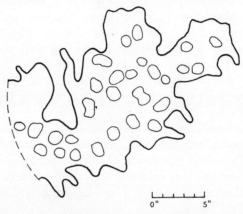

Fig. 3.2. Umbral granulation pattern. The map shows the 28 umbral granules visible on a good-quality fivefold overexposure of the spot illustrated in Plate 3.18 (*a*). Note how the umbral granules form a cellular pattern similar to that of the photospheric granules. The heavy line represents the apparent outline of the umbra.

now appear brighter but the exposure is still insufficient to reveal any detail in the left-hand portion.

The cellular appearance of the umbral granulation is even more evident in Fig. 3.2, which is a map of the 28 umbral granules visible on a very good-quality photograph of the same spot taken with a fivefold overexposure. (Note: some of the granules near the edge of the umbra, drawn in Fig. 3.2, cannot be distinguished on Plate 3.18 owing to the difficulty of recording objects of different brightness with high contrast on a single reproduction.) The *cell size* of the umbral granulation pattern was derived from measurements of this map (Loughhead and Bray, 1960a). The distribution of the 52 intergranular distances measured is given in Fig. 3.3, which also shows the cell size distribution of the photospheric granulation (cf. Section 3.2.3), derived from a good-quality photograph taken with the same instrument. It is evident that the cell size of the umbral granulation is significantly less than that of the photospheric granu-

[14] de Jager (1959: cf. p. 151) has incorrectly implied that the umbral granules tend to occur as isolated points.

lation, the mean cell sizes being 2″3 and 2″9 of arc respectively. Concordant
values for the mean cell size of the umbral granulation were obtained from
measurements made on the smaller groups of granules visible in the umbrae
of the spots illustrated in Plates 3.15 (b) and 3.16 (c) (Bray and Loughhead,

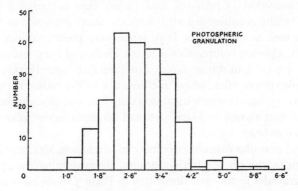

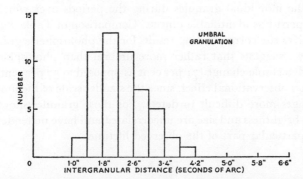

Fig. 3.3. Cell sizes of the umbral and photospheric granulation patterns. The histograms
give the distributions of the distances between the centres of adjacent granules (corrected
for foreshortening), measured for groups of 28 umbral and 92 photospheric gránules.
It is evident that the cell size of the umbral granulation is significantly less than that of
the photospheric granulation; the mean cell sizes are 2″3 and 2″9 of arc respectively.

1959). In the case of both the photospheric and umbral granulation, observations of higher resolution would probably yield more symmetrical cell size
distributions, with some separations of less than 1″.

A determination of the *lifetime* of the umbral granules was made by the present
authors using two consecutive sequences of photographs of the spot shown in
Plate 3.18, each covering a period of nearly an hour (Loughhead and Bray,
1960a). The umbral granules appear quite clearly on both sequences, one of
threefold overexposures and the other of fivefold overexposures. Together, the
two sequences allowed the granules to be studied over a total period of 2 hours

11 minutes. Some of the brighter granules are also visible on a 10-minute sequence of twofold overexposures taken shortly before. The results of the measurements, which refer to 48 granules in all, are as follows: 67 per cent of the granules had lifetimes exceeding 15 minutes, 38 per cent exceeded 30 minutes, 27 per cent exceeded 45 minutes, and 10 per cent exceeded 2 hours. Three particularly bright granules are visible on the short sequence of twofold over-exposures as well as on the two long sequences; these granules lasted for at least 2½ hours, a period comparable with the lifetime of the penumbral filaments (cf. Section 3.5.4). Umbral granules are therefore considerably longer-lived than photospheric granules, whose lifetimes are of the order of 10 minutes (cf. Section 3.2.7). In agreement with this result, four out of eight granules in the umbra of the spot shown in Plate 3.16 had lifetimes in excess of half an hour (Bray and Loughhead, 1959).

The umbral granules exhibit a diversity in brightness, size, and shape similar to that shown by the photospheric granules. By examining the appearances of the umbral granules in the spot illustrated in Plate 3.18 on each photograph in the two sequences referred to above, an attempt was made to study the *evolution* of the individual granules. Table 3.3 gives the results of the attempt to detect changes in the individual granules during the periods over which they were observed to persist as identifiable objects. Comparison of Table 3.3 with Table 3.2, which gives the corresponding results for the photospheric granulation (cf. Section 3.2.7), suggests that rather more umbral than photospheric granules undergo no detectable change (71 per cent as opposed to 57 per cent). However, this may be an observational effect, since the smaller scale of the umbral pattern renders changes more difficult to detect. For those granules showing change, variations in brightness and size are uncorrelated and have no tendency to occur during any particular part of the observed lifetime.

TABLE 3.3

Changes in Umbral Granules

TYPE OF CHANGE		NO. OF GRANULES
No change		34
Brightness	increase	4
	decrease	4
Size	increase	1
	decrease	4
Change of shape		5

Plate 3.19 illustrates the general *stability* over a period of nearly two hours of the umbral granulation in the spot shown in Plate 3.18. Although there are apparent differences from one photograph to another due to seeing, some granules can easily be followed over the entire sequence. (The 'tails' attached to some of the granules at 1^h 18^m 15^s have no real existence but are due to poor seeing on this photograph.) Note the granule which becomes detached from

the upper prong of the 'fork' structure, also the very bright granule which ultimately attaches itself to the lower prong.

Changes in the umbral granules are more difficult to detect during their periods of formation and decay, when the granules cannot easily be identified as such. However, a few cases of well-defined births and deaths were recorded. Some granules are preceded by a diffuse patch of bright material, which is difficult to distinguish from a granule smeared by seeing. Others originate by detachment from the inner ends of penumbral filaments or from the tip of a fork structure (cf. Plate 3.19). Dissolution sometimes occurs by a granule fading into diffuse bright material, which may then become dark. Sometimes a granule loses its identity by coalescing with a neighbouring granule, penumbral filament, or fork structure (cf. Plate 3.19). Some of these modes of formation and decay are similar to those of the photospheric granules, described in Section 3.2.7.

Some of the brighter granules in the umbrae of spots quite close to the limb are frequently visible on negatives taken with a normal exposure. In one case umbral granules were clearly visible in a spot about 25″ of arc from the limb; however, the distance from the limb at which the umbral granules finally become invisible is not yet known. The photospheric granules are observed to persist to within 5″ of the limb (cf. Section 3.2.5) and the facular granules to within 1–2″ (Bray and Loughhead, 1961).

Although no measurements have yet been made of the velocities of umbral granules, the general similarity of the umbral granulation to the photospheric granulation strongly suggests that it also has a *convective origin* (Section 3.2.8). If so, it follows that even the strong magnetic fields present in the umbrae of large spots fail to suppress the convective motions. On the other hand, the influence of the spot magnetic field on the convection currents within its umbra may well be responsible for the smaller cell size actually observed, though it is not clear how the magnetic field would affect the lifetime. The difference in properties between the umbral and photospheric granules supplies direct evidence of an *interaction between the field and the convection currents*. This raises the question of the way in which the convection currents themselves influence the field. For example, do they assist internal solar magnetic fields to penetrate the visible surface layers? Unfortunately, the theory of the origin of sunspots (discussed in Chapter 8) has not yet advanced to the stage of answering this question.

3.7 Light-Bridges

3.7.1   INTRODUCTION

The structure of the umbrae of most sunspots is complicated by the presence of *light-bridges*, which show a great diversity in shape, size, and brightness. The most impressive light-bridges take the form of large masses of bright material extending from one side of the umbra to the other and covering an appreciable part of its area. On the other hand, some bridges take the form of thin *streamers*, often only about 1″ of arc in width, whose continued existence for periods of hours or even days implies a remarkable degree of thermal and dynamical stability. Some bridges and streamers are as bright as, or even slightly brighter than, the surrounding photosphere; others, on the other hand, are so faint that they can be detected only on overexposures. As a general rule, a light-bridge develops slowly and may take several days to extend across an umbra from one side to the other. The process of development is described in Section 3.7.2.

The properties of light-bridges were studied in some detail by Chevalier (1916), who noticed that a light-bridge frequently divides the umbra of a spot into two parts of very different brightness and that the intensity of the umbra is raised in the neighbourhood of a bridge. Since a superficial, overlying structure would not be expected to influence the umbral structure of a spot as a whole, Chevalier concluded that light-bridges are integral parts of the umbral structure. His results are in complete accord with modern observations of the detail in sunspot umbrae which, as we have seen in Section 3.6.3, indicate that the presence of a light-bridge corresponds to the existence of a 'saddle' in the umbral intensity isophotes separating two points of minimum intensity. The intimate relationship of light-bridges to the general structure of the umbra probably explains why they are often remarkably stable and long-lived (cf. Section 3.7.2). On the other hand, the appearance of a light-bridge in the umbra of a spot during the later part of its life is frequently a sign of impending division or final dissolution. The role of light-bridges in the evolution of spots is described in Section 3.7.4.

3.7.2 DEVELOPMENT OF LIGHT-BRIDGES

A good example of the gradual development of a light-bridge across the umbra of a large spot is illustrated in Plates 3.20 and 3.21. The first photograph (Plate 3.20a) shows the appearance of the spot and of other members of the group, photographed fairly close to the east limb on October 15, 1957. The umbra of the main spot is crossed by several long bright streamers which, in places, show a segmented appearance. There are also several bright streamers crossing two smaller spots to the right of the large one.

The second photograph (Plate 3.20b) illustrates the appearance of the same spot on October 17—some 52 hours later. (The apparent increase in area is probably largely due to reduced foreshortening as the spot moves further in from the limb.) A large area of bright material has now grown out from the penumbra at the top of the main umbra. This incipient light-bridge is linked to the penumbra at several places by faint streamers (better seen on the original negative) and, on the lower right-hand side, has a fringe of material of *penumbral* brightness. This fringe is made up of thin threads running out from the light-bridge into the umbra and terminating there; they appear to be examples of a rare type of light-bridge structure which Chevalier (1916: cf. p. B8) referred to as 'filtrations'. Chevalier observed this structure visually but was not able to record it photographically.

Plate 3.21 shows the appearance of the same spot four days later, on October 21. The light-bridge visible in Plate 3.20 (b) has now increased in size and extends practically right across the umbra. It is brighter than the photosphere outside the spot and shows a mottled structure similar to the photospheric granulation. The bridge is linked to the penumbra on the right- and left-hand sides of the spot by lanes of bright material. The thin bright streamer on the right-hand side persisted practically without change for a period of at least $1\frac{1}{2}$ hours. The region between the light-bridge and the penumbra on the lower, left-hand side is crossed by a number of fainter streamers closely resembling ordinary penumbral filaments.

According to Chevalier the mode of development described above is typical. The process is ordinarily very slow and, as in the case described above, may take several days to complete. The bright material normally appears first on one side of the umbra–penumbra boundary and then slowly extends across the umbra, although Chevalier remarks that sometimes the development takes place simultaneously from two opposing parts of the umbra–penumbra boundary.

As a consequence of their slow rate of development, light-bridges often show little change over periods of many hours. A good illustration of this is given in Plate 3.12. Although the definition of the second photograph ($3^h 1^m$) is somewhat inferior to that of the first ($10^h 21^m$), the main light-bridge across the spot shows little perceptible change over the period of 4 hours 40 minutes between the two photographs, although the streamer across the lower, left-hand portion of the umbra appears to undergo some change. The general appearance of a light-bridge may not, in fact, alter radically over a period of 24 hours. This is illustrated by Plates 3.22 (a) and (b). The upper photograph (Plate 3.22a) shows a fairly thick light-bridge dividing the umbra of a medium-sized spot: it extends out from a large blob of bright material on the right-hand side and joins a similar blob of bright material on the lower, left-hand side. Note the incipient development of a second bridge just above the first, and above this a faint streamer running into the upper part of the umbra. The lower photograph (Plate 3.22b) shows the appearance of the light-bridge some 24 hours later: despite changes, it retains an obvious similarity to its appearance of the previous day. The blob of bright material on the right-hand side has increased in size and intrudes further into the umbra. The incipient bridge of the previous day has now developed, so that there are two close parallel bridges crossing the umbra, both showing a segmented appearance. Even the thin streamer in the upper portion of the umbra is still present after the lapse of 24 hours, although its shape has altered somewhat.

Most light-bridges have *lifetimes* of at least several days; in fact, Chevalier (1916: p. B7) found that bridges often date from the period of formation of a spot and, in some cases, persist throughout its entire lifetime.

3.7.3 THE STRUCTURE OF LIGHT-BRIDGES

The bright feature crossing the umbra of the large spot shown in Plate 3.21 is an outstanding example of a large, well-developed light-bridge. It is linked to the penumbra at several places by thin streamers and shows a coarse granular appearance. This *granular structure* was first observed by Janssen (1896: cf. p. 111) and is a characteristic feature. It is clearly evident, for example, in the bridge crossing the umbra of the spot illustrated in Plate 3.15 (*a*). Other examples of light-bridges showing a granular structure are illustrated in Plate 3.9 (*b*) and the Frontispiece.

Light-bridges having the form of thin *streamers* are found in many spots. Very frequently the streamers show a segmented appearance. This is illustrated in Plates 3.11 (*f*) and (*g*): the first photograph shows a number of bright streamers linking the penumbra on the upper part of the spot with a large invading tongue of bright material, which is almost indistinguishable from ordinary photosphere. The second photograph shows a single streamer, clearly resolved into four bright blobs, crossing the lower, left-hand portion of the umbra. Occasionally the brightness of one particular segment in a narrow bridge or streamer is found to greatly exceed that of neighbouring segments.

Light-bridges display a great diversity in *brightness*: the intensity of a light-bridge or streamer can range from a value somewhat greater than that of the photosphere to a value so low that the bridge is visible only on an overexposure (cf. Chevalier, 1916: p. B8). An example of a very bright light-bridge is shown in Plate 3.21; both the bridge and some of the associated streamers are brighter than the surrounding photosphere. In contrast to this example, Plate 3.16 (*b*) illustrates a narrow light-bridge crossing the centre of a spot umbra, which is so faint that it does not appear on a normal photograph of the spot (Plate 3.16*a*). Nevertheless, it is clearly visible in Plate 3.16 (*b*), taken with twice the exposure; it consists of a number of bright segments, an elongated one near the centre being brighter than the others. The remaining photograph (Plate 3.16*c*), taken with three times the exposure of Plate 3.16 (*a*), shows the umbral structure in the neighbourhood of the light-bridge. The centre of the bridge is now overexposed, and neighbouring portions of the umbra begin to appear. It is clear that the portion of the umbra just above the bridge is brighter than the rest: in fact, several umbral granules are now visible in this region although they fail to appear elsewhere. This illustrates the fact, pointed out in Section 3.6.3, that the presence of a light-bridge corresponds to the existence of a 'saddle' in the umbral intensity isophotes.

Light-bridges are even more prominent when seen in spots near the limb, where their brightness may exceed that of the neighbouring white-light faculae. An outstanding example of this is shown in the Frontispiece: the umbra of the spot furthest from the limb is crossed by an extensive light-bridge. It shows evidence of some granular structure and is actually brighter than the faculae outside the spot at the same heliocentric angle. The spots closer to the limb contain large regions of bright material, some closely resembling light-bridges. Near the top, some of the bright material appears to link up with faculae outside the group. This apparent identification of light-bridges with photospheric faculae was also noted by Chevalier (1916: cf. p. B8). Taken in conjunction with the conclusion that light-bridges are intimately related to the general umbral structure, this implies that, in addition to the well-known statistical connection, there may be a fairly close physical relationship between sunspots and faculae. This topic is discussed more fully in Section 7.2.

3.7.4 THE ROLE OF LIGHT-BRIDGES IN THE EVOLUTION OF SUNSPOTS

According to Chevalier (1916: p. B8), light-bridges play an important role in the final stages of the evolution of sunspots, when the appearance of a light-bridge can be a sign of impending division or final dissolution. The process of division is illustrated in Plates 3.22 (c) and (d). The upper photograph shows the follower spot of a typical bipolar group, in which a complicated light-bridge can be seen crossing the upper part of the spot umbra. The light-bridge merges into the photosphere on the right-hand side of the spot and has a granular structure. The lower photograph shows the appearance of the same spot some 26 hours later; it has now divided into two parts, well separated by a wide strip of photospheric granulation which has taken the place of the light-bridge of the previous day. Moreover, the left-hand component of the original spot is now invaded by new light-bridges linked to the photosphere on the right-hand side. In fact, all the penumbra on the right-hand side of this component has dissolved and the large tongue of umbral material intruding into the penumbra in this region on the previous day has completely disappeared. Sometimes an irregular spot is transformed into several smaller spots of more regular shape as a result of divisions occurring in this way (cf. Chevalier, 1916: p. B8). Another good illustration of the role played by light-bridges in the dissolution of a spot is contained in a series of photographs published by Suzuki (1961: cf. Figs. 19–22).

3.8 Appearance of Sunspots near the Limb

3.8.1 INTRODUCTION

The description of the fine structure of sunspots given in the preceding sections refers basically to their appearance in the central region of the solar disk. How does the appearance of a spot change as it moves towards the limb? Besides a general foreshortening, many solar observers have noticed that the width of the penumbra on the side of a spot remote from the limb decreases at a greater rate than that on the limb side (for an account of the earlier work on this topic, see Abetti, 1957: p. 65; Chistyakov, 1961). This phenomenon was discovered by A. Wilson in 1769 and is termed the *Wilson effect*. The work of subsequent observers has in general confirmed the reality of the effect, although the observed magnitude has decreased with the application of more modern techniques. Discrepancies among the results of the early observers can be attributed mainly to the smallness of the quantities to be measured. Near the limb the width of the highly-foreshortened penumbra may be reduced to 2″ of arc or less; this can be measured only with the aid of a high-resolution instrument under conditions of good atmospheric seeing. In addition, by confining their measures to the limb, many early workers failed to take account of any possible asymmetry in the spot which, if large enough, can mask the Wilson effect altogether, or even give the impression of the reverse effect. To avoid systematic error due to this cause, it is necessary to make measurements of a *regular* spot at *both* limbs, and also near the centre of the disk, where the foreshortening is negligible and the true penumbral widths can be determined. This procedure has been used by the authors (Loughhead and Bray, 1958) to demonstrate the reality of the Wilson effect in a small, fairly regular, stable sunspot which was photographed on all but one day of its passage across the disk; an account of these observations is given in Section 3.8.2. In the case of a regular spot, the Wilson effect becomes more and more exaggerated as the limb is approached until, finally, no penumbra at all is visible on the side away from the limb.

Sunspots near the limb often show a variation in the sharpness of the umbra-penumbra boundary, the boundary being quite sharp on the side near the limb but diffuse on the other side. This phenomenon, which is described in Section 3.8.3, is often seen in spots showing a marked Wilson effect.

Until recently it has generally been assumed that sunspots are shallow, saucer-shaped depressions in the photosphere (for references, see Loughhead and Bray, 1958). This interpretation was supported by the observations of early visual observers who reported that large sunspots, when right at the limb, appeared as distinct depressions (cf. Secchi, 1875: p. 75; Young, 1895: p. 128). However, this conclusion is not confirmed by modern photographic observations; a description of the appearance of spots at the extreme limb is given in Section 3.8.4.

3.8.2 THE WILSON EFFECT IN SUNSPOTS

The Wilson effect is very clearly illustrated in Plate 3.23. This shows the changing appearance of a small, fairly regular, stable sunspot during its passage across the disk. Each photograph is orientated so that the direction of greatest foreshortening is horizontal. It is evident that, although the spot is fairly regular when seen near the centre of the disk, at both the east and west limbs the width of the penumbra on the side of the spot nearer the limb is significantly greater than that on the other side. Other spots showing the Wilson effect are illustrated in Plates 3.24 (*a*) and (*b*).

The *magnitude* of the Wilson effect in the spot shown in Plate 3.23 was determined by the authors (Loughhead and Bray, 1958) from measurements of 38 good-quality photographs selected from some 24,000 taken with the Sydney 5-inch photoheliograph during the passage of the spot across the disk. The quantities required are the apparent widths of the penumbra AN and A'N' (cf. Fig. 3.4) measured along the line of greatest foreshortening on the sides of the spot directed towards and away from the solar limb respectively. This line lies along the arc of the great circle whose plane contains the solar radius and the line-of-sight to the centre of the spot.

The ratio $f = $ AN/A'N', which varies with the heliocentric angle θ of the spot, is a quantitative measure of the Wilson effect. A minor difficulty arises in that the spot, as seen from the Earth, appears to rotate about the direction of greatest foreshortening as it moves across the disk. Measurements made on

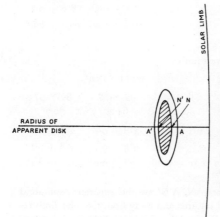

Fig. 3.4. Measurement of the Wilson effect: notation. AA' is the arc intercepted by the spot on the great circle whose plane contains the solar radius and the line-of-sight to the centre of the spot. AN and A'N' are the apparent widths of the penumbra on the sides of the spot towards and remote from the solar limb respectively.

successive days along this line therefore refer to different lines in the spot itself. However, for the three days during which the spot is close to the limb, the apparent rotation is so small that it may safely be ignored. Measurements of f were first made for the three days during which the spot was close to the east limb, and the results compared with measurements for the two days during which the spot was near the central meridian, where the foreshortening is negligible. The line in the spot corresponding to the limb measurements was transferred to the centre photographs with the aid of Stonyhurst disks. All measurements were therefore made along the same line in the spot. A similar set of measurements was made for the west limb and the results also compared with centre measurements. For the reasons mentioned above the east and west limb measurements were made along different lines in the spot, the angle between the lines being approximately 7°. Errors due to seeing distortions were reduced by taking the mean of measures on five photographs for each day, except May 19, when only three of sufficient quality were available.

Tables 3.4 (a) and (b) give the results of the measurements (expressed in seconds of arc) for the east and west limbs respectively, as well as the

TABLE 3.4

The Wilson Effect in a Small Regular Sunspot:
Penumbral Measurements[1]

(*a*) East limb

	EAST LIMB			CENTRE	
Date	19.v.57	20.v.57	21.v.57	24.v.57	25.v.57
θ	$73 \cdot 1°$	$59 \cdot 3°$	$45 \cdot 2°$	$8 \cdot 6°$	$12 \cdot 1°$
AN	$4''1 \pm 0 \cdot 2$	$5''6 \pm 0 \cdot 3$	$6''4 \pm 0 \cdot 3$	$9''3 \pm 0 \cdot 3$	$7''9 \pm 0 \cdot 6$
A′N′	$2''2 \pm 0 \cdot 2$	$3''5 \pm 0 \cdot 3$	$5''4 \pm 0 \cdot 2$	$8''1 \pm 0 \cdot 3$	$8''6 \pm 0 \cdot 3$
f	$1 \cdot 88 \pm 0 \cdot 15$	$1 \cdot 57 \pm 0 \cdot 15$	$1 \cdot 18 \pm 0 \cdot 07$	$1 \cdot 15 \pm 0 \cdot 06$	$0 \cdot 91 \pm 0 \cdot 08$

(*b*) West limb

	CENTRE		WEST LIMB		
Date	24.v.57	25.v.57	28.v.57	29.v.57	30.v.57
θ	$8 \cdot 6°$	$12 \cdot 1°$	$52 \cdot 2°$	$64 \cdot 2°$	$76 \cdot 7°$
AN	$7''9 \pm 0 \cdot 2$	$9''0 \pm 0 \cdot 2$	$6''1 \pm 0 \cdot 2$	$4''8 \pm 0 \cdot 1$	$2''7 \pm 0 \cdot 1$
A′N′	$9''7 \pm 0 \cdot 4$	$7''4 \pm 0 \cdot 2$	$5''5 \pm 0 \cdot 2$	$3''5 \pm 0 \cdot 1$	$1''8 \pm 0 \cdot 1$
f	$0 \cdot 82 \pm 0 \cdot 04$	$1 \cdot 22 \pm 0 \cdot 04$	$1 \cdot 10 \pm 0 \cdot 06$	$1 \cdot 38 \pm 0 \cdot 07$	$1 \cdot 51 \pm 0 \cdot 11$

[1] θ is the heliocentric angle of the spot; AN, A′N′ are the apparent penumbral widths on the sides of the spot directed towards and away from the solar limb respectively; and $f = $ AN/A′N′. It should be noted that the AN of Table 3.4 (*a*) corresponds to the A′N′ of Table 3.4 (*b*), and vice versa. The r.m.s. errors of the mean values AN and A′N′ are calculated from the formula $s = \sqrt{[\Sigma(\text{residuals})^2/\{n(n-1)\}]}$, where *n* is the number of determinations; the corresponding errors in *f* are derived from these errors in the usual way. The east and west limb measurements were made along different lines in the spot, the angle between the lines being approximately $7°$.

corresponding values of the ratio *f*. The r.m.s. errors of the mean values are also given, although these represent only the internal consistency of the measurements. It is evident that the values of *f* steadily increase towards either limb by amounts much greater than the r.m.s. errors. For this particular spot, the Wilson effect appears to be somewhat smaller at the west limb than at the east limb; this may reflect the fact that the penumbral widths in the two cases were measured along somewhat different lines in the spot which, however, differed only by about $7°$. On the other hand it may be connected with the gradual decay of the spot which occurred during its passage across the disk.

The Wilson effect becomes more and more exaggerated the nearer a spot approaches the limb until, finally, little or no penumbra may be visible on the side away from the limb. This is illustrated in Plates 3.24 (*c*) and (*d*); in the latter case the penumbra on the side of the spot remote from the limb has almost disappeared.

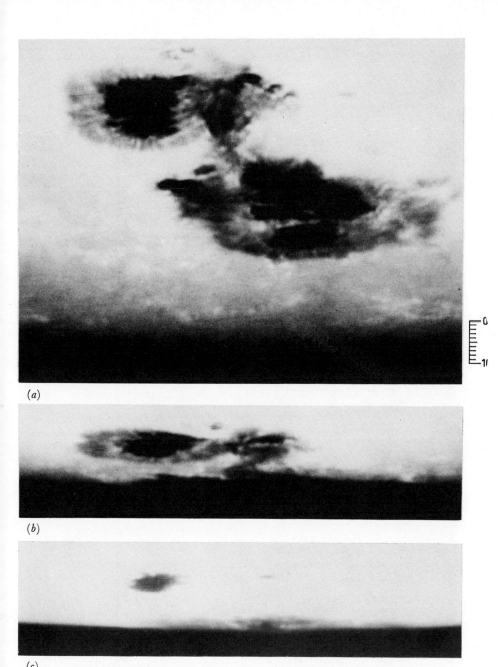

(a)

(b)

(c)

PLATE 3.25. Sunspot at the extreme limb.

(a) Large complicated double spot, some distance from the limb (December 5, 1957).

(b)–(c) Same spot some 24 hours later; the leading edge of the penumbra is now almost at the limb. On (b), a normal reproduction, the spot appears as a depression, but on (c)—printed from the same negative to show detail at the extreme limb—this impression is seen to be illusory. The distinction between umbra and penumbra is almost lost in the leading component. (The slight hump in the limb adjacent to the spot is merely a local seeing distortion.)

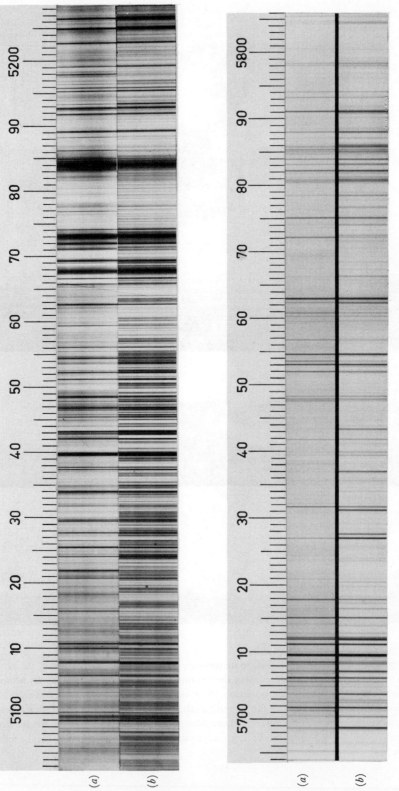

PLATE 4.1. Two sections of the Mt Wilson photographic map of the sunspot spectrum. (a) Disk centre. (b) Sunspot. (By courtesy of Mt Wilson Observatory.)

3.8.3 DIFFUSENESS OF THE UMBRA–PENUMBRA BOUNDARY ON SIDE
REMOTE FROM LIMB

Examination of films taken with the Sydney 5-inch photoheliograph has shown
that the umbra–penumbra boundary of spots near the limb, while quite sharp
on the limb side, is often *diffuse* on the other side (cf. Loughhead and Bray,
1958). This phenomenon is clearly evident in the spot shown in Plate 3.23.
Other illustrations, on a larger scale, are given in Plates 3.24 (*a*) and (*b*):
in both these spots the diffuseness of the umbra–penumbra boundary on the side
remote from the limb is associated with an abnormal *darkening* of the penumbra
on this side. The effect has been seen in a number of spots near the limb,
including one too asymmetrical to show the Wilson effect. Most spots showing
the Wilson effect also have a diffuse umbra–penumbra boundary on the side
remote from the limb, although one exceptional case has been noted.

3.8.4 APPEARANCE OF SUNSPOTS AT THE EXTREME LIMB

Plate 3.25 illustrates the appearance of a large complicated double spot on the
last two days of its passage to the limb. The top photograph (Plate 3.25*a*) shows
the appearance of the spot when it is still some distance from the limb; the umbra
of the leading component is crossed by a light-bridge and several faint streamers.
Plates 3.25 (*b*) and (*c*) are different enlargements of the same original
negative showing the spot some 24 hours later when, according to calculation,
the leading edge of the penumbra was almost at the limb. On a normal print
(Plate 3.25*b*) the leading component appears as a depression in the limb some 3″
of arc deep. However, when the same negative is printed to show detail at the
extreme limb (Plate 3.25*c*), it becomes evident that this impression is illusory
and that in reality spot material remains visible right up to the extreme limb.
In the leading component the contrast of the spot detail is much reduced and
the distinction between umbra and penumbra is largely lost; however, some
bright material can still be distinguished only 1–2″ from the limb. (Note: the
slight hump in the limb adjacent to the spot is merely a local seeing distortion,
and has no real existence.)

Another striking illustration of the appearance of a spot at the extreme limb
is given in Plate 3.24 (*c*), which was taken with twice the exposure normally
given for a limb photograph. Here the distinction between umbra and penumbra
is maintained, despite the close proximity of the spot to the limb. Particularly
noteworthy is the large blob of bright material in the spot, just above the umbra,
which extends almost to the limb. However, there is no evidence of any bright
material associated with the spot projecting *beyond* the limb.

Plate 3.24 (*d*) shows another highly-foreshortened spot at the extreme limb.
Once again, the distinction between umbra and penumbra is retained, although
the penumbra on the side remote from the limb has disappeared almost completely owing to exaggerated Wilson effect. Plate 3.24 (*e*) shows the same spot
some 70 minutes later: the spot is now crossing the limb and only a slight trace

of its umbra remains visible. As before, there is no depression at the limb nor
is there any evidence of any bright material in the spot projecting beyond the
limb. Plate 3.24 (e) was taken with twice the exposure of Plate 3.24 (d).

Two conclusions emerge from the observations described in this section.
In the first place there is no evidence of a depression when a spot goes around the
limb; spot material continues to be visible right up to the extreme limb, although
the contrast of the spot detail becomes greatly reduced. Secondly, the observations indicate that the bright material sometimes present in spots does not
project beyond the limb.

3.8.5 INTERPRETATION OF LIMB OBSERVATIONS OF SUNSPOTS

The whole question of the changing appearance of a sunspot during its passage
across the disk is essentially a problem in the theory of radiative transfer (Jensen
and Ringnes, 1957a; Loughhead and Bray, 1958; Chistyakov, 1962). Theoretically, a knowledge of the emission and absorption coefficients in the photosphere, umbra, and penumbra would yield the intensity profile of the spot along
the line of greatest foreshortening, and hence the Wilson effect (Section 3.8.2), the
variation in the sharpness of the umbra–penumbra boundary (Section 3.8.3),
and the expected appearance of the spot at the extreme limb (Section 3.8.4).
Conversely, measurements of the intensity profile at different heliocentric angles
might provide some information about the emission and absorption coefficients
in the spot and their variation with depth. However, any attempt at a detailed
comparison of theory and observation would face serious difficulties, both
observational and theoretical. The observations would require correction for
scattered light and for distortion by the combined instrumental profile of telescope and atmosphere. These corrections become very large near the limb,
where a resolving limit of 1 ″ of arc may become quite inadequate owing to the
gross foreshortening of the spot. On the theoretical side, as we shall see in
Section 4.2.4, there is the difficulty of matching the scales of optical depth in
the spot and photosphere.

However, in the case of the Wilson effect, some insight into the problem can
be obtained by considering a simple model in which a regular sunspot near the
limb is represented as a cylindrical structure extending through the photo-
sphere, and in which the absorption coefficients κ_1, κ_2, κ_3 of the photosphere,
penumbra, and umbra respectively are independent of depth (Fig. 3.5). The
photosphere and the spot are assumed to have a sharp upper boundary. Making
the reasonable assumption that $\kappa_1 > \kappa_2 > \kappa_3$ and taking the directions of the
apparent photosphere–penumbra and umbra–penumbra boundaries to be such
that *the corresponding lines-of-sight pass through unit optical thickness*, it is easy to show
by simple geometry that the apparent width of the penumbra on the side of the
spot remote from the limb decreases more rapidly than a geometrical foreshortening law would imply, while on the other side the reverse applies. This is
the Wilson effect.

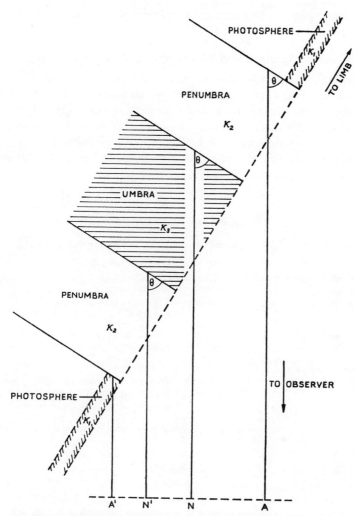

Fig. 3.5. A simple sunspot model: explanation of the Wilson effect. The figure shows a section of the spot by the plane of the great circle containing the solar radius and the line-of-sight to the spot. κ_1, κ_2, and κ_3 are the absorption coefficients of the photosphere, penumbra, and umbra respectively (assumed to be independent of depth).

In view of the crudity of this model it is pointless to attempt to derive values of κ_2 and κ_3 from the Wilson effect observations described in Section 3.8.2. However, the observations do indicate that, in the case of this particular spot, the Wilson effect is due mainly to reduced foreshortening on the limb side of the spot and not to increased foreshortening on the other side. In conjunction with calculations based on the simple model, this implies that the dominant factor is the *great transparency* of the umbra, rather than that of the penumbra, compared with the photosphere. The transparency of sunspot umbrae is further discussed in Section 4.2.4.

3.9 Lifetimes of Sunspot and Photospheric Fine Structures: Summary

To conclude this chapter we summarize in Table 3.5 the available data on the lifetimes of the various sunspot and photospheric fine structures. The table lists for each feature the lifetime, the source reference, and the section in the chapter where the feature is described. Structures whose lifetimes are unknown or for which only a very crude lower limit to the lifetime is available, are not included. It is evident that the fine structures of sunspots, as well as the facular granules, are much longer-lived than the photospheric granules.

TABLE 3.5

Lifetimes of Sunspot and Photospheric Fine Structures

FEATURE	LIFETIME	REFERENCE	SECTION
Penumbral filaments	~ 2 hours	Bray and Loughhead (1958a)	3.5.4
Penumbral bright regions:			
Type 1	30 minutes	Macris (1953)	3.5.6
Umbral granules[1]	15–30 minutes	Bray and Loughhead (1959)	3.6.4
		Loughhead and Bray (1960a)	
Facular granules	2 hours	Waldmeier (1940)	—
	2 hours	Bray and Loughhead (1961)	
Photospheric granules	7–8 minutes	Macris (1953)	3.2.7
	~ 10 minutes	Bray and Loughhead (1958b)	
	10 minutes	Rösch and Hugon (1959)	
	8·6 minutes	Bahng and Schwarzschild (1961)	

[1] Some umbral granules are much longer-lived: 10 per cent last for more than 2 hours.

REFERENCES

ABETTI, G. [1957] *The Sun*. London, Faber and Faber.

ANANTHAKRISHNAN, R. [1951] 'Intensity variation in sunspots', *Nature* **168,** 291.

BAHNG, J. D. R. [1958] 'Lifetime of photospheric pores', *Astrophys. J.* **128,** 145.

BAHNG, J. D. R., and SCHWARZSCHILD, M. [1961] 'Lifetime of solar granules', *Astrophys. J.* **134,** 312.

BERNIÈRE, G., MICHARD, R., and RIGAL, G. [1962] 'Étude statistique des fluctuations locales de brillance et de vitesse dans la photosphère', *Ann. Astrophys.* **25,** 279.

BLACKWELL, D. E., DEWHIRST, D. W., and DOLLFUS, A. [1959] 'The observation of solar granulation from a manned balloon. I. Observational data and measurement of contrast', *Mon. Not. R.A.S.* **119,** 98.

BRAY, R. J., and LOUGHHEAD, R. E. [1957] 'Bright regions in sunspots', *Observatory* **77,** 201.

BRAY, R. J., and LOUGHHEAD, R. E. [1958a] 'The lifetime of sunspot penumbra filaments', *Aust. J. Phys.* **11,** 185.

BRAY, R. J., and LOUGHHEAD, R. E. [1958b] 'Observations of changes in the photospheric granules', *Aust. J. Phys.* **11,** 507.

BRAY, R. J., and LOUGHHEAD, R. E. [1959] 'High resolution observations of the granular structure of sunspot umbrae', *Aust. J. Phys.* **12,** 320.

BRAY, R. J., and LOUGHHEAD, R. E. [1961] 'Facular granule lifetimes determined with a seeing-monitored photoheliograph', *Aust. J. Phys.* **14**, 14.

CHEVALIER, S. [1916] 'Étude photographique des taches solaires', *Ann. Obs. Zô-Sè* **9**, B1.

CHISTYAKOV, V. F. [1961] 'A study of the Wilson effect in sunspots', *Astron. J. U.S.S.R.* **38**, 617 (*Sov. Astron. AJ* **5**, 471).

CHISTYAKOV, V. F. [1962] 'On the observed depths of sunspots', *Astron. J. U.S.S.R.* **39**, 459 (*Sov. Astron. AJ* **6**, 363).

DANIELSON, R. E. [1961a] 'The structure of sunspot penumbras. I. Observations', *Astrophys. J.* **134**, 275.

DANIELSON, R. E. [1961b] 'The structure of sunspot penumbras. II. Theoretical', *Astrophys. J.* **134**, 289.

DAS, A. K., and RAMANATHAN, A. S. [1953] 'Distribution of radiation flux across a sunspot', *Z. Astrophys.* **32**, 91.

EDMONDS, F. N. [1960] 'On solar granulation', *Astrophys. J.* **131**, 57.

EDMONDS, F. N. [1962a] 'A statistical photometric analysis of granulation across the solar disk', *Astrophys. J. Supp.* **6**, 357.

EDMONDS, F. N. [1962b] 'A coherence analysis of Fraunhofer line fine structure and continuum brightness fluctuations near the centre of the solar disk', *Astrophys. J.* **136**, 507.

EDWARDS, A. W. F. [1957] 'The proportion of umbra in large sunspots, 1878–1954', *Observatory* **77**, 69.

EVANS, J. W., MAIN, P., MICHARD, R., and SERVAJEAN, R. [1962] 'Correlations in the time variations of macroscopic inhomogeneities in the solar atmosphere', *Astrophys. J.* **136**, 682.

EVANS, J. W., and MICHARD, R. [1962a] 'Observational study of macroscopic inhomogeneities in the solar atmosphere. II. Brightness fluctuations in Fraunhofer lines and the continuum', *Astrophys. J.* **136**, 487.

EVANS, J. W., and MICHARD, R. [1962b] 'Observational study of macroscopic inhomogeneities in the solar atmosphere. III. Vertical oscillatory motions in the solar photosphere', *Astrophys. J.* **136**, 493.

HALE, G. E., and ELLERMAN, F. [1906] 'The five-foot spectroheliograph of the solar observatory', *Astrophys. J.* **23**, 54.

HART, A. B. [1956] 'Motions in the Sun at the photospheric level. VI. Large scale motions in the equatorial region', *Mon. Not. R.A.S.* **116**, 38.

HOWARD, R. [1962] 'Preliminary solar magnetograph observations with small apertures', *Astrophys. J.* **136**, 211.

JAGER, C. DE [1959] 'Structure and dynamics of the solar atmosphere', *Handbuch der Physik*, ed. S. FLÜGGE, vol. 52, p. 80. Berlin, Springer.

JANSSEN, J. [1896] 'Mémoire sur la photographie solaire', *Ann. Obs. Meudon* **1**, 91.

JENSEN, E., NORDØ, J., and RINGNES, T. S. [1955] 'Variations in the structure of sunspots in relation to the sunspot cycle', *Astrophysica Norvegica* **5**, 167.

JENSEN, E., and RINGNES, T. S. [1957a] 'The variation of the penumbra–umbra ratio of regular sunspots across the solar disk', *Astrophysica Norvegica* **5**, 259.

JENSEN, E., and RINGNES, T. S. [1957b] 'The variation of the penumbra–umbra ratio of regular sunspots with heliocentric latitude', *Astrophysica Norvegica* **6**, 33.

KEENAN, P. C. [1938] 'Dimensions of the solar granules', *Astrophys. J.* **88**, 360.

KIEPENHEUER, K. O. [1960] 'Über die Struktur der gestörten Sonnenatmosphäre', *Z. Astrophys.* **49**, 73.

LEIGHTON, R. B. [1957] 'Some observations of solar granulation', *Pub. Astron. Soc. Pac.* **69**, 497.

LEIGHTON, R. B., NOYES, R. W., and SIMON, G. W. [1962] 'Velocity fields in the solar atmosphere. I. Preliminary report', *Astrophys. J.* **135**, 474.

LOUGHHEAD, R. E., and BRAY, R. J. [1958] 'The Wilson effect in sunspots', *Aust. J. Phys.* **11**, 177.

LOUGHHEAD, R. E., and BRAY, R. J. [1959] ' "Turbulence" and the photospheric granulation', *Nature* **183**, 240.

LOUGHHEAD, R. E., and BRAY, R. J. [1960a] 'The lifetime and cell size of the granulation in sunspot umbrae', *Aust. J. Phys.* **13**, 139.

LOUGHHEAD, R. E., and BRAY, R. J. [1960b] 'Granulation near the extreme solar limb', *Aust. J. Phys.* **13**, 738.

LOUGHHEAD, R. E., and BRAY, R. J. [1961] 'Phenomena accompanying the birth of sunspot pores', *Aust. J. Phys.* **14**, 347.

MACRIS, C. [1953] 'Recherches sur la granulation photosphérique', *Ann. Astrophys.* **16**, 19.

MACRIS, C. J., and BANOS, G. J. [1961] 'Mean distance between photospheric granules and its change with the solar activity', *Mem. Nat. Obs. Athens: Series 1*, No. 8.

MACRIS, C. J., and PROKAKIS, T. J. [1962] 'Sur une différence des dimensions des granules photosphériques au voisinage et loin de la pénombre des taches solaires', *C.R. Acad. Sci.* **255**, 1862.

MALTBY, P. [1960] 'Note on the Evershed effect in sunspots', *Ann. Astrophys.* **23**, 983.

MILLER, R. A. [1960a] 'Filamentary structure between sunspots photographed in integrated light', *J. Brit. Astron. Assoc.* **70**, 100.

MILLER, R. A. [1960b] 'Observations on photospheric brightness surrounding sunspots', *J. Brit. Astron. Assoc.* **70**, 146.

NICHOLSON, S. B. [1933] 'The area of a sunspot and the intensity of its magnetic field', *Pub. Astron. Soc. Pac.* **45**, 51.

OSTER, L. [1957] 'Über die Abkühlung stellarer Materie durch thermische Emission und das Abklingen der Sonneneruptionen', *Z. Astrophys.* **44**, 26.

PECKER, J. C. [1959] 'La zone convective des étoiles: rapport introductif', 9th Coll. Internat. d'Astrophys., Liège. *Mem. Soc. Roy. des Sci. de Liège: Series 5*, vol. 3, p. 343.

PLASKETT, H. H. [1954] 'Motions in the Sun at the photospheric level. V. Velocities of granules and other localized regions', *Mon. Not. R.A.S.* **114**, 251.

PLASKETT, H. H. [1955] 'Physical conditions in the solar photosphere', *Vistas in Astronomy*, ed. A. BEER, vol. 1, p. 637. London, Pergamon.

PLASKETT, H. H. [1959] 'Motions in the Sun at the photospheric level. VIII. Solar rotation and photospheric circulation', *Mon. Not. R.A.S.* **119**, 197.

RICHARDSON, R. S., and SCHWARZSCHILD, M. [1950] 'On the turbulent velocities of solar granules', *Astrophys. J.* **111**, 351.

RÖSCH, J. [1955] 'Données morphologiques nouvelles sur la granulation photosphérique solaire', *C.R. Acad. Sci.* **240**, 1630.

RÖSCH, J. [1956] 'Photographie à cadence rapide de la photosphère et des taches solaires', *C.R. Acad. Sci.* **243**, 478.

RÖSCH, J. [1957] 'Photographies de la photosphère et des taches solaires', *L'Astronomie* **71**, 129.

RÖSCH, J. [1959] 'Observations sur la photosphère solaire. II. Numération et photométrie photographique des granules dans le domaine spectral 5900–6000Å', *Ann. Astrophys.* **22**, 584.

RÖSCH, J., and HUGON, M. [1959] 'Sur l'évolution dans le temps de la granulation photosphérique', *C.R. Acad. Sci.* **249**, 625.

ROYDS, T. [1925] 'Note on spectroheliograms taken with different parts of the Hα-line', *Mon. Not. R.A.S.* **85**, 464.

SCHRÖTER, E. H. [1962] 'Einige Beobachtungen und Messungen an Stratoskop I-Negativen', *Z. Astrophys.* **56**, 183.

SCHWARZSCHILD, M. [1959] 'Photographs of the solar granulation taken from the stratosphere', *Astrophys. J.* **130**, 345.

SECCHI, A. [1875] *Le Soleil*, vol. 1. Paris, Gauthier-Villars.

SEMEL, M. [1962] 'Sur la détermination du champ magnétique de la granulation solaire', *C.R. Acad. Sci.* **254**, 3978.

SERVAJEAN, R. [1961] 'Contribution à l'étude de la cinématique de la matière dans les taches et la granulation solaire', *Ann. Astrophys.* **24,** 1.

SPIEGEL, E. A. [1957] 'The smoothing of temperature fluctuations by radiative transfer', *Astrophys. J.* **126,** 202.

STEPANOV, V. E. [1961] 'On motions in different levels of the solar atmosphere', *Izv. Crim. Astrophys. Obs.* **25,** 154.

STESHENKO, N. V. [1960] 'On the determination of magnetic fields of solar granulation', *Izv. Crim. Astrophys. Obs.* **22,** 49.

STREBEL, H. [1932a] 'Sonnenphotographische Dokumente', *Z. Astrophys.* **5,** 36.

STREBEL, H. [1932b] 'Die innere helle Rand der Penumbra von Sonnenflecken', *Z. Astrophys.* **5,** 96.

STREBEL, H. [1933] 'Beitrag zum Problem der Sonnengranulation', *Z. Astrophys.* **6,** 313.

STUART, F. E., and RUSH, J. H. [1954] 'Correlation analyses of turbulent velocities and brightness of the photospheric granulation', *Astrophys. J.* **120,** 245.

SUZUKI, Y. [1961] 'The fine structures of sunspots', *Bull. Kyoto Gakugei Univ., Series B,* No. 19.

SWEET, P. A. [1955] 'The structure of sunspots', *Vistas in Astronomy,* ed. A. BEER, vol. 1, p. 675. London, Pergamon.

TANDBERG-HANSSEN, E. [1956] 'A study of the penumbra–umbra ratio of sunspot pairs', *Astrophysica Norvegica* **5,** 207.

THIESSEN, G. [1950] 'The structure of the sunspot-umbra', *Observatory* **70,** 234.

WALDMEIER, M. [1939] 'Über die Struktur der Sonnenflecken', *Astron. Mitt. Zürich,* No. 138, p. 439.

WALDMEIER, M. [1940] 'Die Feinstruktur der Sonnenoberfläche', *Helv. Phys. Act.* **13,** 13.

YOUNG, C. A. [1895] *The Sun.* London, Kegan Paul.

CHAPTER 4

Physical Conditions in Sunspots

4.1 Introduction

In this chapter we shall try to give a comprehensive account of the techniques and results of investigations into the physical conditions in sunspot umbrae and penumbrae. As in the previous chapter, we shall be concerned with sunspots as individual entities, i.e. we shall ignore the fact that they occur in groups. In contrast to the previous chapter, on the other hand, we shall have little to say about the *fine detail* in sunspots, since most of our knowledge of the physical conditions is based on spectroscopic observations made with a spatial resolution much lower than the best attained in direct photography (cf. Section 2.4.1). In fact, many of the observations probably have an effective resolution as low as 5″ of arc, or even worse. Consequently the derived physical quantities represent only the *average conditions* in regions some 5″ or more in diameter. However, in a few cases we shall find that our knowledge of the fine detail described in the previous chapter provides an illuminating commentary on the lower-resolution work described here.

The most reliable information about physical conditions in the umbra comes from measurements made in the continuous spectrum, which are accordingly described first (Section 4.2). Photometry in integrated light leads to a value for the effective temperature (Section 4.2.1), while photometry at selected wavelengths leads to a value for the radiation temperature at unit optical depth (Section 4.2.2). The temperature at unit optical depth in a big spot may be more than 2000° below that at unit optical depth in the photosphere. As a result of the large number of independent measurements which have been made on spots of different sizes, it is possible to derive a statistical relationship between temperature and umbral diameter (Fig. 4.1 and Table 4.2).

The observed wavelength dependence of the umbra/photosphere intensity ratio is described in Section 4.2.3. Comparison with theory shows that the temperature distribution in the umbra approximates to that which would obtain in radiative equilibrium.

The dependence of physical conditions (temperature, pressure, electron pressure, etc.) on optical depth is best derived from the observed variation of the umbra/photosphere intensity ratio with $\cos \theta$, where θ is the heliocentric angle of the spot. In Section 4.2.4 two umbral models are discussed in some detail, namely Michard's (Table 4.3) and Mattig's (Table 4.5). According to Michard the umbra is more transparent than the photosphere and the pressure is substantially lower, but Mattig disagrees with these conclusions. Unfortu-nately, these contradictions cannot be resolved until more is known of the role played by magnetic forces in maintaining the equilibrium of sunspots; however, Michard's model receives some independent support from observations of the Wilson effect (Section 3.8.5).

The electrical conductivity of a spot umbra is calculated in Section 4.2.5. In the lower layers of the umbra the conductivity is one to two orders of magni-tude lower than that at equivalent optical depths in the photosphere.

In most of the chapter it is tacitly assumed that the umbra possesses a uni-form brightness, so that it is permissible to regard a single effective temperature or a single umbral model as being applicable to all points of an umbra of a given size. This is an over-simplification, however, and in Section 4.2.7 two-dimen-sional photometric maps are given which show that the intensity contours of both umbra and penumbra are actually quite complex. Of particular interest is the presence of a small umbral 'core', some 2–3″ of arc in diameter, which in one case was located not at the centre of the umbra but near one end. The core is much darker than the rest of the umbra, the temperature being lower by some 500°.

The Fraunhofer spectrum of the umbra (Section 4.3) not only provides valuable confirmatory information but in addition is potentially capable of supplying data not available from the continuous spectrum. However, it has not been exploited as fully as the continuous spectrum and much work remains to be done, particularly on the molecular lines.

One source of additional information provided by the Fraunhofer spectrum is the sunspot curve of growth (Section 4.3.2). Excitation temperatures derived from the curve of growth are in good agreement with temperatures derived from measurements made in the continuous spectrum. It is interesting to note that the 'non-thermal' velocities in the umbra appear to be significantly larger than those in the photosphere. The effect of the umbral magnetic field on the equivalent widths of the atomic lines is considered in Section 4.3.3: this is a topic about which, owing to theoretical difficulties, there are currently great uncertainties.

The molecular spectrum of the umbra is discussed in Section 4.3.4. Table 4.10 lists the molecular bands identified in the umbra, and Table 4.11 gives 'rotational' temperatures for the umbra derived from measurements of the band-line intensities. In spite of the rather large observational errors, the temperatures are in good agreement with those derived by other means. The fact that the molecular lines are free from the disturbing influence of the Zeeman effect makes their study of great potential usefulness, and further work—both observational and theoretical—is very desirable.

Less is known about the physical conditions in the penumbra (Section 4.4). However, information is available about the effective temperature and the dependence of temperature on optical depth; this is given in Section 4.4.1.

One of the most interesting properties of the penumbra is the occurrence of material motions: this phenomenon, called the Evershed effect, has been known and studied for more than half a century. The results obtained by the different observers are described in Section 4.4.2, while in Section 4.4.3 the observations are discussed in the light of the hypothesis that the observed outflow consists of a laminar motion along the penumbral filaments. Further progress in understanding the Evershed effect depends on the possibility of obtaining velocity measurements with a spatial resolution comparable to that of the photographs of the previous chapter.

The chapter concludes with a summary of the available quantitative information about the physical conditions in the umbra and penumbra (Section 4.5).

4.2 The Continuous Spectrum of Sunspot Umbrae; Physical Conditions in the Umbra

4.2.1 PHOTOMETRY IN INTEGRATED LIGHT: THE EFFECTIVE TEMPERATURE OF THE UMBRA

One of the simplest methods of deriving the temperature of a spot umbra is to compare the total radiation, i.e., integrated over all wavelengths, with the corresponding figure for the undisturbed photosphere. By definition, the temperature obtained in this way is the *effective* temperature and, if I^* and I are the measured radiation fluxes, then from Stefan's law

$$T_{\text{eff}}^* = T_{\text{eff}} \sqrt[4]{(I^*/I)}, \qquad (4.1)$$

where T_{eff}^* and T_{eff} are the effective temperatures of umbra and photosphere respectively. In passing, it may be noted that for an atmosphere in radiative equilibrium the effective temperature is equal to the actual temperature at optical depth 0·67 (see equation 4.6).

Measurements of I^*/I have been made by Pettit and Nicholson (1930), Wormell (1936), and Sitnik (1940), as well as by several earlier workers. As an example of the technique we shall describe the careful work carried out by Wormell at the Cambridge Observatories. The image of a sunspot was formed on a screen pierced by a small aperture, behind which a microscope objective re-imaged the light from the region of the spot over the aperture onto the blackened junction of a vacuum thermocouple. The thermocouple was connected to a micro-galvanometer, whose deflection could be recorded on bromide paper. Slow motions applied to the coelostat allowed the spot and the photosphere outside it to drift across the aperture. The measured intensity ratios were corrected for parasitic light due to seeing by Wanders' method (Section 4.2.6), the measured values of Wanders' parameter a ranging from 0·006 to 0·070. The corrected values of I^*/I for the spots measured ranged from 0·27 to 0·77. Although the corrections were sometimes quite small, e.g., a measured value of 0·31 might only be reduced to 0·30, occasionally they were quite large: e.g., from 0·49 (uncorrected) to 0·28 (corrected).

Wormell derived the figure 0·30 as the most reliable value for fairly large spots, corrected for scattered light. However, a final correction must be applied: this arises from *selective attenuation* in the telescope and atmosphere and is a consequence of the fact that the energy/wavelength curves of the spot and the photosphere are not the same. This correction reduced the value to 0·27 which, with a value for T_{eff} of 5785°K (Allen, 1955), gives

$$T_{eff}^* = 5785 \sqrt[4]{0\cdot27} = 4160°K,$$

indicating a temperature difference between photosphere and umbra of 1625°.

In subsequent sections we shall often find it convenient to make use of the parameter $\Theta = 5040/T$, rather than T itself. If we define

$$\Delta\Theta \equiv 5040\left(\frac{1}{T^*} - \frac{1}{T}\right), \tag{4.2}$$

then Wormell's observations give $\Delta\Theta_{eff} = 0\cdot34$, a figure which, as we shall see in the next section, is consistent with $\Delta\Theta$-values derived from measurements of umbral intensities made at selected wavelengths.

Pettit and Nicholson (1930) had earlier employed a similar technique. These workers obtained values of I^*/I ranging from 0·43 to 0·52 for 6 spots; the mean value was 0·47—a figure considerably greater than Wormell's mean value. Most of the discrepancy is probably due to the difficulty of applying the various corrections, although part may be due to real differences in temperature among the spots measured in the two investigations.

4.2.2 PHOTOMETRY AT SELECTED WAVELENGTHS; DEPENDENCE OF TEMPERATURE ON UMBRAL DIAMETER

Values for the quantity $\Delta\Theta$ defined by equation (4.2)—and hence values for the umbral temperature—can also be derived from measurement of the umbra/photosphere intensity ratio at individual wavelengths. A large number of independent measurements for spots of different sizes have been carried out by different observers, enabling us to derive a statistical relation between temperature and umbral diameter. The measurements described in this section, taken as a whole, provide the most reliable estimates of umbral temperatures at present available.

The theory behind the method (Unsöld, 1955: cf. p. 561) is very simple: in the visible and ultra-violet regions of the spectrum, and at the temperatures with which we are concerned, we may with sufficient accuracy use Wien's approximate form of Planck's radiation law. Hence, for the umbra and photosphere respectively we have

$$I^*(\lambda) \propto \lambda^{-5} e^{-c_2/\lambda T^*}; \; I(\lambda) \propto \lambda^{-5} e^{-c_2/\lambda T}, \tag{4.3}$$

where $c_2 = 1\cdot439$ cm deg and λ is expressed in cm. Taking ratios and introducing $\Delta\Theta$, we obtain

$$\log_e \frac{I^*(\lambda)}{I(\lambda)} = -\frac{c_2}{\lambda \cdot 5040} \cdot \Delta\Theta. \tag{4.4}$$

Consequently, with the aid of (4.4), measurement of $I^*(\lambda)/I(\lambda)$ at any wavelength λ yields a value for $\Delta\Theta$; measurements of the same spot made at different wavelengths should lead to the same value of $\Delta\Theta$.

It is evident that in making use of (4.3) and (4.4) we are ignoring the effect of *stratification* in the spot and photosphere. As we shall see in Section 4.2.4, the validity of this procedure can be tested on the basis of observational models of the umbra. When $\Delta\Theta$ is calculated from the tabulated umbral and photospheric temperatures at the same *optical* depths,[1] it is found that it does in fact remain remarkably constant over a large range of optical depths. Hence, although the value of $\Delta\Theta$ derived from (4.4) refers to an optical depth in spot and photosphere of approximately unity, the same value should obtain at other optical depths throughout the spot model. This circumstance makes the quantity $\Delta\Theta$ of great utility.

The ratio $I^*(\lambda)/I(\lambda)$ has been measured by a large number of workers. Most have confined their measurements to only two or three wavelengths, but a few have made measurements over a wide range of wavelengths. The work of the former is of great value as it yields a large number of independent determinations—by a variety of observational techniques—of the quantity $\Delta\Theta$. The work of the latter is even more valuable since, in addition to providing values of $\Delta\Theta$, it allows a comparison to be made between the observed wavelength dependence of the ratio and one calculated on the basis of radiative equilibrium (cf. Section 4.2.3).

The various techniques which have been employed in measuring the ratio may be broadly classified as follows:

(1) *photographic:* in this method direct photographs of the spot are taken in light of known effective wavelength and are then microphotometered at leisure. This method has the advantage that owing to the short exposure times possible in direct photography the correction of the measured intensity ratios for seeing (Section 4.2.6) is likely to be small. In fact, if the photographs are of sufficient quality, no correction at all may be required. This method has been used by the following workers: Richardson (1933b); Korn (1940); Bray and Loughhead (1962);

(2) *spectrographic:* spectrograms of the spot are obtained in the normal way. This technique has the advantage that the measurements can be made at precisely defined 'windows' in the continuum, often distributed over a wide spectral region. In addition, a simultaneous study of the Fraunhofer spectrum can be made for the same spot. It suffers from the disadvantage that owing to the longer exposure times necessary, the seeing corrections are often very large. The following workers have used the spectrographic technique: Pettit and Nicholson (1930); Wanders (1935); Abetti and Colacevich (1939); Richardson (1939); Krat (1948); Das and Ramanathan (1953); Michard (1953); Ramanathan (1954); Stepanov (1957); Stumpff (1961);

(3) *thermocouple or photomultiplier:* Pettit and Nicholson (1930) used the 150-

[1] Not necessarily the same *geometrical* depths.

foot tower telescope at Mt Wilson, a monochromator, and a thermocouple to obtain $I^*(\lambda)/I(\lambda)$ at wavelengths ranging from 4000 to 22,000Å. (In addition, they covered the region 3000–4000Å by the spectrographic method.) Their work represents the most extensive determination of the wavelength dependence of the umbra/photosphere intensity ratio yet made; their curve is discussed in Section 4.2.3. A more sensitive technique has been employed by Makita and Morimoto (1960), who placed three photomultipliers in the focal plane of a spectrograph at three 'windows' in the visible spectrum. A rapidly-rotating

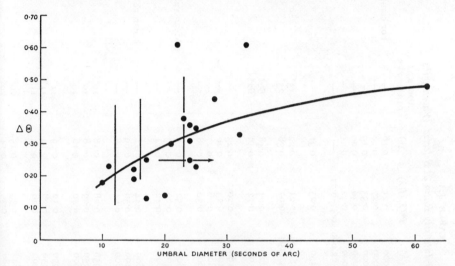

Fig. 4.1. Temperature of the umbra as a function of diameter. The dots represent observed values of $\Delta\Theta$, derived with the aid of (4.4). (The vertical lines and the arrow represent Makita and Morimoto's and Michard's results respectively.) Knowing $\Delta\Theta$, the umbral temperature at unit optical depth (T^*) can be obtained from the relation $\Delta\Theta = 5040(1/T^* - 1/T)$, where T = photospheric temperature at unit optical depth. Representative values of T^* are given in Table 4.2.

prism outside the spectrograph moved the sunspot across a small pinhole, 2″ of arc in diameter, which replaced the normal spectrograph slit. 290 spot profiles were obtained at each of the three wavelengths. The effective exposure times were much smaller than the several seconds usually required for spectrograms.

Table 4.1 summarizes the results of all of these investigations. The first two columns give the source reference and the observatory at which the work was carried out; the next two give the wavelength in Ångstroms and the measured value of $I^*(\lambda)/I(\lambda)$; the fifth column gives the corresponding value of $\Delta\Theta$ calculated from (4.4); the sixth gives the diameter of the umbra, when this is known, calculated from the published data (all umbrae are assumed circular; 1″ of arc = 725 km).

TABLE 4.1
Umbra/Photosphere Intensity Ratio at Selected Wavelengths

REFERENCE	OBSERVATORY	λ(Å)	$I^*(\lambda)/I(\lambda)$	ΔΘ	UMBRAL DIAMETER	REMARKS
Pettit and Nicholson (1930)	Mt Wilson	3000	0·21	0·16	—	For complete curve, see Fig. 4.2.
		4000	0·27	0·18	—	
		5000	0·33	0·19	—	
		6000	0·38	0·20	—	
		7000	0·42	0·21	—	
		8000–22,000	0·47–0·77	—	—	
Richardson (1933b)	Mt Wilson	6450	0·25	0·31	—	
Wanders (1935)	Utrecht	4027; 4194	0·21	0·22	15″	
		4027; 4194	0·07	0·38	23″	
Abetti and Colacevich (1939)	Arcetri	4227	0·04	0·48	62″	
		6568	0·12	0·49	62″	
Richardson (1939)	Mt Wilson	4100	0·20	0·23	—	
		5100	0·23	0·26	—	
		5800	0·26	0·27	—	
		6600	0·30	0·28	—	
Korn (1940)	Munich	3800	0·32	0·15	—	Mean values for 'large' umbrae.
		4400	0·29	0·19	—	
		4600	0·42	0·14	—	
		3800	0·44	0·11	—	Mean values for 'small' umbrae.
		4400	0·56	0·09	—	
		4600	0·61	0·08	—	
Krat (1948)	Pulkovo	3950	0·28	0·18	10″	
		4250	0·29	0·18	10″	
		4900	0·40	0·16	10″	
		3950	0·12	0·29	11″	
		4250	0·28	0·19	11″	
		4900	0·28	0·22	11″	
Das and Ramanathan (1953)	Kodaikanal	3843	0·34	0·15	20″	
		3888	0·35	0·14	20″	

Reference	Location	Wavelength				Notes
Michard (1953)	Meudon	3300	0·13	0·24	>19″	Mean values for spots whose umbral diameters exceed 19″.
		3659	0·15	0·24	>19″	
		4043	0·16	0·26	>19″	
		4795	0·22	0·25	>19″	
		5452	0·26	0·26	>19″	
		6040	0·31	0·25	>19″	
Ramanathan (1954)	Kodaikanal	4000	0·19	0·23	25″	
		5000	0·27	0·23	25″	
		6500	0·36	0·23	25″	
		4000	0·39	0·13	17″	
		5000	0·47	0·13	17″	
		6500	0·55	0·14	17″	
		4000	0·26	0·19	15″	
		5000	0·34	0·19	15″	
		6500	0·44	0·19	15″	
Makita and Morimoto (1960)	Tokyo	4795	0·04–0·23	0·54–0·25	23″	Range of values for 'large' umbrae.
		5451	0·07–0·36	0·51–0·19	23″	
		6040	0·10–0·31	0·49–0·25	23″	
		4795	0·07–0·32	0·45–0·19	16″	Range of values for 'medium' umbrae.
		5451	0·09–0·37	0·46–0·19	16″	
		6040	0·15–0·38	0·40–0·20	16″	
		4795	0·09–0·53	0·41–0·11	12″	Range of values for 'small' umbrae.
		5451	0·07–0·53	0·51–0·12	12″	
		6040	0·16–0·62	0·39–0·10	12″	
Stumpff (1961)	Göttingen	4000–8600	—	0·61	33″	$\Delta\Theta$-values calculated from the published temperatures; see original paper for intensity ratios.
		4000–8600	—	0·33	32″	
		4000–8600	—	0·44	28″	
		4000–8600	—	0·35	25″	
		4000–8600	—	0·31	24″	
		4000–8600	—	0·61	22″	
		4000–8600	—	0·30	21″	
		4000–8600	—	0·25	17″	
Bray and Loughhead (1962)	Sydney	5400	0·26	0·25	24″	Over most of inner umbra.
		5400	0·15	0·36	24″	At darkest point.

Note: extensive umbral intensity measurements have also been published by Stepanov (1957); however, the data necessary for calculating the umbral diameters are not given in his paper.

It will be noticed from Table 4.1 that in almost all cases $I^*(\lambda)/I(\lambda)$ increases with λ. However, in accordance with expectation, values of $\Delta\Theta$ for a given spot are independent of λ. (The only exceptions are the values derived from the observations of Pettit and Nicholson (1930) and Richardson (1939), where a slight apparent increase with λ occurs.) It is accordingly permissible to *average* the values of $\Delta\Theta$ obtained at different wavelengths for a given spot by each investigator. These average values are plotted as a function of umbral diameter in Fig. 4.1 above, where the points represent observed values, the horizontal arrow represents Michard's mean value for umbral diameters exceeding 19″ of arc, and the vertical lines represent the ranges of values found by Makita and Morimoto for umbrae having the mean diameters indicated in Table 4.1.

Figure 4.1 shows that large umbrae are darker than small ones; a smooth curve has been drawn through the observed values in an attempt to represent this variation. This fact has been pointed out by a number of workers, including Wanders (1935), Wormell (1936), Michard (1953), and Makita and Morimoto (1960).[2]

TABLE 4.2

Umbral Temperatures at Unit Optical Depth as a Function of
Umbral Diameter

UMBRAL DIAMETER

sec of arc	km	$\Delta\Theta = 5040\left(\dfrac{1}{T^*} - \dfrac{1}{T}\right)_{T^*=\tau=1}$	$T^*_{\tau^*=1}$	$\Delta T = T_{\tau=1} - T^*_{\tau^*=1}$
10	7,250	0·18	5200°	1200°
20	14,500	0·29⁵	4700	1700
30	21,700	0·37	4300	2100
40	29,000	0·42	4200	2200
50	36,200	0·45⁵	4100	2300
60	43,500	0·48	4000	2400

Figure 4.1 also shows that for spots of a given size there is considerable scatter in the values of $\Delta\Theta$. While some of the scatter may be due to systematic differences from one observer to another, the results of some of the individual observers (for example, those of Makita and Morimoto) also show considerable scatter. There is little doubt that it is real, individual umbrae of a given size differing in temperature, on occasion, by as much as 1000°.

Despite the scatter, the smooth curve in Fig. 4.1 provides a useful means of summarizing the available observational data, and values of $\Delta\Theta$ derived from it (for umbral diameters ranging from 10″ to 60″ of arc) are given in Table 4.2. As explained earlier, the temperatures T^* and T in (4.3) and (4.4) refer to the

[2] Sitnik (1940) considers that the observed variation is only apparent and is a consequence of the fact that the parasitic light due to seeing increases with decreasing spot area. Although this is true, it should be pointed out that the method of correction (cf. Section 4.2.6) is such that an *over*-correction is just as likely as an *under*-correction (Michard, for example, actually obtained some negative values for the corrected umbral intensity). Therefore, the effect of the corrections is to increase the scatter of the results rather than to introduce a spurious trend.

values at approximately unit optical depths in the umbra and photosphere respectively. The values of T^* are given in the fourth column of Table 4.2, while the fifth gives ΔT, the difference between the temperature at unit optical depth in the photosphere (taken to be 6400° from Plaskett, 1955: cf. Fig. 2) and the temperature at unit optical depth in the umbra. In a statistical sense the umbral temperatures in Table 4.2 should be very reliable, since they are based on a large number of completely independent determinations.

4.2.3 WAVELENGTH DEPENDENCE OF UMBRA/PHOTOSPHERE INTENSITY RATIO

The wavelength dependence of the ratio $I^*(\lambda)/I(\lambda)$ has been measured over more or less limited spectral regions by a number of workers (see Table 4.1).

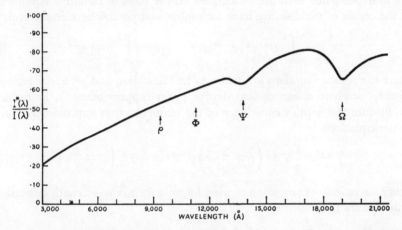

Fig. 4.2. Wavelength dependence of umbra/photosphere intensity ratio, according to Pettit and Nicholson (1930). The Greek letters indicate the positions of atmospheric water-vapour bands.

The most extensive observations, however, are those of Pettit and Nicholson (1930) who, using techniques described in the previous section, obtained measurements of the ratio over wavelengths ranging from 3000 to 22,000Å. Pettit and Nicholson's curve is shown in Fig. 4.2.

Between 3000 and 17,000Å the curve is nearly linear, the ratio increasing from 0·21 at 3000Å to 0·80 at 17,000Å. The positions of four infra-red atmospheric water-vapour bands are indicated by their conventional symbols: it will be noticed that rather marked dips occur at Ψ and Ω, although none occur at the positions corresponding to the other two bands. It is not known whether these dips are a real feature of the umbra/photosphere intensity ratio curve or whether they are merely indirect effects of these bands on the measurement of rather small quantities. It is interesting to note that Stumpff (1961) has detected a *discontinuity* in the curve in the region between 4500 and 5000Å;

its magnitude is about 12 per cent. The discontinuity does not appear in Pettit and Nicholson's curve, and its origin is obscure.

Although Fig. 4.2 probably represents the *variation* of $I*(\lambda)/I(\lambda)$ with λ quite well, some caution should be exercised in making use of the actual values since Pettit and Nicholson failed to make adequate correction for parasitic light. For this reason it is not surprising that $\Delta\Theta$-values calculated from their data (see Table 4.1) are not only rather small but, in addition, fail to show the expected constancy with λ. However, the wavelength dependence found by other observers over more limited regions in general agrees with Pettit and Nicholson's curve where the spectral regions coincide. In the absence of more modern observations the curve of Fig. 4.2 represents the best data available.

Minnaert and Wanders (1932) were the first to compare the observed wavelength dependence with one calculated on the basis of radiative equilibrium. At the centre of the disk, we have for umbra and photosphere respectively

$$I*(\lambda) = \int_0^\infty B*(\tau*)\,e^{-\tau*}\,d\tau*; \quad I(\lambda) = \int_0^\infty B(\tau)\,e^{-\tau}\,d\tau, \qquad (4.5)$$

where the source functions are taken to be Planckian, and $\tau*$ and τ represent suitably weighted mean optical depths ('grey' approximation). In radiative equilibrium the depth dependence of the temperature is approximately given by the equations

$$T*^4 = \frac{1}{2}\,T_{\text{eff}}^{*4}\left(1 + \frac{3}{2}\,\tau*\right); \quad T^4 = \frac{1}{2}\,T_{\text{eff}}^4\left(1 + \frac{3}{2}\,\tau\right). \qquad (4.6)$$

Hence we readily obtain the following formula for the wavelength dependence of the umbra/photosphere intensity ratio:

$$\frac{I*(\lambda)}{I(\lambda)} = \int_0^\infty \frac{e^{-\tau*}\,d\tau*}{e^{\frac{c_2}{\lambda T_{\text{eff}}^*}\left(\frac{1}{2} + \frac{3}{4}\tau*\right)^{-\frac{1}{4}}} - 1} \bigg/ \int_0^\infty \frac{e^{-\tau}\,d\tau}{e^{\frac{c_2}{\lambda T_{\text{eff}}}\left(\frac{1}{2} + \frac{3}{4}\tau\right)^{-\frac{1}{4}}} - 1}. \qquad (4.7)$$

There is only one disposable parameter in (4.7), namely T_{eff}^*, the effective temperature of the umbra: a single value of T_{eff}^* should therefore lead to the correct values of $I*(\lambda)/I(\lambda)$ at all wavelengths, if the assumption of radiative equilibrium is valid.

Minnaert and Wanders and, more recently, Unsöld (1955: cf. Fig. 192) compared (4.7) with Pettit and Nicholson's curve. Unsöld 'corrected' the latter curve by subtracting 0·04 units from the original values, and over the range 3000–10,000Å obtained good agreement between theory and observation with values for T_{eff}^* and T_{eff} of 4480° and 5740° respectively. (This leads to a value for $\Delta\Theta_{\text{eff}}$ of 0·25.) However, beyond 10,000Å the theoretical curve falls below the observed one. A similar procedure was later used by Stumpff (1961), who compared his own observations of $I*(\lambda)/I(\lambda)$ over a wavelength range of 4000–8600Å with values calculated on the basis of radiative equilibrium. Stumpff used Mark's exact solution for the depth dependence of the temperature, rather than (4.6); once again good agreement between theory and observation was obtained. By comparing the theoretical and observed curves he derived values of T_{eff}^* for several spots. The corresponding $\Delta\Theta$-values have already been given in Table 4.1.

The good agreement between the observed and calculated curves from the near ultra-violet to the near infra-red shows that the temperature distribution in sunspot umbrae must approximate to that which would obtain in radiative equilibrium. (The difference between observation and theory for $\lambda > 10,000$Å may possibly indicate departures from the radiative distribution, but it may be due merely to errors of observation.) This conclusion is perhaps somewhat surprising in view of the existence of umbral granules, which almost certainly represent convection currents carrying part of the outward energy flux (cf. Section 3.6.4). How is this apparent discrepancy to be explained? Perhaps the answer may lie in modern theoretical studies of fluid convection which take account of its intrinsically non-linear character. Such an investigation has been carried out by Kuo (1961), who found that the temperature distribution in a convecting fluid is nearly the same as that in the *absence* of convection except in relatively narrow regions at the top and bottom of the convection zone.

It is interesting to note that Michard's *observational* model (discussed in the next section), which has a steeper temperature gradient at small optical depths than the radiative model, gives good agreement with Pettit and Nicholson's observations. Nevertheless, more measurements of $I^*(\lambda)/I(\lambda)$, particularly in the infra-red, would undoubtedly contribute further to our understanding of the radiative properties of the umbra.

4.2.4   COS θ-DEPENDENCE OF UMBRA/PHOTOSPHERE INTENSITY RATIO: PHYSICAL CONDITIONS AS A FUNCTION OF DEPTH

In the same paper as that in which they compared the observed wavelength dependence of the umbra/photosphere intensity ratio with the predictions of radiative equilibrium, Minnaert and Wanders (1932) also calculated the dependence of the ratio on $\cos\theta$ (θ = heliocentric angle). They found that in radiative equilibrium the ratio is nearly independent of $\cos\theta$, and subsequent observations, both in integrated radiation (Wormell, 1936; Sitnik, 1940) and at selected wavelengths (Richardson, 1933b, 1939; Wanders, 1935; Michard, 1953; Makita and Morimoto, 1960), were in agreement with this prediction within the limits of observational error.[3] These observations therefore provided additional evidence that the temperature distribution in sunspot umbrae is approximately the same as that which would obtain in radiative equilibrium.

The good agreement between observation and theory led Waldmeier (1942) to construct a *radiative* model of the umbra, the temperature distribution with depth being given by (4.6). A similar procedure was later followed by Berdichevskaya (1954), but most subsequent workers (Michard, 1953; Mattig,

---

[3] Michard found, however, that in the ultra-violet the ratio decreases towards the limb.

1958a; Makita and Morimoto, 1960) have preferred to use the observed $\cos\theta$-dependence of $I^*(\lambda)/I(\lambda)$ to construct *observational* models of the umbra. Such a procedure, which has proved very successful in the case of the photosphere (see, for example, Plaskett, 1955), has the advantage of involving the fewest possible assumptions. These models are described below.

(1) *Michard's model.* Observations were made at Meudon of 32 spots, using a Hilger quartz spectrograph and a coelostat; many of the spots were observed at two or more values of $\cos\theta$. Measurements were made at six wavelengths little affected by Fraunhofer lines: $\lambda\lambda 6040, 5452, 4795, 4043, 3659$, and 3300. Corrections were made for scattered light and seeing according to the methods described in Section 4.2.6. In view of the magnitude of the corrections found to be necessary for the smaller spots, all deductions are based on the results for large spots only. Although Michard takes his model as referring to a spot with an umbral diameter of $24''$ of arc ($17{,}600$ km), it should be pointed out that his $\Delta\Theta$-values, shown in the fourth column of Table 4.3, are more appropriate to a spot with an umbral diameter of $15''$ (cf. Fig. 4.1).

TABLE 4.3

Comparison between Empirical Models of Photosphere and Umbra (Michard, 1953)

$\tau_{5000}, \tau^*_{5000}$	T	T^*	$\Delta\Theta$	p dynes/cm²	p^* dynes/cm²	p_e dynes/cm²	p_e^* dynes/cm²	h km	h^* km
0	4270°	3550°	0·24	—	—	—	—	—	—
0·02	4520	3720	0·24	1·29 × 10⁴	6·17 × 10³	0·89	0·13	0	0
0·05	4760	3880	0·24	2·34 × 10⁴	9·12 × 10³	1·70	0·25	77	413
0·10	5010	4060	0·23	3·47 × 10⁴	1·32 × 10⁴	2·88	0·45	132	733
0·15	5200	4180	0·23	4·37 × 10⁴	1·59 × 10⁴	4·17	0·63	167	927
0·25	5490	4340	0·24	5·89 × 10⁴	1·91 × 10⁴	7·76	0·93	206	1190
0·50	5930	4590	0·25	7·76 × 10⁴	(2·51 × 10⁴)	19·1	(1·59)	256	(1620)
1·00	6450	4870	0·25	9·77 × 10⁴	(3·55 × 10⁴)	58·2	(2·69)	300	(2120)
2·00	7100	5250	0·25	1·15 × 10⁵	(5·25 × 10⁴)	214	(5·43)	332	(2660)
3·00	(7620)	(5490)	(0·26)	(1·23 × 10⁵)	(6·17 × 10⁴)	(562)	(8·32)	(345)	(2975)
4·00	(8080)	(5720)	(0·26)	(1·26 × 10⁵)	(6·76 × 10⁴)	(1230)	(12·9)	(353)	(3190)

Note: less reliable values are given in parentheses. Michard takes his model as referring to an umbral diameter of $24''$ of arc; however, in view of the rather low values of $\Delta\Theta$ it is more appropriate to a spot with an umbral diameter of $15''$ (cf. Fig. 4.1).

For each of the six wavelengths, the measured $\cos\theta$-dependence of $I^*(\lambda)/I(\lambda)$ was represented by a straight line. Once the $\cos\theta$-dependence is known, the dependence of the source function on optical depth τ_λ^* may readily be determined by any one of the methods commonly used in analysing the photospheric limb darkening. If the source function is identified with Planck's function, the dependence of T^* on τ_λ^* may then be established. The resulting temperature distribution is given in the third column of Table 4.3, while the second column gives Michard's photospheric temperature distribution; the optical depths in the first column refer to a wavelength of 5000Å. The $\Delta\Theta$-values, calculated from the formula $\Delta\Theta = 5040\left(\dfrac{1}{T^*} - \dfrac{1}{T}\right)_{\tau^* = \tau}$, are shown in the fourth column.

The next step was to determine p_e^*, the electron pressure in the umbra. This was accomplished by re-discussing ten Bruggencate and von Klüber's measurements of the equivalent widths of certain iron and titanium lines (these observations are described in Section 4.3.2). The equivalent width of a weak Fraunhofer line formed by pure absorption depends in a fairly simple way on the electron pressure, and from the measurements Michard derived the values of p_e^* at optical depths 0·08 and 0·15. The value of p_e^* at $\tau^* = 4$ was derived from the measured magnitude of the Balmer discontinuity. The run of p_e^* with optical depth finally adopted is given in the eighth column of Table 4.3.

The electron pressure p_e^* depends on the gas pressure p^*, the temperature T^*, and the ratio of easily-ionized metals to hydrogen $1/A$. It follows that if p_e^*, T^*, and A are known, p^* can be calculated: this was the procedure followed by Michard. Strömgren's ionization tables were employed and the value of A was taken to be 10^4; the helium abundance was assumed to be negligible. The values of p^* are given in the sixth column of Table 4.3.

Finally, the geometrical depth in kilometres h^* corresponding to optical depth τ^* was calculated from the equation

$$h^* = 4 \cdot 16 \times 10^6 \int_{\tau_0^*}^{\tau^*} \frac{d\tau^*}{\kappa^* \Theta^* p^*}. \tag{4.8}$$

(4.8) is readily derived by integrating the defining relation $d\tau^* = \kappa^* \rho^* \, dh^*$, the perfect gas laws being assumed (κ^* is the continuous absorption coefficient due to H^- and ρ^* is the gas density). The origin of the h-scales for both spot and photosphere was arbitrarily taken to be $\tau_0^* = \tau_0 = 0 \cdot 02$.[4] The values of h^* are given in the last column of Table 4.3.

It is important to realize that apart from the assumption concerning the matching of the zeros of optical depth, Michard's model is entirely empirical: it supposes neither radiative nor hydrostatic equilibrium.

When the sunspot and photospheric models are compared, two interesting conclusions immediately emerge: (a) the umbra is much more *transparent* than the photosphere, optical depth unity, for example, corresponding to 2120 km in the umbra but only to 300 km in the photosphere; (b) in the umbra *the gas pressure is less* than in the photosphere by a factor of two to three. To explain this discrepancy, Michard suggested that magnetic forces may play a role in maintaining the equilibrium of the spot. Both of these conclusions have been contradicted by Mattig (1958a), whose sunspot model is described below.

Michard has compared his empirical sunspot model with a model in radiative and hydrostatic equilibrium (Michard, 1953: Fig. 34): the observed pressures are some five times smaller than those corresponding to hydrostatic equilibrium. In addition, the empirical model shows a steeper temperature gradient at small optical depths than the theoretical model (cf. Section 4.2.3).

(2) *Mattig's model.* Mattig employed the same temperature distribution as

---

[4] Present observations unfortunately do not permit a more satisfactory determination of the level at which equality occurs. However, owing to the relatively steep pressure gradient the effect of any error in the matching of the zeros on the model itself is small. An accurate matching is impossible until we have a more certain knowledge of the relative roles played by hydrodynamic and magnetic forces in maintaining the quasi-equilibrium of the umbra.

Michard, and graphically extrapolated Michard's curve to $\tau^* = 0.001$. However, although the two models proceed from the same temperature distribution, there are gross differences in the derived gas and electron pressures and in the geometrical depths. These differences are due firstly to a difference in the method of calculating the gas pressure and secondly to a difference in the adopted chemical abundances. In contradiction to Michard, Mattig calculated the gas pressure on the basis of *hydrostatic equilibrium*. (Michard's method of calculating the gas pressure is described above.) He justifies this procedure on the grounds that since observation shows the magnetic lines of force to be perpendicular to the solar surface, magnetic forces cannot contribute to the support of the umbral material (see, however, Section 5.4.8). With this assumption, $dp = g\rho \, dh$ and, since $d\tau = \kappa\rho \, dh$,

$$\frac{dp}{d\tau} = \frac{g}{\kappa}. \qquad (4.9)^5$$

The hydrogen/helium ratio was taken to be 5, so that a volume of gas containing one neutral H atom also contains 0.2 He atoms, $1/A$ metal atoms, and $\phi(T)/p_e$ H$^+$ ions, where $\phi(T)$ is the temperature-dependent part of Saha's equation, i.e.

$$\log \phi(T) = -0.48 - 13.53 \times \frac{5040}{T} + \frac{5}{2} \log T. \qquad (4.10)$$

(The figure 13.53 volts represents the ionization potential of hydrogen.) The value of A was taken to be 2×10^4, in contradiction to Michard's value of 10^4.

Formulae for p and p_e were derived as follows: the total mass of a volume of gas containing 1 neutral H atom and 0.2 He atoms is $m_H + 0.2 \times 4 \, m_H = 1.8 \, m_H$ (m_H = mass of H atom), where the mass of the small number of H$^+$ and H$^-$ ions and metal atoms is neglected. Since κ is the absorption coefficient per gram, we have

$$\kappa = \frac{1}{1.8 m_H} (\kappa_H m_H + k_{H^-} p_e), \qquad (4.11)$$

where κ_H is the absorption coefficient per gram of H and k_{H^-} is the absorption coefficient of H$^-$ per neutral hydrogen atom per unit electron pressure. Both κ_H and k_{H^-} have been extensively tabulated as functions of λ and T.

A further relation is obtained by introducing x_i, the degree of ionization of the metals. The total number of electrons in the volume element is $\phi(T)/p_e + x_i/A$ and, since the total number of particles is approximately 1.2, we have

$$\frac{p_e}{p} = \frac{\phi(T)/p_e + x_i/A}{1.2}. \qquad (4.12)$$

---

[5] In the theoretical discussion which follows we shall, for convenience, omit the asterisks on p^*, p_e^*, etc.

From (4.9–4.12) the following formulae for p and p_e are readily derived:

$$p\frac{\mathrm{d}p}{\mathrm{d}\tau} = \frac{1.8g}{\dfrac{\kappa_\mathrm{H}}{p} + \dfrac{k_{\mathrm{H}-}}{1\cdot2\,m_\mathrm{H}}\left[\dfrac{\phi(T)}{p_e} + \dfrac{x_i}{A}\right]}$$
$$p_e = \frac{px_i}{2\cdot4A} + \left[\left(\frac{px_i}{2\cdot4A}\right)^2 + \frac{\phi(T)p}{1\cdot2}\right]^{\frac{1}{2}}$$
$$\text{(4.13)}$$

Besides the unknowns p and p_e, (4.13) contains a further unknown, namely x_i, the degree of ionization of the metals. The calculation of p and p_e from (4.13) therefore had to proceed by successive approximations; at each approximation x_i was calculated anew, using as a basis the metallic abundances published by Weidemann. The values of x_i for the umbra are given in the third and sixth columns of Table 4.4, while the photospheric values are shown in the second and fifth columns.

TABLE 4.4

Degree of Metal Ionization in Photosphere and Umbra (Mattig, 1958a)

τ, τ^*	x_i	$x_i{}^*$	τ, τ^*	x_i	$x_i{}^*$
0·001	0·64	0·11	0·2	0·86	0·33
0·002	0·67	0·12	0·3	0·89	0·39
0·005	0·75	0·13	0·5	0·92	0·47
0·01	0·82	0·14	0·7	0·93	0·54
0·02	0·82	0·16	1·0	0·94	0·59
0·05	0·82	0·19	1·5	0·95	0·66
0·1	0·83	0·25	2·0	0·96	0·72

Mattig's final values for p^*, $p_e{}^*$, and h^*, are given in Table 4.5 which, for comparison, also shows Michard's values for these quantities. The gross differences in the values of p^*, $p_e{}^*$, and h^* given by the two models are well-marked. In fact, we have approximately

$$\frac{p^* \text{ (Matt.)}}{p^* \text{ (Mich.)}} \simeq 7; \qquad \frac{p_e{}^* \text{ (Matt.)}}{p_e{}^* \text{ (Mich.)}} \simeq 2; \qquad \frac{h^* \text{ (Matt.)}}{h^* \text{ (Mich.)}} \simeq \tfrac{1}{4} \text{ to } \tfrac{1}{8}.$$

Particularly serious is the large difference between the height scales, Mattig finding the umbra to be somewhat *less* transparent than the photosphere, whereas Michard finds it to be considerably *more* transparent. A similar discrepancy is present in the values of p^*, Mattig disagreeing with Michard's conclusion that the pressure in the umbra is substantially lower than that in the photosphere.

The validity of Mattig's model depends on his assumption that hydrostatic equilibrium obtains in the umbra. However, this assumption may be incorrect if the umbral magnetic field contains, in addition to the vertical component, a significant transverse component. But this still leaves open the question as to whether the field is 'force-free' or not (see Section 8.3.3). At the moment, therefore, it is not possible to adjudicate between Michard's and Mattig's models on the basis of theoretical considerations alone.

However, a certain amount of help comes from a relatively unexpected quarter, namely, observations of the Wilson effect. The authors have shown, on the basis of a simple geometrical model, that the Wilson effect receives a ready explanation if it is assumed that the umbra is more transparent than the photosphere (see Section 3.8.5). If this interpretation of the Wilson effect is correct, then Michard's model is to be preferred.[6]

TABLE 4.5

Comparison between Michard's (1953) and Mattig's (1958a) Umbral Models

τ^*_{5000}	T^*	p^* dynes/cm² Michard	Mattig	p_e^* dynes/cm² Michard	Mattig	h^* km Michard	Mattig
0	(3100°)	—	—	—	—	—	—
0·001	(3330)	—	$8·13 \times 10^3$	—	0·037	—	0
0·002	(3400)	—	$1·18 \times 10^4$	—	0·056	—	25
0·005	(3520)	—	$1·91 \times 10^4$	—	0·10	—	57
0·01	(3620)	—	$2·75 \times 10^4$	—	0·16	—	83
0·02	3720	$6·17 \times 10^3$	$3·89 \times 10^4$	0·13	0·26	0	110
0·05	3880	$9·12 \times 10^3$	$6·17 \times 10^4$	0·25	0·49	413	144
0·10	4060	$1·32 \times 10^4$	$8·71 \times 10^4$	0·45	0·89	733	171
0·20	4270	$1·76 \times 10^4$	$1·23 \times 10^5$	0·79	1·70	1070	206
0·30	4400	$2·03 \times 10^4$	$1·48 \times 10^5$	1·08	2·40	1290	224
0·50	4590	$2·51 \times 10^4$	$1·91 \times 10^5$	1·59	3·80	1620	246
0·70	4720	$2·94 \times 10^4$	$2·24 \times 10^5$	2·04	5·01	1840	261
1·00	4870	$3·55 \times 10^4$	$2·63 \times 10^5$	2·69	6·76	2120	278
1·50	5080	$4·45 \times 10^4$	$3·31 \times 10^5$	3·88	9·77	2440	299
2·00	5250	$5·25 \times 10^4$	$3·80 \times 10^5$	5·43	12·9	2660	314

(3) *Makita and Morimoto's temperature distribution.* These workers have derived only a temperature distribution, not a complete model, from their observations. The observations, which were made at three wavelengths ($\lambda\lambda 4795$, 5451, and 6040), have already been described (Section 4.2.2). The temperature distribution was determined separately for 'large' and 'medium' umbrae (mean umbral diameters 23″ and 16″ of arc respectively) and also for the penumbra (Section 4.4.1), using a procedure similar to Michard's. The results are given in Table 4.6, which also shows Michard's umbral temperature distribution; the second column gives Michard's photospheric temperature distribution.

Except for the rather unreliable surface values, Makita and Morimoto's temperatures for a spot of umbral diameter 23″ lie considerably below Michard's

---

[6] van't Veer (1961) has derived values for p^* intermediate to those of Michard and Mattig from his own observations of the wing strengths of certain iron and magnesium lines. His observations, however, are not adequate for the determination of an independent model umbra: in fact, van't Veer's model assumes the value of $d(\log p^*)/d(\log \tau^*)$, as well as the temperature distribution, found by Michard.

values for a spot of nearly identical size; the difference increases with optical depth and amounts to over 700° at $\tau^* = 2\cdot0$. The lower temperatures found by Makita and Morimoto are reflected in their rather high $\Delta\Theta$-values, shown in the last two columns of Table 4.6. They find representative values for spots of umbral diameter 23″ and 16″ to be about 0·38 and 0·29 respectively: these figures are higher than the values 0·32 and 0·25 given by the smooth curve of

TABLE 4.6

Umbral Temperature Distributions
(*Michard*, 1953; *Makita and Morimoto*, 1960)

$$\Delta\Theta = 5040\left(\frac{1}{T^*} - \frac{1}{T}\right)$$

$\tau_{5000}, \tau^*_{5000}$	T (Michard)	T^* Michard $d = 24''$	Makita and Morimoto $d = 23''$	$d = 16''$	Michard $d = 24''$	Makita and Morimoto $d = 23''$	$d = 16''$
0	4270°	3550°	(3910°)	(3910°)	0·24	(0·11)	(0·11)
0·05	4760	3880	(3960)	(3960)	0·24	(0·21)	(0·21)
0·1	5010	4060	(4000)	(4010)	0·23	(0·25)	(0·25)
0·2	5370	4280	(4080)	(4110)	0·24	(0·30)	(0·29)
0·3	5620	4400	4140	4200	0·25	0·32	0·30
0·4	5790	4510	4160	4280	0·25	0·34	0·31
0·5	5930	4590	4190	4350	0·25	0·35	0·31
0·6	6060	4670	4220	4420	0·25	0·36	0·31
0·7	6170	4730	4250	4490	0·25	0·37	0·31
0·8	6270	4780	4270	4550	0·25	0·38	0·30
0·9	6360	4830	4300	4610	0·25	0·38	0·30
1·0	6450	4870	4330	4670	0·25	0·38	0·30
1·2	6620	4960	4370	4770	0·25	0·39	0·30
1·4	6770	5030	4420	4870	0·26	0·40	0·29
1·6	6880	5110	4470	4960	0·25	0·40	0·28
1·8	7010	5180	4500	5050	0·25	0·40	0·28
2·0	7100	5250	4540	5150	0·25	0·40	0·27

Note: less reliable values are given in parentheses; d = diameter of umbra in seconds of arc ($1'' = 725$ km).

Fig. 4.1 for umbrae of these dimensions. This discrepancy does not necessarily imply observational error in Makita and Morimoto's work since, as we have seen in Section 4.2.2, the temperatures of individual umbrae of a given size show considerable scatter. However, it may be that Makita and Morimoto's umbral temperatures are a bit on the low side, whereas Michard's are too high.

To complete this discussion of umbral models it is instructive to consider de Jager's suggestion (1959: p. 161) that since $I^*(\lambda)/I(\lambda)$ is observed to be nearly independent of $\cos\theta$, the umbral temperature distribution can be obtained simply by adding the observed $\Delta\Theta$ to the photospheric Θ-values, Michard's more refined procedure being hardly justified in view of the limited accuracy of the observations. The theoretical justification for this suggestion is as follows: for any given θ the value of the ratio $I^*(\lambda)/I(\lambda)$ is determined by the equation

$$\left[\frac{I^*(\lambda)}{I(\lambda)}\right]_\theta = \frac{\int_0^\infty B^*(t)\, e^{-t\sec\theta}\sec\theta\, dt}{\int_0^\infty B(t)\, e^{-t\sec\theta}\sec\theta\, dt}, \tag{4.14}$$

where B^* and B are the umbral and photospheric source functions respectively. If now $I^*(\lambda)/I(\lambda)$ is *independent of* θ, so that

$$\frac{I^*(\lambda)}{I(\lambda)} = A(\lambda), \quad \text{say}, \tag{4.15}$$

it follows from (4.14) and (4.15) that

$$\int_0^\infty [B^*(t) - A(\lambda) \cdot B(t)]\, e^{-t\sec\theta}\sec\theta\, dt \equiv 0.$$

This identity can be satisfied if, and only if,

$$\frac{B^*(t)}{B(t)} \equiv A(\lambda),$$

showing that the ratio of the source functions at equal optical depths in the umbra and photosphere is independent of optical depth. Hence from (4.15)

$$\frac{I^*(\lambda)}{I(\lambda)} = \frac{B^*}{B}. \tag{4.16}$$

In the visible and ultra-violet regions of the spectrum, and at the temperatures concerned, it is permissible to use Wien's approximate form of Planck's radiation law to calculate B^* and B. With the aid of this law we obtain from (4.16) the relation

$$\frac{I^*(\lambda)}{I(\lambda)} = e^{-\frac{c_2}{\lambda.5040}\cdot\Delta\Theta} \qquad (c_2 = \text{const.}), \tag{4.17}$$

showing that $\Delta\Theta$ is also independent of optical depth. It follows that the umbral temperature distribution can be obtained from the photospheric temperature distribution simply by adding $\Delta\Theta$—calculated from (4.17) or, more conveniently, from (4.4)—to the photospheric Θ-values.

Turning now to the observations, we see from Table 4.6 that Michard's $\Delta\Theta$-values are indeed remarkably constant, as are Makita and Morimoto's values for spots of mean umbral diameter $16''$. However, the values given by the latter authors for spots of mean umbral diameter $23''$ show a steady increase of $\Delta\Theta$ with τ^*, indicating that for this group of spots the measured values of $I^*(\lambda)/I(\lambda)$ are *not* independent of $\cos\theta$. In the general case, therefore, it is preferable to use Michard's procedure. On the other hand, temperature distributions have so far been derived for only three values of the umbral diameter (16″, 23″, and 24″): should one require a model for an umbra of *any* selected diameter, the best procedure—in the absence of measurements of the cos θ-dependence of the intensity ratio—would be to follow de Jager's suggestion, using a $\Delta\Theta$-value appropriate to the chosen diameter derived from Fig. 4.1 or Table 4.2.

Finally, we should point out that in view of the large discrepancies between Michard's and Mattig's results, the present position with regard to sunspot models is far from satisfactory. There is, in fact, a great need for further theoretical work on the subject. On the observational side, it is unfortunate that most of the existing observational data concerning the cos θ-dependence of $I*(\lambda)/I(\lambda)$ consist of measurements of a number of *different* spots at various heliocentric angles. Ideally, measurements should be made of a *single, stable* spot throughout its passage across the solar disk. Such observations would yield more refined umbral temperature distributions than those described above, although they would not solve the problem of the discrepancies in the other physical quantities.

4.2.5 ELECTRICAL CONDUCTIVITY OF THE UMBRA

Having derived numerical values for the umbral gas pressure, electron pressure, and temperature as functions of optical depth, we are now in a position to calculate the electrical conductivity σ of the umbral material. As we shall see in Chapter 8, this quantity occupies an important place in the magnetohydrodynamic theory of sunspots since it determines, among other things, the rate at which electric currents flowing in the spot can be dissipated by Joule heating. This in turn leads to certain basic conclusions concerning the possible modes of formation and disappearance of sunspot magnetic fields. We shall find in this section that σ is considerably smaller in the umbra than in the photosphere, although the umbra remains a good conductor in the astrophysical sense.

In calculating σ for the case of a partially ionized gas such as a spot umbra, the photosphere, or the lower chromosphere, one must not only take account of collisions of charged particles with one another but also collisions between charged particles and neutral hydrogen atoms. It is true that the collisional cross-sections for collisions of the latter type are much smaller but, on the other hand, neutral hydrogen atoms vastly outnumber any other particles present. Equations for the electrical conductivity of such a gas have been obtained by a number of authors, including Nagasawa (1955). Nagasawa's formulae are somewhat complicated for the purpose of computation, but Kopecky (1957, 1958) has shown that under photospheric conditions a considerable simplification is possible. Kopecky gives the following approximate formula for the conductivity in e.s.u.:

$$\sigma = 0 \cdot 26 \, \frac{e^2}{(3km_e)^{\frac{1}{2}}} \cdot \frac{1}{S} \cdot \frac{p_e}{pT^{\frac{1}{2}}}, \qquad (4.18)$$

where m_e and e are the electron mass and charge, k is Boltzmann's constant, and S is the effective collisional cross-section.

The validity of (4.18) depends on (a) the relatively low ionization which obtains in the photosphere and sunspot umbrae, and (b) the assumption that $S \geqslant 10^{-15}$ cm². The actual value of S is somewhat uncertain: Alfvén has quoted a value of the order of 10^{-15} cm² for collisions involving electrons whose energy does not exceed a few hundred electron volts (this is certainly the case for the visible layers of the photosphere and umbra), while some more recent estimates

TABLE 4.7

Electrical Conductivity of Photosphere and Umbra

τ, τ^*	σ (e.s.u.)	σ^* (e.s.u.)	σ^*/σ
0·02	$1\cdot0 \times 10^{11}$	$3\cdot4 \times 10^{10}$	0·34
0·05	$1\cdot0 \times 10^{11}$	$4\cdot3 \times 10^{10}$	0·42
0·10	$1\cdot1 \times 10^{11}$	$5\cdot2 \times 10^{10}$	0·46
0·15	$1\cdot3 \times 10^{11}$	$6\cdot0 \times 10^{10}$	0·46
0·25	$1\cdot7 \times 10^{11}$	$7\cdot2 \times 10^{10}$	0·42
0·50	$3\cdot1 \times 10^{11}$	$(9\cdot2 \times 10^{10})$	(0·29)
1·00	$7\cdot3 \times 10^{11}$	$(1\cdot1 \times 10^{11})$	(0·15)
2·00	$2\cdot2 \times 10^{12}$	$(1\cdot4 \times 10^{11})$	(0·06)
3·00	$(5\cdot1 \times 10^{12})$	$(1\cdot8 \times 10^{11})$	(0·03)
4·00	$(1\cdot1 \times 10^{13})$	$(2\cdot5 \times 10^{11})$	(0·02)

Note: values in parentheses are less reliable owing to uncertainties in Michard's umbral and photospheric models.

give $S \simeq 10^{-14}$ cm² (see Kopecky, 1957). Kopecky has compared (4.18) with the more exact formulae of Nagasawa. Under photospheric conditions the differences are very small, particularly at small optical depths, say $\tau < 1$. At larger values of τ slight differences between the approximate and exact formulae begin to occur, but they remain less than errors due to uncertainty in the photospheric model itself. The errors are due to the steadily increasing value of p_e/p, which has the effect of gradually invalidating one of the approximations made.[7]

If we can assume that under the conditions which obtain in sunspots the conductivity is, to a first approximation, independent of the strength of the

---

[7] In calculating the average mass of the ions present, Kopecky incorrectly assumes that hydrogen ionization can be neglected, and he accordingly takes this quantity to be $40m_{\rm H}$. While it is true that metal ions are more abundant than hydrogen ions at very small optical depths, hydrogen ions quickly become predominant at greater depths. This may readily be shown by calculating the degree of hydrogen ionization $x_{\rm H} = \phi(T)/p_e$ on the basis of the photospheric model given in Table 4.3 ($\phi(T)$ is given by equation 4.10). $x_{\rm H}$ remains less than x_i, the degree of metal ionization, but hydrogen atoms are about 10^4 times as abundant as metal atoms and hydrogen ions become predominant as the depth increases. However, this error does not invalidate the derivation of (4.18), because the relevant terms involving the mean ionic mass conveniently disappear when the main approximation is made.

In equation (19) of Kopecky (1957), $3/(8\pi)$ should read $3/(8\sqrt{\pi})$.

magnetic field, then (4.18) can also be applied to sunspot umbrae. (For a discussion of the effect of a magnetic field on the conductivity of a partially ionized gas, see Cowling, 1962: p. 260.) Table 4.7 gives values of σ, σ^*, and σ^*/σ calculated according to Michard's photospheric and umbral models (Table 4.3). Values of σ^* calculated according to Mattig's umbral model are about one-third of the values given in the third column of Table 4.7. However, the greatest uncertainty in the values of σ and σ^* is due to uncertainty in the value of S, the effective collisional cross-section, which may be in error by an order of magnitude (this, on the other hand, does not affect σ^*/σ).

It will be noticed that between $\tau = 0.02$ and $\tau = 4.0$ the photospheric conductivity increases by two orders of magnitude, whereas the umbral conductivity increases by less than an order of magnitude over the same range of

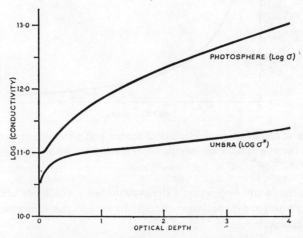

Fig. 4.3. Electrical conductivities of photosphere and umbra as functions of optical depth. Apart from the region $\tau^* < 1$, the electrical conductivity in the umbra is one to two orders of magnitude smaller than that at equivalent optical depths in the photosphere.

optical depths. The run of log σ and log σ^* with τ and τ^* is shown in Fig. 4.3, where the umbral conductivity is seen to increase with depth at a much slower rate than the photospheric conductivity.

Finally, in Fig. 4.4 the ratio of the umbral and photospheric conductivities (calculated at equal optical depths) is plotted as a function of optical depth. Between $\tau = \tau^* = 0.02$ and $\tau = \tau^* = 0.15$, σ^*/σ increases slightly (from 0.34 to 0.46). Thereafter, however, there is a steady decline until at $\tau = \tau^* = 4.0$, σ^* is only 2% of σ. Hence, apart from the region $\tau^* < 1$, *the electrical conductivity in the umbra is one to two orders of magnitude smaller than that at equivalent optical depths in the undisturbed photosphere.* The disparity is even more marked when comparison is made between values calculated at equal *geometrical* depths since, as we have seen in the previous section, the umbra is more transparent than the photosphere.

It is of interest to compare the conductivities of the photosphere and umbra with the conductivities of some familiar terrestrial substances. This comparison is shown in Table 4.8. The photospheric and umbral gases are evidently not 'good' conductors in the commonly-used sense, but rather have a conductivity similar to that of a semi-metal such as tellurium. The conductivity of the umbra is approximately the same as that of a saturated salt solution. However, when

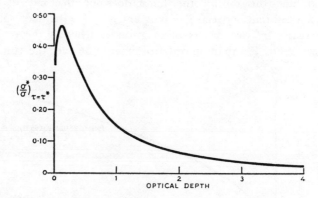

Fig. 4.4. Ratio of electrical conductivities of umbra and photosphere as a function of optical depth. Note: σ^* and σ are calculated at *equal optical depths*.

account is taken of the *large scale* of the systems with which we are concerned, these conductivities are quite large enough to ensure the importance of magnetohydrodynamic forces.

TABLE 4.8

Comparison between Conductivities of Photosphere, Umbra, and some Familiar Substances

SUBSTANCE	σ (e.s.u.)
Copper	$5\cdot3 \times 10^{17}$
Carbon	$2\cdot6 \times 10^{14}$
Tellurium	$4\cdot5 \times 10^{12}$
Photosphere	$10^{11}\text{--}10^{13}$
Sunspot umbra	$10^{10}\text{--}10^{11}$
Saturated solution of NaCl (15°C)	$1\cdot8 \times 10^{11}$
Commercial plate glass	$4\cdot5 \times 10^{-2}$
Shellac	$9\cdot0 \times 10^{-5}$

Note: 'terrestrial' data from *Handbook of Chemistry and Physics*, 28th ed., 1944: Chemical Rubber Publishing Co., Cleveland, Ohio.

4.2.6 CORRECTION OF SUNSPOT INTENSITY MEASUREMENTS FOR PARASITIC LIGHT

The reliability of many of the umbral intensity measurements described in Sections 4.2.1 and 4.2.2—and hence the reliability of the temperatures and models derived from them—depends to a certain degree upon the accuracy with which the measured intensities can be corrected for parasitic light from the surrounding photosphere. Accordingly, it is appropriate to include in this chapter a brief discussion of the methods of correction commonly used.

The parasitic light may be split into two components, which originate in different ways and require different methods for their correction:

(1) a component due to *scattered light*, which originates in the Earth's atmosphere and at one or more of the optical components of the telescope. Insofar as the condition of the telescope and atmosphere remains constant, the scattering function (i.e. the curve giving the scattered light intensity as a function of distance, measured in the image plane, from its source) is independent of time. It should also be reasonably independent of position in the image plane. The scattering function can be determined by measuring the distribution of scattered light beyond the solar limb. Kinman (1953) has devised a convenient graphical method of correcting for scattered light; Kinman's method has been fully described by Goldberg and Pierce (1959: p. 10), who also give references to a number of other determinations of scattered light intensity in various solar telescopes.

As we have seen in Section 2.4.4 the scattered light intensity tends to be rather high in the conventional, all-mirror, tower-type of solar telescope but is much lower in the photoheliograph-type instrument, particularly when a suitable diaphragm is placed at the primary focal plane;

(2) a second component due to the *combined instrumental profile of telescope and atmosphere*. The most widely-used method of correcting for this component is that of Wanders (1934). Many of the measurements described in Sections 4.2.1 and 4.2.2 were corrected by this method, which may be described as follows:

Let $O(x,y)$ = the observed intensity profile of a sunspot;
$\quad T(x,y)$ = its true profile;
$\quad A(x,y)$ = the combined instrumental profile of telescope and atmosphere.

Then the integral equation of obliteration can be written

$$O(x,y) = \int_{-\infty}^{+\infty} \int_{-\infty}^{+\infty} T(x',y') \cdot A(x-x', y-y') \, dx' \, dy'. \qquad (4.19)$$

Let us now assume that the instrumental profile is a Gaussian function with radial symmetry, so that

$$A(r) = \frac{a}{\pi} e^{-ar^2}, \qquad (4.20)$$

where $r = (x^2 + y^2)^{\frac{1}{2}}$ and a is a constant (the factor a/π ensures that the integral of $A(r)$ over the whole plane equals unity).

The value of a is determined by measuring the profile of the solar limb and comparing the observed curve with a set of curves *computed* for various values of the parameter (see Wanders, 1934: Fig. 2). Knowing a, and hence $A(r)$, the true spot profile can be obtained from the observed profile by solving (4.19) by any one of the methods available for solving integral equations of this type (see, for example, Goldberg and Pierce, 1959: p. 9). In solving the integral equation, most observers assume that the spot possesses circular symmetry, but Stumpff (1961) takes exact account of the spot geometry.

In many cases the atmosphere (i.e. the seeing), rather than the telescope, dominates the instrumental profile. The validity of Wanders' method then depends on the correctness of the following two assumptions: (*a*) the seeing is the same at different parts of the image, and (*b*) the seeing is independent of time. In the case of short exposures, e.g. ~ 1 millisecond, both of these assumptions are false, as we have seen in Section 2.2.4. In the case of fairly long exposures, e.g. ~ 1 second or more, an averaging effect appears and the assumptions are probably closer to the truth.

In conclusion, it is probably true to say that those umbral intensity measurements which refer to fairly large spots are reliable; but for small spots, where large corrections must be applied, the corrected values may be subject to considerable uncertainty (we recall that Michard (1953) actually obtained *negative* corrected values for some of his smaller spots). Penumbral measurements should be much more reliable, and it is noteworthy that the penumbral $\Delta\Theta$-values given in Table 4.12 are in good agreement among themselves. Umbral observations made with an effective resolution of $1-2''$ of arc should require no correction for instrumental profile.

4.2.7   ISOPHOTAL CONTOUR MAPS

In the previous sections of this chapter we have tacitly assumed that the umbra possesses a uniform brightness, so that it is permissible to regard a single effective temperature or a single umbral model as being applicable to all points of an umbra of given size. And in fact, several authors have remarked that although photometric traces across a sunspot often show a smooth transition from photosphere to penumbra and penumbra to umbra, the transition becomes much steeper and the intensities within the umbra and penumbra much more uniform when proper corrections are made for parasitic light (see, for example, Wormell, 1936; Sweet, 1955: p. 676; de Jager, 1959: p. 157). However, isolated photometric traces simply cannot suffice to give an accurate idea of the intensity distribution throughout the spot as a whole. This is clearly evident from the qualitative description of the detail found in sunspot umbrae given in Section 3.6.3 (cf. Plates 3.15–3.18). Sunspot umbrae, in reality, show a very complex structure: a large umbra, for example, may possess two, three, or even more points of minimum intensity. There are often large differences in intensity between different regions of an umbra, and even an apparently regular sunspot *pore*, such as the one illustrated in Plate 3.7 (*c–d*), can show a marked difference in brightness from one side to the other.

A far more adequate method of investigating the intensity distribution inside a sunspot is to prepare a complete *two-dimensional photometric map*. This procedure has been carried out by the present authors (Bray and Loughhead, 1962) using two films of two large spots (mean umbral diameters 45″ and 24″) taken on January 12 and May 11, 1959 respectively. Photographs of these spots are reproduced in Plates 3.17 and 3.18. The film of May 11 contains, in addition to normal exposures of the spot, sequences of twofold, threefold, and fivefold overexposures. The latter two sequences, which served to bring up the detail in the umbra, were taken through a small diaphragm placed at the prime focus (cf. Section 3.6.2). This eliminated scattered light within the telescope and no corrections for parasitic light, whether due to scattering or seeing, were necessary. (In the film of January 12, diaphragmed exposures of up to sevenfold normal were available.)

Microphotometry was carried out with a circular scanning aperture of effective diameter 2″ of arc. This value was deliberately chosen so as to wash out the fine detail present on the negatives: with a smaller scanning aperture, the photometric traces would have been impossibly confused in both umbra and penumbra (the umbral granulation, for example, has a mean cell size of 2″.3), making the subsequent drawing of the isophotes very difficult. The level of the photospheric intensity was established by making traces both near to and well away from the spot; no significant variation with distance from the spot could be detected (cf. Section 3.3.3). The sunspot on each of the selected exposures was then microphotometered, the line of scan being displaced vertically 2″ after each run, so that every point of the spot was ultimately measured. On the overexposures the photosphere is of course 'burnt out', so in order to express the umbral intensities in terms of the photospheric intensity, a step-by-step procedure had to be adopted. This consisted of matching, by virtue of their size and shape, isophotes derived from negatives with different degrees of exposure, the lightest giving intensities directly in photospheric units. There is some possibility of error here, since a given isophote could conceivably alter in the interval between successive sequences of exposures, which sometimes extended for an hour or more; and indeed, significant changes were occasionally noticed. However, no real difficulty was experienced in applying the matching process.

The resulting isophotal contour maps for the spot of May 11 are shown in Figs. 4.5 (*a*) (penumbra and outer umbra) and 4.5 (*b*) (umbra). With regard to the penumbra the following points are noteworthy:

(1) the isophotes are very irregular and reflect the presence of both bright and dark regions, which are also evident on a photograph of the spot (Plate 3.18*a*);

(2) there is a gradual decrease in intensity from the photosphere (1·00) to the inner edge of the penumbra ($\sim$0·70), although this is not readily apparent on

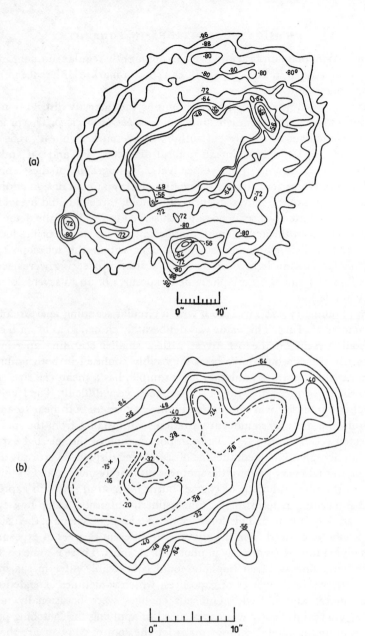

(a)

(b)

Fig. 4.5. Isophotal contour maps (spot of May 11, 1959).

(a) Penumbra and outer umbra.

(b) Umbra. The darkest region is located well towards the left-hand end of the umbra; its temperature is some 500° below that of the rest of the umbra. For other noteworthy features, see text.

Note: some of the isophotes are incomplete, owing to truncation by the focal-plane diaphragm used in taking the original photographs.

the photograph. However, over quite a large part of the penumbra the intensity lies in the range 0·72–0·80;

(3) the umbra–penumbra boundary is characterized by a sharp increase in the intensity gradient and a marked simplification in the shapes of the isophotes.

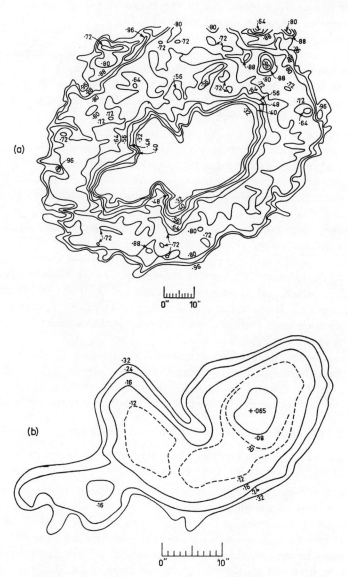

Fig. 4.6. Isophotal contour maps (spot of January 12, 1959).

(a) Penumbra and outer umbra.

(b) Umbra. In this spot the dark umbral core is located more centrally than in the spot of Fig. 4.5.

The intensity distribution within the umbra has the following noteworthy features:

(1) the sharp gradient at the boundary continues for a short distance into the umbra, after which the intensity becomes more uniform; over a large part of the inner umbra the intensity lies in the range 0·24–0·28, giving a $\Delta\Theta$-value lying somewhat below the smooth curve of Fig. 4.1;

(2) the isophotes are still irregular in shape;

(3) there are three intensity minima, one of which has an intensity of only 0·15 (this leads to a $\Delta\Theta$-value lying *above* the curve of Fig. 4.1);

(4) this extra-dark region, which we shall call the 'core' of the umbra, is only a few seconds of arc in diameter *and is located not at the centre of the umbra but near one end*, to the left of a 'fork' structure (cf. Plate 3.18b–c).

The larger spot of Figs. 4.6 (a) and (b) leads to similar conclusions. The only noteworthy difference is that there is no extensive region of more or less constant intensity in the umbra (the spot of Fig. 4.5 is *not* typical in this regard), although the gradient decreases as the central regions are approached. Figure 4.6 contains very useful information about the morphology of the umbral and penumbral isophotes of this large spot, but for various photometric reasons (cf. Bray and Loughhead, 1962) not too much reliance can be placed on the absolute values of the intensities. They have not been included in Table 4.1.

One of the most important results of the work described above is the demonstration of the existence of a 'core' in the umbra. The core is only a few seconds of arc in diameter, and its temperature is as much as 500° less than that of the rest of the umbra. It is interesting to note that as early as 1916, Chevalier hinted at the existence of an umbral core; it is probable that some of the early visual observers also noticed this feature.

The results described in this section raise the question of what relationship, if any, the complex intensity structure of the umbra bears to the magnetic field distribution. For example, is there a concentration of the field in the umbral core? These interesting and fundamental questions are discussed in Section 5.4.9.

4.3 The Fraunhofer Spectrum of Sunspot Umbrae

4.3.1 GENERAL CHARACTERISTICS OF THE SUNSPOT FRAUNHOFER SPECTRUM; EQUIVALENT SPECTRAL CLASS

As a direct consequence of the umbra's lower temperature, its spectral class is later than that of the photosphere—dKo as compared to dGo–2 for the photosphere. It is interesting to see whether the $\Delta\Theta$-value calculated from the observed spectral class agrees with the value obtained from intensity measurements made in the continuous spectrum: taking the effective temperatures of the photosphere and a dKo star (Ambartsumyan, 1958: Table 11, p. 252) to be 5785° and 4910° respectively, we obtain

$$\Delta\Theta_{\text{eff}} = 5040 \left(\frac{1}{4910} - \frac{1}{5785} \right) = 0\cdot16.$$

In comparison with the $\Delta\Theta$-values given in Table 4.2, this figure is very low. It follows that the spectral class of the umbra is decidedly *earlier* than the temperature derived from intensity measurements made in the continuous spectrum would lead one to expect. The origin of this discrepancy is unknown; it may be partly due to a difference in stratification between a sunspot umbra and the atmosphere of a dK0 star.

Continuing the analogy between the umbral spectrum and that of a dK0 star, let us compare the electron pressures obtained by Michard and Mattig with the value expected for such a star. According to Ambartsumyan (1958: Table 8, p. 221) the latter value is 1·2 dynes/cm², while from Table 4.5 we see that the electron pressure in the umbra attains this value at $\tau^* \simeq$ 0·3 (Michard) or 0·1 (Mattig). Hence the agreement is good.

However, the analogy with stellar spectra should not be pushed too far. Van Dijke (1946) has directly compared the sunspot curve of growth (cf. Section 4.3.2) with that of 70 *Ophiuchi* A, a K0 dwarf, with the result that although the temperatures agree very well and the gas pressures agree fairly well, the relative numbers of absorbing atoms show systematic differences. In fact, she found that $\log (N_{spot}/N_{star})$ was positive for all neutral atoms but negative for all ionized atoms—a result which, like the discrepancy in the $\Delta\Theta$-values pointed out above, may be due to a difference in stratification of the two atmospheres.

The substantial temperature difference between photosphere and umbra, which as we have seen can exceed 2000° for a large or medium-sized spot, results in a substantial difference in the intensities of individual spectral lines and groups of lines. It is convenient to discuss the behaviour of *atomic* and *molecular* lines separately. In the case of atomic lines, some are weakened in the sunspot spectrum, others strengthened. Among the principal Fraunhofer lines, for example, the lines of the Balmer series are weakened whereas $\lambda 4227$ of Ca I and the H and K lines of Ca II are strengthened. Lines due to neutral atoms of low ionization potential present in the umbral spectrum tend to be absent from the photospheric spectrum owing to the onset of ionization. Examples are the leading lines of Li, Rb, and In, which were first detected as 'spot' lines since they occur only as mere traces in the photospheric spectrum (Moore, 1953). On the other hand, many photospheric lines are completely absent from the spot spectrum. Plate 4.1 shows two of the twenty-six 100Å sections which comprise the Mt Wilson photographic map of the sunspot spectrum,[8] together with identical wavelength regions of the photospheric spectrum; marked differences between the two spectra are very apparent.

The sunspot spectrum displays hundreds of molecular lines absent from the photospheric spectrum. In fact, according to H. D. Babcock (1945), under the best conditions extensive regions of the sunspot spectrum are found to show no

---

[8] Hale and Ellerman (1920). An accompanying catalogue of atomic lines in the sunspot spectrum has been published by Moore (1933). Work on the atomic and molecular spectra of sunspots carried out at Mt Wilson in the early years of the century by Hale, Adams, and their collaborators has been summarized by Goldberg (1953).

true continuum at all! Instead, there are innumerable tenuous features due to the superposition of lines too faint to be seen separately, arising from overlapping molecular bands. Unidentified molecular bands in the sunspot spectrum are far more numerous and extensive than identified ones. As an example of the complexity of the spectrum, there are nearly 5000 lines of just a single identified compound, titanium oxide, between 3900 and 7000Å (Abetti, 1929: p. 134). The complexity is very well illustrated by a high-dispersion spectrogram of the region in the neighbourhood of the Mg b lines published by ten Bruggencate and von Klüber (1939: Fig. 12). Numerous molecular lines, most of which are due to magnesium hydride, almost obliterate the continuum of the sunspot spectrum in this region, although the photospheric continuum is relatively undisturbed. Fortunately, the Zeeman effect due to the spot magnetic field provides an excellent means of distinguishing molecular from atomic lines (Moore, 1953): in spectra obtained with a Nicol prism and a compound quarter-wave plate most atomic lines display the characteristic zig-zag pattern due to magnetic splitting (cf. Section 5.3.2), but this is absent in the molecular lines (see, for example, Russell, Dugan, and Stewart, 1927: Fig. 201). The identification of molecular compounds in the umbra and the determination of rotational temperatures from band-line intensities are discussed in Section 4.3.4.

To complete this short review of the general characteristics of the Fraunhofer spectrum of the umbra, we must draw attention to the very useful compilation by Merrill (1956) of the behaviour of the lines of the chemical elements and their compounds in astronomical spectra. With the aid of this treatise one can quickly ascertain the characteristic behaviour of the lines of any given element or compound occurring in various astronomical sources, including sunspots and the solar photosphere. Merrill gives copious references to the literature.

4.3.2 SUNSPOT CURVES OF GROWTH: EXCITATION TEMPERATURE, ELECTRON PRESSURE, AND NON-THERMAL VELOCITY

Our principal source of information about the physical conditions in sunspots consists of intensity measurements in the continuous spectrum, which have been extensively discussed in Section 4.2. Nevertheless, it is important to obtain supporting data from other sources whenever this is possible. One such source is the *sunspot curve of growth*, obtained by measuring the equivalent widths W of a number of multiplet lines and plotting log (W/λ) against the quantity log X given by the relation

$$\log X = \log gf + \log \lambda - \frac{5040}{T_{\text{ex}}^*} \chi, \qquad (4.21)$$

where g = statistical weight of the lower level of the transition; f = oscillator strength; χ = excitation potential of the lower level; T_{ex}^* = excitation temperature of the umbra.

The excitation temperature is found by plotting curves of growth for a number of values of T_{ex}^* and determining which value best represents the observations (see, for example, Abhyankar and Ramanathan, 1955; Howard, 1958; Kornilov, 1961). Secondly, if curves of growth are constructed for successive ionization states of a given element, it is possible—with the aid of Saha's equation—to obtain the electron pressure p_e^*. Finally, from the height of the flat, transitional region of the curve of growth (AB in Fig. 4.7) it is possible (knowing T_{ex}^*) to determine the non-thermal component ξ^* of the atomic velocities.[9] In this section we shall assume that the umbral magnetic field has

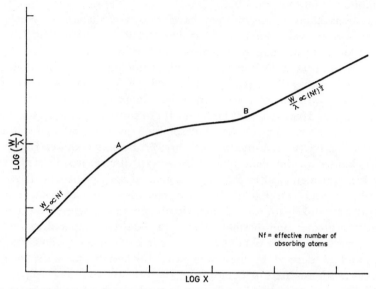

Fig. 4.7. Schematic curve of growth. W is the equivalent width, λ the wavelength; $\log X$ is given by (4.21).

no effect on the equivalent widths although, as we shall see in the next section, there is currently some uncertainty on this point.

ten Bruggencate and von Klüber (1939, 1944) were the first to obtain sunspot curves of growth from photometric measurements of the equivalent widths of multiplet lines. In all, 32 lines of Fe I, 40 lines of Ti I, and 28 lines of Ti II were measured in both sunspot and photosphere. Excitation temperatures of 3800–3900° and 5040° were obtained for the umbra and photosphere respectively and, from the relative numbers of Ti I and Ti II atoms in the ground state in spot and photosphere, p_e^*/p_e was found to be 1/27. However, this value for the ratio is much smaller than the value given by Michard's model (Table 4.3) at the appropriate optical depth. The gas pressures in umbra and photosphere, derived by applying Strömgren's method of model atmospheres, were found to be very nearly equal—again in contradiction to Michard.

[9] We prefer to use the term 'non-thermal' rather than the more usual word 'turbulent' to describe this component: as in the case of the photosphere, the observations provide no evidence for the existence of turbulence in the strict aerodynamic sense (cf. Loughhead and Bray, 1959).

ten Bruggencate and von Klüber were also the first to notice that the transition region of the sunspot curve of growth (AB in Fig. 4.7) lies above the corresponding region of the photospheric curve. They attributed this to *magnetic intensification* (cf. Section 4.3.3); however, this interpretation was later criticized by a number of workers, including Michard (1953). Michard carried out a complete re-discussion of ten Bruggencate and von Klüber's measurements and derived entirely different values for $p_e{}^*$, which he incorporated into his umbral model (Section 4.2.4).

The observations of later workers have broadly confirmed ten Bruggencate and von Klüber's value for T_{ex}^*. Abhyankar and Ramanathan (1955), working at Kodaikanal, measured the equivalent widths of 82 Cr I, 70 Fe I, and 74 Ti I lines in the umbrae of four spots and found values of T_{ex}^* for the three groups of lines of $4030 \pm 80°$, $4200 \pm 150°$, and $3800 \pm 100°$. Howard (1958) obtained a mean value of $4100 \pm 29°$ from two groups of Fe I lines. Like ten Bruggencate and von Klüber, he found the transition region of the curve of growth to lie above that of the photospheric curve. He attributed this to a greater value for the non-thermal velocity, ξ^* being 2·9 km/sec compared to $\xi = 1·7$ km/sec for the photosphere. Zwaan (1959) measured lines of Ti I and Fe I; his equivalent widths are in good general agreement with those of ten Bruggencate and von Klüber but disagree with those of Abhyankar and Ramanathan. Like Howard, Zwaan attributed the raising of the transition region to a greater non-thermal velocity (2·4 km/sec compared to 1·2 km/sec for the photosphere). He obtained a value for T_{ex}^* of $3900 \pm 200°$. The values of T_{ex}^* and ξ^* derived by these and other workers by the curve of growth method are summarized in Table 4.9.

All but one of the values for ξ^* given in the last column of Table 4.9 exceed the value $\xi = 1·7$ km/sec for the photosphere (at the centre of the disk) given by Allen (1955). Although it is not certain how much significance should be attached to these figures, they do provide additional evidence for the view that the magnetic field in the umbra fails to suppress motions of the umbral material— a conclusion to which observations of the umbral granulation have already led us (Section 3.6.4).

In the fourth column of Table 4.9 we have given the value of $\Delta\Theta_{ex}$, calculated on the assumption that the excitation temperature of the photosphere is 5000° (this figure is itself subject to considerable uncertainty). Apart from a very low value from Zhuravlev's Ti I measurements, these values are in good agreement with the $\Delta\Theta$-values derived from intensity measurements made in the continuous spectrum (cf. Fig. 4.1 and Table 4.2). Hence in spite of the many uncertainties inherent in the curve of growth method (for example, the rather crude theory of line formation upon which it is based, the assumption that all lines are formed at the same optical depth, uncertainties in the gf-values, etc.) we can conclude that the Fraunhofer line spectrum and the continuous spectrum give satisfactorily concordant results.

Finally, we must mention T. V. Krat's interesting method (see Ambartsumyan,

TABLE 4.9

Excitation Temperatures and Non-Thermal Velocities in the Umbra

REFERENCE	ELEMENT	T_{ex}^*	$\Delta\Theta_{ex}$	ξ^*(km/sec)
ten Bruggencate and	Ti I, Fe I	3800°	0·32	—
von Klüber (1939, 1944)..	Ti I, Ti II	3900	0·28	—
Abhyankar and	Ti I	3800 ± 100°	0·32	—
Ramanathan (1955)......	Fe I	4200 ± 150	0·19	—
	Cr I	4030 ± 80	0·24	—
Zhuravlev (1957)[1]	Ti I	4670 ± 40	0·07	2·1
	Fe I	3840 ± 200	0·30	1·2
Howard (1958)	Fe I	4100 ± 29	0·22	2·9
Mattig (1958a)[2]	—	—	—	2·8, 3·7
Stepanov (1958)[3]	—	—	—	1·8
Zwaan (1959)	Ti I, Fe I	3900 ± 200	0·28	2·4
Kornilov (1961)	Ti I	4100	0·22	—
	Ti II	3600	0·39	—
	Fe I	4000	0·25	—
	Fe II	3600	0·39	—
	V I	3700	0·35	—

[1] See Zwaan (1959).
[2] This author made use of ten Bruggencate and von Klüber's data.
[3] Derived from the observed profile of $\lambda6173$ of Fe I.

1958: p. 295) of determining p_e^*/p_e from *strong* lines *without* making use of the curve of growth. We write Saha's equation logarithmically for spot and photosphere and subtract; then

$$\log\frac{p_e^*}{p_e} = \log\frac{N_1}{N_1^*} + \log\frac{N_0^*}{N_0} + \frac{5}{2}\log\frac{T^*}{T} - 5040\chi'\left(\frac{1}{T^*} - \frac{1}{T}\right), \quad (4.22)$$

where N_0 and N_1 are the numbers of neutral and singly-ionized atoms, respectively, of a given element—both in the ground state—and χ' is the ionization potential. The ratios N_1/N_1^* and N_0^*/N_0 can be determined if the neutral and singly-ionized states are represented in both spot and photosphere by strong resonance lines. This is the case for the line $\lambda4227$ of Ca I and the H and K lines of Ca II, and for these lines Krat assumed that the effective numbers of atoms in spot and photosphere were proportional to the squares of the measured equivalent widths (cf. Fig. 4.7). Strictly speaking, T and T^* in (4.22) represent *ionization* temperatures: however, Krat assumed that they are equal to the corresponding *excitation* temperatures which, following ten Bruggencate and von Klüber, he took to be 5040 and 3800°. He derived a value for p_e^*/p_e of 1/40, a figure which once again is substantially lower than the value given by Michard's model. Krat's method suffers from both observational and theoretical uncertainties, and therefore cannot be regarded as being particularly reliable. Nevertheless, it is one of the few observational methods of determining the electron pressure in a sunspot umbra.

4.3.3 MAGNETIC INTENSIFICATION OF FRAUNHOFER LINES

In attempting to derive the physical conditions in sunspot umbrae by the analysis of equivalent widths or line profiles, we encounter a difficulty which we do not normally have to contend with in astrophysics—namely the effect of the magnetic field on the line absorption or scattering coefficient, with a consequent effect on the line itself. Unfortunately, different workers are not yet in complete agreement about the magnitude of this effect. Hence in this section we shall merely try to indicate the factors involved and give references to the more important investigations.[10]

Following Ambartsumyan (1958: p. 293) let us first consider the case of a faint line. The equivalent width of such a line is proportional to N, the number of absorbing atoms (cf. Fig. 4.7). The effect of magnetic splitting is to produce a number of Zeeman components, the equivalent width of each component being proportional to the number of atoms producing it, which is less than N. However, since the total number of atoms remains the same, it is evident that the combined equivalent width is the *same* as in the absence of the field. Now let us consider a very strong line. In this case even fields of the order of several thousand gauss produce a splitting which is considerably less than the extent of the wings. Since the wings provide the main contribution to the equivalent width of such a line, it is clear that once again there is no change in the equivalent width. Finally, let us consider a line of medium strength, lying on the flat, transition region of the curve of growth (AB in Fig. 4.7). As before, the number of atoms producing each Zeeman component is less than the number producing the original line; however, for a line on the flat portion of the curve of growth a large change in the number of absorbing atoms produces only a small change in the equivalent width, so each component has nearly the same equivalent width as the original line. The net result for a medium line is therefore an increase in the equivalent width. Taking all these results together, it is clear that the effect of the magnetic field on the curve of growth is to *raise the transition region*, as originally suggested by ten Bruggencate and von Klüber.

A more refined discussion (Unno, 1956; Stepanov, 1958) shows that a normal Zeeman triplet undergoes no magnetic intensification in the case of a longitudinal field, but this does not apply to lines with anomalous splitting. Since the Zeeman pattern varies from line to line, one evidently should not combine different lines into a single curve of growth without regard to their magnetic splitting patterns (cf. Aller, 1953: p. 347; Warwick, 1955).

However, there is still some doubt concerning the importance of the magnetic intensification effect for fields of the magnitude of those occurring in sunspot

[10] The effect is also important in the case of magnetic stars (Section 8.5).

umbrae, although the theory of the effect has been studied in detail by a number of workers, including Hubenet (1954); Unno (1956, 1959); Mattig (1958a); Stepanov (1958, 1960); Zwaan (1959); and Boyarchuk, Efimov, and Stepanov (1960).[11] Mattig and Zwaan concluded that magnetic intensification was *not* sufficient to explain the observed raising of the transition region of the sunspot curve of growth, which instead they attributed to a greater non-thermal velocity; their derived velocities have already been given in Table 4.9.

Stepanov considered the case of a magnetic field with an *arbitrary* inclination to the line-of-sight, and his theory has been used by Boyarchuk *et al.* in the most complete discussion of magnetic intensification which has so far been published. This discussion is particularly valuable for its analysis of the various factors involved. The authors begin by defining a quantity $q = \log (W^*/W)$ as a measure of the magnetic intensification, where W^* and W are the equivalent widths in the presence and in the absence of the field respectively. They then proceed to investigate how q depends on various relevant factors and obtain the following results:

(1) q increases with γ, the angle between the magnetic field and the line-of-sight, but beyond $\gamma = 55°$ it remains practically constant: this confirms the conclusion reached by Unno (1956);

(2) in the case of *anomalous* splitting in a longitudinal field ($\gamma = 0$), q is proportional to $Hn\delta$, H being the magnitude of the field, n the number of subcomponents, and δ their separation;

(3) q rises to a maximum and then decreases to zero as x_0, the concentration of absorbing atoms, increases;

(4) for all γ and x_0, q decreases as the damping constant increases, the rate of decrease increasing with increasing x_0;

(5) magnetic intensification leads to a significant upward shift of the curve of growth, in disagreement with the conclusions of Mattig (1958a) and Zwaan (1959); it also increases the scatter of the points on the curve of growth.

In addition to its effect on the equivalent width, the magnetic field also affects the line profile. This latter aspect of magnetic intensification is of crucial importance in interpreting measurements of the magnitude and direction of sunspot magnetic fields, and is discussed in Chapter 5.

4.3.4   THE MOLECULAR SPECTRUM OF THE UMBRA

The general characteristics of the molecular spectrum of the umbra have already been described (Section 4.3.1). In this section we shall discuss two additional topics, namely (a) the identification of molecular compounds in the umbra, and (b) the determination of the rotational temperature, T_{rot}^*, from band-line intensities.

The first list of fairly reliable identifications was published by Richardson (1931) and is reproduced in Table 4.10. This table is of particular interest as it

---

[11] Similar calculations have also been made by Warwick (1955, 1957), but Unno (1959) and Stepanov (1960) have shown that his basic equations are incorrect.

includes comparative data for stars and comets. An even more extensive study was later undertaken by Babcock (1945). In general, Babcock confirmed Richardson's identifications; however, in view of the complexity of the molecular spectrum of the umbra it is not surprising that there are some differences (see footnote (2), Table 4.10). A new result due to Babcock was the recognition in the disk spectrum of all the *identified* compounds found in spots; only about one-third of these had been recognized in the disk spectrum before. However, numerous weak bands due to *unidentified* compounds have so far been seen only in spots, and in fact unidentified bands in the sunspot spectrum are more numerous than those now accounted for. It is evident that much work remains to be done.

TABLE 4.10

Molecular Bands Definitely or Tentatively Identified in the Spectra of Sunspot Umbrae, Stars, and Comets (Richardson, 1931)[1,2]

NAME OF BAND	IDENTIFICATION	OCCURRENCE	
		Stars and Comets	*Sun*
Aluminium hydride[2]	AlH	—	Disk?; spot?
Calcium hydride.............	CaH	—	Spot
G-band	CH	F5, G, K, M, N, R; comets	Disk; spot
Raffety bands	CH?	Comets	—
	CN?	Comets	—
Magnesium hydride..........	MgH	o Ceti?	Disk; spot
Nitrogen hydride	NH	—	Disk; spot
Hydrogen monoxide	OH	—	Disk; spot
Silicon hydride	SiH	—	Spot
Aluminium oxide	AlO	o Ceti	Spot
Boron monoxide[2]	BO	—	Spot
Comet tail	CO+	Comets	—
Scandium oxide	ScO	o Ceti	—
Titanium oxide[3]	TiO	M, S	Spot
Yttrium oxide	YO?	o Ceti?	—
Zirconium oxide	ZrO?	S	Spot
Swan[3].....................	C_2	R, N; comets	Disk; spot
Cyanogen	CN	G, K, M, N, R; comets	Disk; spot
Nitrogen	$N_2{}^+$	Comets	—
Hydrogen[2].................	H_2	—	Spot?
Silicon fluoride[2]	SiF?	—	Disk; spot

[1] For the wavelengths of prominent band heads and literature references, see original paper.

[2] Babcock (1945) fails to find AlH, BO, H_2, or SiF in the sunspot or disk spectra, but instead gives the following newly-identified compounds: YO, ScO, O_2, MgO, BH, MgF, SrF. See also Merrill's publication, referred to in Section 4.3.1.

[3] Richardson comments on the interesting fact that the bands of TiO and C_2 in the umbral spectrum are of about equal intensity, whereas in stellar spectra they are invariably mutually exclusive. This shows, once again, that the umbral spectrum cannot be identified with any one spectral class.

Babcock occasionally noticed a marked difference in the intensity of a given molecular band from spot to spot, even when the atomic lines appeared much the same. This is further evidence for the fact that spot umbrae vary in temperature among themselves; Babcock calculated that a difference of only 200–300° should be detectable by this means.

Actual measurement of the band-line intensities yields a value for the temperature of the layer of formation of the lines. The principle behind the method (cf. Waldmeier, 1955: p. 189) is as follows: the various lines in a given molecular band differ from one another only in their different rotational quantum numbers J. The quantum theory gives the following expression for the rotational energy of a molecule in terms of J and the moment of inertia A:

$$E_J = \frac{h^2}{8\pi^2 A} \cdot J(J + 1),$$

where h is Planck's constant. The 'intensity' I_J of a given line is proportional to the number of molecules in the initial state of the transition; this in turn is proportional to the *a priori* probability of the state, $2J + 1$, and to a Boltzmann factor $e^{-E_J/kT_{\rm rot}}$, $T_{\rm rot}$ being the rotational temperature. Hence

$$I_J \propto (2J + 1)\, e^{-J(J+1)h^2/8\pi^2 A k T_{\rm rot}}. \tag{4.23}$$

There are various ways of using (4.23) to find $T_{\rm rot}$. One method involves determining the most intense line in the band: if $J_{\rm max}$ is the quantum number of this line, then it is easy to show that

$$T_{\rm rot} = \frac{h^2(2J_{\rm max} + 1)^2}{16\pi^2 A k}.$$

A may be found from laboratory data.

Values of the umbral and photospheric rotational temperatures obtained by Richardson (1931) and Laborde (1961) from the band spectra of various molecules are given in Table 4.11. The large probable errors show that not too much reliance can be placed on these results; nevertheless, in the last column we have given the corresponding $\Delta\Theta$-values, calculated in the usual way. Richardson's values appear to be systematically too low, but Laborde's are in good agreement with values given by the continuous spectrum (cf. Fig. 4.1 and Table 4.2). However, although the last five of Laborde's values refer to a single spot, they show a rather wide dispersion. In its present form this method of determining umbral temperatures is clearly rather inaccurate.

Both Richardson (1933a) and Laborde (1961) attempted to use their measurements of band-line intensities to find p^*/p, the ratio of the gas pressures in umbra and photosphere. Richardson made use of an equation similar to (4.23), together with the equation of dissociation equilibrium for molecules. From measurements of the $\lambda 4300$ CH band he obtained $p^*/p = 0.5$. Laborde, on the other hand, used Minnaert's formula for the equivalent width of a weak line; he concluded from his measurements of the C_2 lines that $p^*/p \simeq 1$ at the optical depth of formation of these lines ($\sim$0·04). However, Laborde considers this result uncertain, the method used being very sensitive to the temperature distribution adopted for the *photosphere*. Richardson's figure is close to the value given by Michard's model (cf. Table 4.3).

TABLE 4.11

Rotational Temperatures of Photosphere and Umbra

REFERENCE	COMPOUND	T_{rot}	T_{rot}^*	$\Delta\Theta_{rot}$
Richardson (1931)	C_2	6000 ± 700°	4900 ± 600°	0·19
	C_2	5700 ± 1600	4900 ± 750	0·14
	C_2	5300 ± 400	4500 ± 400	0·17
Laborde (1961)	MgH	4640 ± 300	3730 ± 300	0·26
	C_2	4860 —	3600 ± 250	0·36
	C_2	4890 ± 250	3640 ± 250	0·35
	CH	4490 ± 250	3850 ± 200	0·19
	CN	4720 ± 240	3900 ± 180	0·22
	NH	4640 ± 300	3600 ± 250	0·31
	OH	4800 ± 200	3880 ± 150	0·25

4.3.5 EXTENSION OF SUNSPOTS INTO THE SOLAR INTERIOR AND CHROMOSPHERE

Hitherto in this chapter we have considered sunspots as more or less *surface* phenomena. This viewpoint has been imposed upon us by the fact that the radiation carrying the information upon which our knowledge of the physical conditions is necessarily based originates in a rather shallow layer. In fact, even when we accept Michard's transparent model (Table 4.3) the thickness of the layer throughout which we have reasonably precise information about the physical conditions is only two or three thousand kilometres, a figure much less than the lateral extent of the spot. In this section we shall discuss how far the sunspot extends (a) below this layer into the solar interior, and (b) above this layer into the chromosphere.

Since the deeper layers are inaccessible to observation we here have to rely on purely theoretical considerations. Cowling (1953: pp. 569–71; 1957: p. 23) has suggested that the coolness of a sunspot cannot extend to a depth of more than a few thousand kilometres owing to the inability of the so-called 'magnetic pressure' $H^2/8\pi$ to exert a very great influence at depths where the gas pressure exceeds, say, 10^8 dynes/cm²—even when allowance is made for a tighter bunching of the lines of force below the surface. However, this argument presupposes that there is no large increase of magnetic field strength with depth, an assumption which must be regarded as purely speculative at the present time. There is no doubt that sunspots owe their origin to internal magnetic fields, but we have no certain knowledge of either the strength or the variation with depth of such fields, which may depend on such complicated factors as (a) the interaction between the internal fields and deep-seated convection currents, and (b) the variation of the rate of solar rotation with distance to the centre of the Sun. In fact, here we encounter fundamental questions involving the very origin of sunspots and the solar cycle, some of which are discussed in Chapter 8.

Turning now to the chromosphere above sunspots, we find that observations made in Fraunhofer lines of chromospheric origin provide clear evidence that sunspots persist as coherent structures for at least several thousand kilometres into the chromosphere—indeed they may penetrate right through the chromosphere. In the first place, spectroheliograms of spot groups (whether taken in $H\alpha$ or in the H or K line of Ca II) often show at least the larger spots quite distinctly (cf. Section 7.3.1). Moreover, in spite of the rather low resolution of such photographs, the distinction between umbra and penumbra is often maintained. Good illustrations of these facts are to be found in several series of photographs published by Abetti (1929: Figs. 119–21): each series shows the appearance of a spot group in white light, hydrogen, and Ca II.

Secondly, Severny and Bumba (1958), by measuring the Zeeman effect in $H\alpha$, $H\beta$, and K_2 of Ca II, have shown that the spot magnetic field penetrates well into the chromosphere. Directly over the spot, field strengths of the order of 300–500 gauss were measured; outside the spot, but still within the group, they were $\sim$ 100–150 gauss. The height of penetration was estimated to be $\sim$ 2000 km.

A third piece of evidence comes from the behaviour of the bright components H_2 and K_2 at the centres of the H and K lines in the sunspot spectrum. In a plage region the centre of the K line, for example, contains two bright components K_2 separated by a dark reversal K_3 (the K_2 components are fainter in a plage-free region), but over a spot the K_2 components merge and the dark gap disappears. Spectrograms illustrating this phenomenon have been obtained by St John (see Abetti, 1929: Fig. 114), Bumba (1960a), and Mustel and Tsap (1960). The interpretation of the phenomenon is complicated by the possible influence of the Zeeman and Evershed effects, but Mustel (1955) has concluded that the absence of H_3 and K_3 over spots is due to the fact that the temperature of the lower layers of the chromosphere above a spot does not increase appreciably with height, in contradiction to its behaviour outside the spot.[12] In addition, Mustel remarks that direct chromospheric observations, both outside of and during an eclipse, testify to an appreciable lowering of the chromosphere above sunspots.

A final piece of evidence is provided by the independent discovery by Mattig (1958b) and by Severny and Bumba (1958: cf. Plate Ic) that the core of the $H\alpha$ line in the spectrum of a spot near the limb is shifted towards the limb, as if the spot had a 2000 km extension into the chromosphere in the form of a dark, absorbing cloud. In addition to a shift in $H\alpha$, Mattig (1959) has observed

[12] It is not known what relationship, if any, the phenomenon bears to the Wilson–Bappu effect.

similar, though smaller, shifts in $H\beta - H\delta$ and in certain metal lines. He has concluded that in the chromosphere, as in the photosphere, the temperature in a spot is less than in the undisturbed surroundings.

4.4 Physical Conditions in the Penumbra

4.4.1 THE TEMPERATURE OF THE PENUMBRA

The temperature of the penumbra can be derived from intensity measurements by methods identical to those used for the umbra (cf. Sections 4.2.1, 4.2.2, and 4.2.4). As Figs. 4.5 (a) and 4.6 (a) demonstrate, the intensity is *not* uniform over the penumbra. In fact, superimposed on numerous irregularities, there is a gradual decline in intensity from the photosphere to the border of the umbra—

TABLE 4.12

Penumbra/Photosphere Intensity Ratio

REFERENCE	$\lambda(\text{Å})$	$I^*(\lambda)/I(\lambda)$	$\Delta\Theta$
Abetti and	4227	0·57	0·083
Colacevich (1939)	6568	0·82	0·046
Korn (1940)	3800	0·72	0·044
	4400	0·72	0·051
	4600	0·77	0·042
Michard (1953)	3300	0·67⁵	0·045
	3659	0·70⁵	0·045
	4043	0·71⁵	0·048
	4795	0·75	0·048
	5452	0·77⁵	0·049
	6040	0·79	0·050
Stepanov (1957)	3400	0·47–0·62	0·090–0·057
	3670	0·56–0·71	0·076–0·044
	4370	0·66–0·81	0·065–0·032
	4825	0·63–0·87	0·078–0·024
	5310	0·62–0·82	0·089–0·038
	6055	0·69–0·82	0·080–0·042
Makita and	4795	0·63–0·85	0·078–0·027
Morimoto (1960)	5451	0·62–0·84	0·091–0·033
	6040	0·69–0·90	0·078–0·022
Bray and Loughhead (1962)	5400	0·76	0·052
Wormell (1936)	Integrated	0·80	0·051

Note: all $\Delta\Theta$-values except Wormell's have been calculated from (4.4) (based on Wien's approximate form of Planck's law); Wormell's value has been calculated from (4.1) (Stefan's law), assuming T_{eff} for the photosphere to be 5785°.

a fact first pointed out by Pettit and Nicholson (1930). However, it is evident from Fig. 4.5 (a) that over quite a large part of the penumbra the intensity lies in the fairly narrow range 0·72–0·80.

From measurements of a large number of spots, Makita and Morimoto (1960) found the penumbra/photosphere intensity ratio to be independent of the size of the spot, although there was a large scatter from spot to spot. This result disagrees with an earlier conclusion of Richardson (1933b) that the ratio decreases as the size of the spot increases according to the relation $I*/I = 1·35 - 0·282 \log A$, A being the total area of the spot in millionths of the visible hemisphere. In agreement with Michard (1953), Makita and Morimoto found little or no variation of $I*/I$ with the heliocentric angle of the spot.

Table 4.12 summarizes (without regard to spot size) the various penumbral intensity measurements which have been made. Each of the workers referred to has already been mentioned in connection with umbral intensity measurements in Sections 4.2.1–4.2.2 and Table 4.1. The $\Delta\Theta$-values, shown in the last column, are in good agreement. The mean value is 0·054. The corresponding temperature at unit optical depth is almost exactly 6000°, assuming the temperature at unit optical depth in the photosphere to be 6400° (Section 4.2.2). The effective temperature, derived from Wormell's measurements, is 5500°, compared to 5785° for the photosphere.

No comprehensive model of the penumbra has so far been published, although Makita and Morimoto (1960) have derived a temperature distribution from their measurements of the $\cos\theta$-dependence of the penumbra/photosphere intensity ratio. Their distribution is given in the third column of Table 4.13;

TABLE 4.13

Penumbral Temperature Distribution (Makita and Morimoto, 1960)

$\tau_{5000}, \tau^*_{5000}$	T (MICHARD)	$T*$ (MAKITA & MORIMOTO)	$\Delta\Theta$
0	4270°	4310°	—
0·05	4760	4710	0·011
0·1	5010	4910	0·021
0·2	5370	5170	0·036
0·3	5620	5350	0·045
0·4	5790	5490	0·048
0·5	5930	5610	0·048
0·6	6060	5710	0·051
0·7	6170	5800	0·052
0·8	6270	5880	0·053
0·9	6360	5950	0·055
1·0	6450	6020	0·056
1·2	6620	6150	0·058
1·4	6770	6260	0·061
1.6	6880	6370	0·059
1·8	7010	6470	0·060
2·0	7100	6560	0·059

the second column gives Michard's (1953) photospheric temperature distribution. The surface temperatures in the penumbra are not very reliable, so little significance should be attached to the first few values of $\Delta\Theta$ in the fourth column of the table.

4.4.2 MOTIONS IN THE PENUMBRA: THE EVERSHED EFFECT

Motions in sunspot penumbrae were discovered spectroscopically in 1909 by J. Evershed at the Kodaikanal Observatory. They were found to consist predominantly of a radial outflow parallel to the solar surface, of magnitude

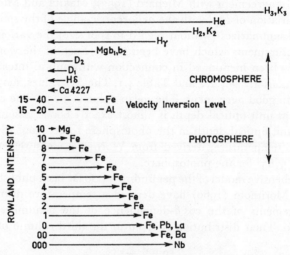

Fig. 4.8. Evershed velocity as a function of line strength (schematic). The lengths of the arrows are proportional to the measured velocities. Arrows to the right indicate an outflow, arrows to the left an inflow (St John, 1913).

~ 2 km/sec. However, in the case of the dark central components H_3 and K_3 of the H and K lines of Ca II (cf. Section 4.3.5) the motion was directed radially *inwards*. A few years later St John, at the Mt Wilson Observatory, measured the Evershed displacements of 506 carefully selected lines, including several which—like the H and K lines—are formed in the chromosphere. Once again the weak lines were found to give a radial outflow while the stronger, chromospheric lines gave a radial inflow (Fig. 4.8), the magnitude of the velocities being about the same in the two cases. Lines of Rowland Intensity $\simeq 20$ showed no Evershed displacement. St John found that for lines of a given intensity the displacements were proportional to wavelength, a fact consistent with the Doppler interpretation. Following St John, the subject was taken up during the sunspot maximum of 1926–32 by G. Abetti at Arcetri. Abetti obtained substantially larger radial velocities than Evershed or St John, ranging from nearly zero to 6 km/sec. In addition, he found evidence for the existence of an irregular *tangential* component of nearly the same magnitude as the radial component. Excellent historical

accounts (with references) of early work on the Evershed effect have been given
by Abetti (1929: pp. 170–9) and Kinman (1952).

One of the most comprehensive investigations of the Evershed effect in
more recent times is that of T. D. Kinman. In his first paper Kinman (1952)
analysed velocity measurements at 237 points in the penumbra of a large regular
spot and its surroundings. No evidence was found for a tangential or vertical
velocity component. The radial component V_r (parallel to the solar surface)
increased from about 1 km/sec at the edge of the umbra to a maximum of about
2 km/sec at the centre of the penumbra; zero velocity was reached well out in

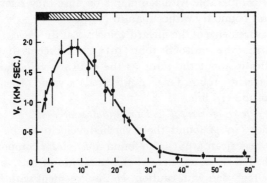

Fig. 4.9. Variation of Evershed velocity with distance from umbral border, according
to Kinman (1952). Distances are given in seconds of arc ($1'' = 725$ km). Vertical lines
represent observational errors.

the photosphere (see Fig. 4.9). In a second paper Kinman (1953) investigated
the dependence of the Evershed velocity on spot size. Results were obtained
for five isolated, regular, stable spots, including one from his previous investigation; all measures were carefully corrected for parasitic light (cf. Section 4.2.6).
The maximum radial velocity, $(V_r)_{max}$, was found to be linearly dependent on
the umbral radius according to the equation $(V_r)_{max} = (0.12 \pm 0.01)a +
(0.56 \pm 0.17)$, a being the umbral radius in seconds of arc ($1'' = 725$ km).
Maximum velocity occurred within the penumbra for the larger spots ($a = 14''.5$,
$14''$, $5''$) but outside it for the smaller spots ($a = 3''.5$, $2''$); in all cases the velocity
field continued beyond the outer edge of the penumbra. One spot showed a
decided asymmetry in V_r, the maximum velocity being bigger on the east side
(2.9 km/sec) than on the west (1.6 km/sec).

The work of later observers has in general confirmed Kinman's results,
although in connection with certain aspects of the Evershed effect there remain
important discrepancies. In the following pages we shall discuss each of these
aspects in turn, and shall try to indicate the extent to which different observers
agree or disagree.

(1) *Magnitude of the radial component V_r.* Holmes (1961) and Servajean (1961)
(who both employed lines free from Zeeman splitting) confirmed Kinman's

figure of 1–2 km/sec as a representative value for V_r. However, Holmes remarks that her value of 0·6 km/sec for a *large* spot is much less than the figure 3·5 km/sec predicted by Kinman's equation for a spot of the relevant size. Abetti (1932; 1957: p. 157) found values ranging from nearly zero to 6 km/sec, and the occurrence of such large velocities was later confirmed by Maltby (1960), who on three occasions found values in the range 6–8 km/sec. (As Maltby points out, these velocities are close to the velocity of sound in the photosphere, 8 km/sec.) Michard (1951) agrees with Kinman that the Evershed velocity increases with spot size. However, his results, based on Abetti's observations, are not directly comparable with Kinman's because they refer to the velocity at the penumbral boundary rather than to $(V_r)_{max}$.

Very few measurements of the inward velocities at the *chromospheric* level have been published since the work of St John (1913), but it seems that the magnitude of these velocities is about the same as the outward velocities at the photospheric level. Severny (1960), for example, finds a value of 1·5–2 km/sec from measurements of the Hβ line.

(2) *Possible existence of vertical and tangential components.* In agreement with Kinman, Holmes (1961) found the Evershed velocity to be predominantly radial, whereas Servajean (1961) also found a *downward* component, of magnitude 0·3 km/sec. However, in view of the smallness of this value Servajean's result should be regarded with caution. In disagreement with Kinman, Abetti (1932; 1957: p. 157) found that a *tangential* component was almost always present. Its mean value was 1 km/sec, the highest value recorded being 3 km/sec; it is rather surprising that a tangential component of this magnitude has escaped detection by later workers.[13] A number of authors have remarked that the Coriolis force due to solar rotation acting on the radial velocity component could be expected to give rise to a tangential component, and have cited the cyclonic or anticyclonic structure observed on Hα spectroheliograms in the neighbourhood of spots (cf. Section 7.3.2) as evidence for this component. However, this structure, when observed, extends far beyond the penumbral boundary—indeed, far beyond the limit of any measured radial velocities. It would therefore seem to be unrelated to the Evershed effect.

(3) *Variation of V_r with distance from centre of spot.* Kinman (1953) found that the Evershed velocity reached a maximum within the penumbra in the case of large spots, as shown in Fig. 4.9, but outside it in the case of small spots. In the spots observed by Holmes (1961) and Servajean (1961) the maximum velocity occurred within the penumbra. On the other hand, Abetti (1957: p. 157) found that the maximum velocity occurred at the *umbral boundary*.

Richardson (1948) noted that the velocity became sharply zero at the outer penumbral boundary, but both Kinman and Holmes found that it continued for a considerable distance into the photosphere, as illustrated in Fig. 4.9. If this were the case then one might expect the granulation to exhibit a very

[13] Evershed occasionally detected a very small tangential component but came to the conclusion that it was not regularly present (see Kinman, 1952).

disturbed appearance in the neighbourhood of spots: yet this is *not* so, as we have emphasized in Section 3.3.2.

Michard (1953) has pointed out that any apparent variation of the Evershed velocity with distance from the centre of the spot may be due to both horizontal and vertical variations, owing to differences in transparency between umbra, penumbra, and photosphere.

(4) *Asymmetry in the Evershed velocity.* Kinman's observation that the magnitude of the radial component is not always the same on opposite sides of the spot has already been mentioned. Abetti (1957: pp. 158–9) noticed a similar asymmetry and, in addition, found that the *sign* of the radial component was sometimes anomalous on one or both sides of the spot, indicating an inflow instead of the outflow usually observed in the weaker lines. Abetti remarks that these abnormal cases usually occur when the umbra is double or complex. Bumba (1960b) found that the *velocities directed away from and towards the observer* on the two sides of the spot could be represented by the empirical relations $V^+ = 2V \sin \theta$, $V^- = V(\sin \theta + \cos \theta)$, where $V \simeq 1$ km/sec and θ is the heliocentric angle of the spot. Thus only at $\theta = 45°$ are the two velocities equal.[14] He explained these relations on the hypothesis that the Evershed flow follows the lines of magnetic force, which fan out from the umbra and therefore present different angles, on the two sides of the spot, to the observer (see Bumba, 1960c: Fig. 1).

(5) *Depth-dependence of V_r.* St John (1913) was the first to establish the fact that the Evershed velocity varies markedly with depth in the photospheric and chromospheric levels of a spot (Fig. 4.8). Unfortunately, in the case of the strong, chromospheric lines, there are very few modern confirmatory measurements. However, Servajean (1961) has measured the velocity for a number of the weaker, photospheric lines originating at different optical depths. The optical depth of formation was calculated from the theory of equivalent widths of weak lines and, although the observations showed some scatter, a definite increase of Evershed velocity with depth was established; the velocity increased from zero at $\tau^* = 0$ (depth = 0 km) to 1·4 km/sec at $\tau^* = 0·1$ (330 km). Since V_r increases with depth, it would be expected to decrease with θ, the heliocentric angle of the spot, because near the limb the effective level of observation is higher. This expectation has been confirmed by Michard (1951), who finds that the observed velocities fit a relation of the form $V_r = V_0 (1 - 0·8 \sin \theta)$.

(6) *The velocity field distribution and its relation to the magnetic field distribution.* Hitherto in this section we have been concerned with velocity measurements either at a handful of points in the penumbra, at a number of points along a single slit position or, in the case of Kinman's (1952) measurements, at five successive slit positions. Of considerable interest, however, are measurements of the *entire velocity field* of a spot and its immediate surroundings. Measurements of this type have been carried out by Bumba (1960c), Semel (1960),

[14] These formulae clearly break down at the centre of the disk ($\theta = 0°$).

Severny (1960), and Servajean (1961). The first three of these obtained, in addition, simultaneous magnetic field measurements. Servajean's velocity map (see his Fig. 8), which refers to a complex group, is very complicated and appears to show no relation whatsoever to the visual aspect of the group. Semel's map is also complex: only around a certain portion of the main spot does it recall the simple configuration expected of an isolated spot. It is very difficult to see how these results can be reconciled with those of other investigations into the Evershed effect described above.

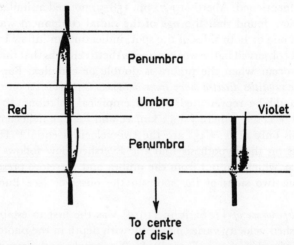

Fig. 4.10. The 'flag' phenomenon in a sunspot spectral line, according to Bumba (1960b). This phenomenon provides support for the hypothesis that the Evershed effect consists of an outflow along the individual penumbral filaments (see Section 4.4.3).

Semel found that the relationship between the velocity field and the magnetic field was not marked, but this he attributed to the fact that only the line-of-sight components of V and H were measured. Bumba (1960c), on the other hand, found a close relation between V and H in the case of single spots; and in fact all his data could be explained on the hypothesis that the motion in the penumbra takes place along the lines of force, which fan out from the umbra. In agreement with Fig. 4.8, he found the velocity at the photosphere-chromosphere boundary to be close to zero. Beneath this region, the velocity is practically parallel to the solar surface, but at greater depths it becomes inclined downward. Severny (1960), from observations of a spot near the limb, showed that the chromospheric inflow seen in $H\beta$ also follows the lines of force, although the magnetic field at this level is only ~ 50 gauss.

(7) *The 'flag' phenomenon.* The application of spectrographs of much higher resolving power than hitherto employed has recently led to the discovery that in the Evershed effect we are concerned not merely with a simple displacement of a spectral line, but also with a broadening, an asymmetry, and even a *doubling* of the line (Bumba, 1960b; Holmes, 1961; Servajean, 1961). The nature of the

phenomenon is, according to Bumba, shown by Fig. 4.10, which is taken from his paper. Bumba finds that generally the line as a whole is *not* shifted, but instead has a strong asymmetry ('flag'), which gives the appearance of a shift (the centre of gravity of the line is shifted). At the point where the Evershed velocity is greatest the flag often becomes a sharp separate line (cf. Fig. 4.10), whose displacement from the centre of the unshifted line often corresponds to a velocity exceeding 5 km/sec. (Holmes gives a figure of 6 km/sec for the extent of the line broadening.) The direction of the flags for fairly strong lines, such as the Na D or the Mg *b* lines, is opposite to that of the weak lines and, in addition, the flags are less conspicuous. Observations of the type described here of the fine structure of Fraunhofer lines in spot penumbrae are still in their infancy.

Despite the various discrepancies noted in the previous pages, one can form the following general, simplified picture of the Evershed effect. In a simple spot, at any rate, the motion at the photospheric level consists of a predominantly radial outflow which is largely confined to the penumbra. The velocity has a magnitude of a few kilometres per second; the magnitude increases with the size of the spot and also, for a given spot, with depth. At the chromospheric level, on the other hand, the motion consists of a radial *inflow* of about the same magnitude as the outflow observed in the photospheric lines.

4.4.3   REMARKS ON THE INTERPRETATION OF THE EVERSHED EFFECT

The suggestion has often been made that the Evershed effect is related in some way to the filamentary structure of the penumbra. For example, in 1958 the present authors remarked: 'The simplest interpretation of the Evershed effect is that it consists of a laminar flow of matter outwards from the umbra along the filaments, which are probably shallow structures of depth comparable with their width'. This hypothesis is attractive in several ways, although it presents at least one serious difficulty, as we shall see below.

In the first place, since there is considerable diversity in the morphology of the penumbral structure from spot to spot (cf. Section 3.5.3) the hypothesis provides a possible explanation for some of the discrepancies in the results of the different observers described in the previous section. As an example, let us consider the spots of Plates 3.10 (*d*) and (*b*). The penumbra of the first of these shows a *blobby* rather than the usual filamentary appearance, whereas in the second the filaments are clearly defined and radial to the umbra, at least over portion of the penumbra. If the hypothesis is correct the first spot would give only a small or zero velocity (the lines might become broadened), whereas the second would show the normal Evershed effect. As a second example, consider the complex spot of Plate 3.9 (*b*). Here, many of the filaments, although clearly defined, run in almost arbitrary directions: in this case, the direction of the Evershed velocity might be expected to be equally arbitrary, depending on which particular part of the penumbra was located on the spectrograph slit. Finally, consider the spot of Plate 3.5 (*c*). In this case some of the filaments are

TABLE 4.14

Physical Conditions in Sunspots: Summary of Data

PHYSICAL QUANTITY	OPTICAL DEPTH	MAGNITUDE Photosphere	MAGNITUDE Umbra	UMBRAL DIAMETER (sec of arc)	SOURCE	SECTION
Effective temperature (T_{eff})......	0·67	5785°K	4160°K	—	Wormell	4.2.1
Radiation temperature (T)	1·0	6400°K	4500°K	25"	Many observers	4.2.2
	1·0	6450°K	4870°K	24"	Michard	4.2.4
	1·0	—	4330°K	23"	Makita and Morimoto	4.2.4
	1·0	—	4670°K	16"	Makita and Morimoto	4.2.4
	0·1	5010°K	4060°K	24"	Michard	4.2.4
	0·1	—	4000°K	23"	Makita and Morimoto	4.2.4
	0·1	—	4010°K	16"	Makita and Morimoto	4.2.4
Excitation temperature (T_{ex})	—	5000°K	3800–4200°K	—	Several observers[1]	4.3.2
Rotational temperature (T_{rot})....	—	4490–6000°K	3600–4900°K	—	Several observers	4.3.4
Gas pressure (p)..............	1·0	$9 \cdot 8 \times 10^4$ dynes/cm²	$3 \cdot 6 \times 10^4$ dynes/cm²	24"	Michard	4.2.4
	1·0	—	$2 \cdot 6 \times 10^5$ dynes/cm²	24"	Mattig	4.2.4
	0·1	$3 \cdot 5 \times 10^4$ dynes/cm²	$1 \cdot 3 \times 10^4$ dynes/cm²	24"	Michard	4.2.4
	0·1	—	$8 \cdot 7 \times 10^4$ dynes/cm²	24"	Mattig	4.2.4
Electron pressure (p_e)	1·0	58 dynes/cm²	2·7 dynes/cm²	24"	Michard	4.2.4
	1·0	—	6·8 dynes/cm²	24"	Mattig	4.2.4
	0·1	2·9 dynes/cm²	0·45 dynes/cm²	24"	Michard	4.2.4
	0·1	—	0·89 dynes/cm²	24"	Mattig	4.2.4

		Photosphere	Penumbra			
Geometrical depth (h)	1·0	300 km	2120 km	24″	Michard	4.2.4
	1·0	—	278 km	24″	Mattig	4.2.4
	0·1	132 km	733 km	24″	Michard	4.2.4
	0·1	—	171 km	24″	Mattig	4.2.4
Degree of hydrogen ionization (x_H)	1·0	$5·1 \times 10^{-4}$	$2·0 \times 10^{-6}$	24″	Michard	4.2.4
	1·0	—	$8·1 \times 10^{-7}$	24″	Mattig	4.2.4
	0·1	$5·0 \times 10^{-6}$	$1·2 \times 10^{-8}$	24″	Michard	4.2.4
	0·1	—	$6·3 \times 10^{-9}$	24″	Mattig	4.2.4
Degree of metal ionization (x_i)	1·0	0·94	0·59	24″	Mattig	4.2.4
	0·1	0·83	0·25	24″	Mattig	4.2.4
Electrical conductivity (σ)	1·0	$7·3 \times 10^{11}$ e.s.u.	$1·1 \times 10^{11}$ e.s.u.	24″	—	4.2.5
	0·1	$1·1 \times 10^{11}$ e.s.u.	$5·2 \times 10^{10}$ e.s.u.	24″	—	4.2.5
Non-thermal velocity (ξ)	—	1·7 km/sec	1·8–3·7 km/sec	—	Several observers[1]	4.3.2
Spectral class	—	dGo-2	dKo	—	van Dijke	4.3.1
Effective temperature (T_{eff})	0·67	5785°K	5500°K	—	Wormell	4.4.1
Radiation temperature (T)	1·0	6400°K	6000°K	—	Several observers	4.4.1
	1·0	6450°K	6020°K	—	Makita and Morimoto	4.4.1
	0·1	5010°K	4910°K	—	Makita and Morimoto	4.4.1
Evershed velocity	—	—	1–2 km/sec	—	Several observers	4.4.2

[1] Note: Zhuravlev's values (cf. Table 4.9) have not been included.

inclined at quite a steep angle to the radial direction and, accordingly, a substantial *tangential* component of the velocity would be expected. It is evident that if the Evershed effect had been investigated in each of these spots, very discrepant results would have been obtained. In particular, three of the spots would have yielded descriptions of the Evershed effect very different from the simplified picture given at the end of the previous section.

The *flag phenomenon* provides additional support for the hypothesis of an outflow along the individual filaments. Here we have clear evidence for the existence of two or even more velocity fields which are superimposed on the spectrograph slit except under the very best seeing conditions. If the hypothesis is correct, then the displaced component of the line corresponds to light from the filaments, whereas the undisplaced component corresponds to light from the darker interfilamentary background, assumed to have only a small (or zero) velocity.

One piece of evidence *against* the hypothesis is the observed variation of the Evershed velocity with height: this would seem to be of sufficient magnitude to prevent the occurrence of a purely laminar flow. Indeed, it would seem that the shearing effect of the vertical velocity gradient would quickly lead to the disintegration of the filaments. Yet individual filaments are observed to persist for periods comparable with the time needed for penumbral material to travel along the entire length of a filament with an Evershed velocity of 1–2 km/sec (Section 3.5.4).

No trustworthy mechanism for the origin of the *driving force* of the Evershed flow has yet been proposed. It is rather interesting to note that at the photospheric level the direction of the motion is *opposed* to the pressure gradient, the pressure in the umbra being less than that in the photosphere (Section 4.2.4). It may be that magnetic forces are involved.

It will be evident from our discussion of the Evershed effect that much observational work remains to be done before its nature is fully understood. Several aspects of the effect remain obscure and, in particular, further observations of the flag phenomenon may bring about considerable revision of our ideas. The most important observational need is for velocity measurements having a spatial resolution of 1″ of arc or better. This is beyond existing technique, but the application of some of the high-resolution observing methods described in Chapter 2 may ultimately bring such observations within the realm of possibility.

4.5 Summary of Data

To complete this chapter, we summarize in Table 4.14 the known quantitative data concerning the physical conditions in sunspot umbrae and penumbrae. This table provides a useful bird's-eye view, but for more comprehensive information the reader should, of course, refer to the appropriate section of the chapter. For convenience, the relevant section numbers are given in the last column of the table.

REFERENCES

ABETTI, G. [1929] 'Solar physics', *Handbuch der Astrophysik*, ed. G. EBERHARD, A. KOHL-SCHÜTTER, and H. LUDENDORFF, vol. 4, p. 57. Berlin, Springer.

ABETTI, G. [1932] 'Moti dei vapori metallici nelle macchie del Sole', *Osserv. e Mem. Arcetri* No. 50, p. 47.

ABETTI, G. [1957] *The Sun*. London, Faber and Faber.

ABETTI, G., and COLACEVICH, A. [1939] 'Temperatura e spettro di una notevole macchia solare', *Osserv. e Mem. Arcetri* No. 58, p. 9.

ABHYANKAR, K. D., and RAMANATHAN, A. S. [1955] 'Equivalent widths of atomic lines in sunspot spectra', *Astrophys. J.* **121**, 739.

ALLEN, C. W. [1955] *Astrophysical Quantities*. University of London Press.

ALLER, L. H. [1953] *Astrophysics: the Atmospheres of the Sun and Stars*. New York, Ronald.

AMBARTSUMYAN, V. A. [1958] *Theoretical Astrophysics*. London, Pergamon.

BABCOCK, H. D. [1945] 'Chemical compounds in the Sun', *Astrophys. J.* **102**, 154.

BERDICHEVSKAYA, V. S. [1954] 'Structure of sunspots and the theory of radiative equilibrium', *Astron. J. U.S.S.R.* **31**, 51.

BOYARCHUK, A. A., EFIMOV, Y. S., and STEPANOV, V. E. [1960] 'The increase in the equivalent widths of absorption lines in a magnetic field', *Astron. J. U.S.S.R.* **37**, 812 (*Sov. Astron. AJ* **4**, 766).

BRAY, R. J., and LOUGHHEAD, R. E. [1958] 'The lifetime of sunspot penumbra filaments', *Aust. J. Phys.* **11**, 185.

BRAY, R. J., and LOUGHHEAD, R. E. [1962] 'Isophotal contour maps of sunspots', *Aust. J. Phys.* **15**, 482.

BRUGGENCATE, P. TEN, and KLÜBER, H. VON [1939] 'Das Spektrum von Sonnenflecken. I. Die Temperatur der Flecken', *Z. Astrophys.* **18**, 284.

BRUGGENCATE, P. TEN, and KLÜBER, H. VON [1944] 'Temperatur und Elektronendruck in Sonnenflecken', *Nachr. Akad. Wiss. Gött., Math.-Phys. Kl.* **1944**, p. 165.

BUMBA, V. [1960a] 'Results of an investigation of the magnetic field of single sunspots', *Izv. Crim. Astrophys. Obs.* **23**, 212.

BUMBA, V. [1960b] 'Results of the study of the Evershed effect in single sunspots', *Izv. Crim. Astrophys. Obs.* **23**, 253.

BUMBA, V. [1960c] 'The connection between the magnetic field and motion in single sunspots', *Izv. Crim. Astrophys. Obs.* **23**, 277.

CHEVALIER, S. [1916] 'Étude photographique des taches solaires', *Ann. Obs. Zô-Sè* **9**, B1.

COWLING, T. G. [1953] 'Solar electrodynamics', *The Sun*, ed. G. KUIPER, p. 532. Univ. Chicago Press.

COWLING, T. G. [1957] *Magnetohydrodynamics*. New York, Interscience.

COWLING, T. G. [1962] 'Magnetohydrodynamics', *Rep. Progress Phys.* **25**, 244.

DAS, A. K., and RAMANATHAN, A. S. [1953] 'Distribution of radiation flux across a sunspot', *Z. Astrophys.* **32**, 91.

DIJKE, S. E. A. VAN [1946] 'A comparative study of the spectra of α Boötis and 70 Ophiuchi A', *Astrophys. J.* **104**, 27.

GOLDBERG, L. [1953] *Introduction to 'The Sun'*, ed. G. KUIPER. Univ. Chicago Press.

GOLDBERG, L., and PIERCE, A. K. [1959] 'The photosphere of the Sun', *Handbuch der Physik*, ed. S. FLÜGGE, vol. 52, p. 1. Berlin, Springer.

HALE, G. E., and ELLERMAN, F. [1920] 'The Mount Wilson photographic map of the sun-spot spectrum', *Pub. Astron. Soc. Pac.* **32**, 272.

HOLMES, J. [1961] 'A study of sunspot velocity fields using a magnetically undisturbed line', *Mon. Not. R.A.S.* **122**, 301.

HOWARD, R. [1958] 'Excitation temperatures and turbulent velocities in sunspots', *Astrophys. J.* **127**, 108.

HUBENET, H. [1954] 'Fraunhofer-Linien im inhomogenen Magnetfeld', *Z. Astrophys.* **34,** 110.

JAGER, C. DE [1959] 'Structure and dynamics of the solar atmosphere', *Handbuch der Physik*, ed. S. FLÜGGE, vol. 52, p. 80. Berlin, Springer.

KINMAN, T. D. [1952] 'Motions in the Sun at the photospheric level. II. The Evershed effect in sunspot Mt Wilson No. 9987', *Mon. Not. R.A.S.* **112,** 425.

KINMAN, T. D. [1953] 'Motions in the Sun at the photospheric level. III. The Evershed effect in sunspots of different sizes', *Mon. Not. R.A.S.* **113,** 613.

KOPECKY, M. [1957] 'Elektrische und Magnetische Erscheinungen in der Sonnenatmosphäre. I. Elektrische Leitfähigkeit in der Sonnenphotosphäre', *Bull. Astron. Inst. Czech.* **8,** 71.

KOPECKY, M. [1958] 'An approximative calculation of electric conductivity in the lower layers of the solar atmosphere', *Electromagnetic Phenomena in Cosmical Physics*, ed. B. LEHNERT, p. 513. Cambridge Univ. Press.

KORN, J. [1940] 'Untersuchungen über die Intensität der Sonnenflecke', *Astron. Nachr.* **270,** 105.

KORNILOV, A. I. [1961] 'Spectrophotometry of sunspots', *Publ. Sternberg Astron. Inst.* No. 117, p. 27.

KRAT, T. V. [1948] 'Temperature and pressure in sunspots', *Izv. Astron. Obs. Pulkovo* **17,** No. 137, p. 1.

KUO, H. L. [1961] 'Solution of the non-linear equations of cellular convection and heat transport', *J. Fluid Mech.* **10,** 611.

LABORDE, G. [1961] 'Étude de la photosphère et des taches solaires à l'aide des bandes moléculaires', *Ann. Astrophys.* **24,** 89.

LOUGHHEAD, R. E., and BRAY, R. J. [1959] ' "Turbulence" and the photospheric granulation', *Nature* **183,** 240.

MAKITA, M., and MORIMOTO, M. [1960] 'Photoelectric study of sunspots', *Pub. Astron. Soc. Jap.* **12,** 63.

MALTBY, P. [1960] 'Note on the Evershed effect in sunspots', *Ann. Astrophys.* **23,** 983.

MATTIG, W. [1958a] 'Zur Linienabsorption im inhomogenen Magnetfeld der Sonnenflecken', *Z. Astrophys.* **44,** 280.

MATTIG, W. [1958b] 'Beobachtungen randnaher Sonnenflecken in Hα', *Naturwiss.* **45,** 104.

MATTIG, W. [1959] 'Beobachtungen randnaher Sonnenflecken in den Balmerlinien Hα bis Hδ', *Monat. der Deut. Akad. Wiss. Berlin* **1,** 65.

MERRILL, P. W. [1956] 'Lines of the chemical elements in astronomical spectra', *Carnegie Inst. Washington Publ.* No. 610.

MICHARD, R. [1951] 'Remarques sur l'effet Evershed', *Ann. Astrophys.* **14,** 101.

MICHARD, R. [1953] 'Contribution à l'étude physique de la photosphère et des taches solaires', *Ann. Astrophys.* **16,** 217.

MINNAERT, M., and WANDERS, A. J. M. [1932] 'Zur Theorie der Sonnenflecke', *Z. Astrophys.* **5,** 297.

MOORE, C. E. [1933] *Atomic Lines in the Sun-spot Spectrum.* Princeton University Observatory.

MOORE, C. E. [1953] 'The identification of solar lines', *The Sun*, ed. G. KUIPER, p. 186. Univ. Chicago Press.

MUSTEL, E. R. [1955] 'On the chromosphere above sunspots', *Izv. Crim. Astrophys. Obs.* **13,** 96.

MUSTEL, E. R., and TSAP, T. T. [1960] 'The spectrophotometry of bright reversals in the H and K lines of the spectrum of spot umbrae', *Izv. Crim. Astrophys. Obs.* **22,** 75.

NAGASAWA, S. [1955] 'Electrical conductivity of the lower chromosphere of the Sun', *Pub. Astron. Soc. Jap.* **7,** 9.

PETTIT, E., and NICHOLSON, S. B. [1930] 'Spectral energy-curve of sun-spots', *Astrophys. J.* **71,** 153.

PLASKETT, H. H. [1955] 'Physical conditions in the solar photosphere', *Vistas in Astronomy*, ed. A. BEER, vol. I, p. 637. London, Pergamon.

RAMANATHAN, A. S. [1954] 'Radiation flux in sunspot umbrae. Paper II', *Z. Astrophys.* **34**, 169.

RICHARDSON, R. S. [1931] 'An investigation of molecular spectra in sunspots', *Astrophys. J.* **73**, 216.

RICHARDSON, R. S. [1933a] 'Hydrocarbon bands in the solar spectrum', *Astrophys. J.* **77**, 195.

RICHARDSON, R. S. [1933b] 'A photometric study of sunspots and faculae', *Astrophys. J.* **78**, 359.

RICHARDSON, R. S. [1939] 'The intensities of sunspots from center to limb in light of different colours', *Astrophys. J.* **90**, 230.

RICHARDSON, R. S. [1948] 'Sunspot groups of irregular magnetic polarity', *Astrophys. J.* **107**, 78.

RUSSELL, H. N., DUGAN, R. S., and STEWART, J. Q. [1927] *Astronomy.* II. *Astrophysics and Stellar Astronomy.* Boston, Ginn and Co.

SEMEL, M. [1960] 'Observations simultanées du champ magnétique et du champ des vitesses dans un grand groupe de taches solaires', *C.R. Acad. Sci.* **251**, 1346.

SERVAJEAN, R. [1961] 'Contribution à l'étude de la cinématique de la matière dans les taches et la granulation solaires', *Ann. Astrophys.* **24**, 1.

SEVERNY, A. B. [1960] 'Some peculiarities of plasma motion in solar magnetic fields', *Izv. Crim. Astrophys. Obs.* **24**, 281.

SEVERNY, A. B., and BUMBA, V. [1958] 'On the penetration of solar magnetic fields into the chromosphere', *Observatory* **78**, 33.

SITNIK, G. F. [1940] 'Dependence of the intensity of sunspots [on] their position on the disk and their geometric area', *Astron. J. U.S.S.R.* **17**, 23.

STEPANOV, V. E. [1957] 'On the problem of determining the temperature of sunspots', *Pub. Sternberg Astron. Inst.* No. 100, p. 3.

STEPANOV, V. E. [1958] 'On the theory of the formation of absorption lines in a magnetic field and the profile of Fe $\lambda6173$Å in the solar sunspot spectrum', *Izv. Crim. Astrophys. Obs.* **19**, 20.

STEPANOV, V. E. [1960] 'The absorption coefficient in the inverse Zeeman effect for arbitrary multiplet splitting and the transfer equation for light with mutually orthogonal polarization', *Astron. J. U.S.S.R.* **37**, 631 (*Sov. Astron: AJ* **4**, 603).

ST JOHN, C. E. [1913] 'Radial motion in sun-spots. I. The distribution of velocities in the solar vortex', *Astrophys. J.* **37**, 322.

STUMPFF, P. [1961] 'Photometrie des Kontinuierlichen Spektrums von Sonnenflecken im Spektralbereich 4000Å–8600Å', *Z. Astrophys.* **52**, 73.

SWEET, P. A. [1955] 'The structure of sunspots', *Vistas in Astronomy*, ed. A. BEER, vol. I, p. 675. London, Pergamon.

UNNO, W. [1956] 'Line formation of a normal Zeeman triplet', *Pub. Astron. Soc. Jap.* **8**, 108.

UNNO, W. [1959] 'Comments on Warwick's paper on the formation of absorption lines in a sunspot', *Ann. Astrophys.* **22**, 430.

UNSÖLD, A. [1955] *Physik der Sternatmosphären*, 2nd ed. Berlin, Springer.

VEER, F. VAN'T [1961] 'Die Bestimmung des Gasdrucks in Sonnenflecken aus Flügelstärken von Fraunhoferlinien', *Z. Astrophys.* **52**, 165; **55**, 208 (1962).

WALDMEIER, M. [1942] 'Der Aufbau der Sonnenatmosphäre', *Helv. Phys. Act.* **15**, 405.

WALDMEIER, M. [1955] *Ergebnisse und Probleme der Sonnenforschung*, 2nd ed. Leipzig, Geest u. Portig.

WANDERS, A. J. M. [1934] 'Die Reduktion für Einstrahlung bei Intensitätsmessungen an Sonnenflecken', *Z. Astrophys.* **8**, 108.

WANDERS, A. J. M. [1935] 'Die Änderung der Sonnenfleckenintensität über die Scheibe', *Z. Astrophys.* **10**, 15.

WARWICK, J. W. [1955] 'Magnetic intensification and the sunspot curve of growth', *Z. Astrophys.* **35,** 245.

WARWICK, J. W. [1957] 'Formation of absorption lines in a sunspot', *Ann. Astrophys.* **20,** 165.

WORMELL, T. W. [1936] 'Observations on the intensity of the total radiation from sunspots and faculae', *Mon. Not. R.A.S.* **96,** 736.

ZWAAN, C. [1959] 'Curves of growth for a large sunspot', *Bull. Astron. Inst. Neth.* **14,** 288.

CHAPTER 5

The Magnetic Field of Individual Sunspots

5.1 Introduction

In this chapter we shall complete our discussion of the physical conditions in individual sunspots by considering their most important property of all: the magnetic field. As we have seen in the previous chapter the most reliable information about the other physical conditions has come from measurements made in the continuous spectrum, while the role of the Fraunhofer spectrum has been largely confined to supplying confirmatory data. However, for the investigation of sunspot magnetic fields we have to depend solely on the observation of the Zeeman effect in specially selected Fraunhofer lines in the sunspot spectrum. The easiest parameter to measure is the *strength* of the field: the measured values range from the order of 100 gauss in the smallest spots or pores to somewhat over 4000 gauss in the largest. However, great difficulties arise when we wish to go further and determine the *direction* and *gradient* of the magnetic field; we then face one of the most difficult observational problems in the whole of solar physics.

The study of sunspot magnetic fields was inaugurated by G. E. Hale in 1908 (cf. Section 1.7), and for the following three decades it remained the almost exclusive preserve of the Mt Wilson Observatory. A number of important discoveries about the magnetic fields of sunspots and sunspot groups were made during this period. In particular, there emerged a picture of the field configuration in individual regular sunspots with the following properties: (a) the field is symmetrical around the axis of the spot; (b) it has its maximum value at the centre of the umbra, the lines of force at this point being perpendicular to the solar surface (longitudinal field); and (c) away from the centre of the umbra the field becomes smaller and inclined to the vertical, an inclination of about 70° being attained at the outer border of the penumbra (transverse field). This picture was later elaborated by workers at the Potsdam Observatory, where an extensive programme of observations of sunspot fields was commenced in 1941, during the early part of the Second World War. However, in the same year Evershed in Great Britain made an observation which, in complete contradiction to the classical picture, indicated the presence of a *transverse* field in the umbra and a *longitudinal* field in the penumbra.

In 1955 observations of sunspot magnetic fields were started with the solar tower telescope of the Crimean Astrophysical Observatory and subsequently a research programme on sunspot fields of considerable importance has developed at this institution. Amongst other advances the Soviet astronomers have extended the use of the solar magnetograph, originally devised by H. W. and H. D. Babcock in 1952 for the measurement of photospheric magnetic fields, to the detailed study of the strong fields in and around sunspots. The presence of transverse fields has been confirmed not only by the Soviet observers but also by Dollfus and Leroy at Meudon, who have used a very sensitive white-light polarimeter to actually measure the consequent plane polarization of the Fraunhofer lines.

Although a great deal is now known about sunspot magnetic fields, the contradiction between the results of modern observations and the classical picture of the field configuration clearly indicates that we do not as yet have an accurate idea of their exact morphology. Three factors have undoubtedly contributed to the present rather unsatisfactory state of this aspect of the subject. In the first place, the observations are intrinsically difficult and are subject to many subtle causes of error, such as scattered light and instrumental polarization. Secondly, in most cases the spatial resolution achieved has not been adequate for studying the field in detail; although a number of new methods have been devised or suggested in recent years, the problem of obtaining magnetic observations with a resolution comparable to that achieved in direct photography has not yet been fully solved. The third difficulty arises in the interpretation of the observations which, in many cases, requires an accurate knowledge of the line profile under the influence of the inverse Zeeman effect. However, such information is available only if the process of line formation in the presence of a magnetic field is properly understood.

We begin (Section 5.2.1) by summarizing the basic spectroscopic principles underlying the direct Zeeman effect observed in laboratory emission spectra. The much more complicated *inverse* Zeeman effect actually observed in solar absorption lines is considered in Sections 5.2.2 and ·5.2.3. A detailed account is given of Unno's calculation of the profile of a line formed by true absorption in the presence of a uniform magnetic field, assuming that the line has a simple triplet splitting. (Because of their simplicity, lines with triplet Zeeman patterns are particularly suitable for magnetic field measurements.) Polarization rules are given for the individual Zeeman components; these rules serve as a basis for determining sunspot polarities. In addition, the relative intensities of the components are computed as a function of the angle between the field and the line-of-sight. Mention is also made of V. E. Stepanov's extensive work on the theory of the formation of an absorption line in the presence of a magnetic field, and the influence of a field varying with depth is briefly considered.

Most methods of measuring sunspot magnetic fields depend on the use of some form of polarizing analyser. Consequently, in Section 5.2.4 we continue our discussion of Unno's analysis in order to elucidate the effect of a quarter- or half-wave plate analyser on the line profile. The selection of Fraunhofer lines suitable for measuring sunspot fields is considered in Section 5.2.5, which is based largely on an exhaustive discussion published by von Klüber in 1948.

The various methods used for measuring the strengths and directions of sunspot fields are described in detail in Section 5.3. We begin with the quarter- and half-wave plate analysers, which provide the standard means of determining the strengths and polarities of sunspot fields on a routine basis. (Routine observations as such are described in Section 5.3.9.) Section 5.3.3 gives an account of two methods based on polarization interferometry, one of which (due to P. J. Treanor) is specifically designed for determining the *directions* of sunspot fields. The principles of the Babcock-type magnetograph are outlined in Section 5.3.4, with particular reference to the magnetograph of the Crimean Astrophysical Observatory. The Crimean magnetograph has been extensively used in the detailed investigation of the fields in and around sunspots.

The application of the magnetograph to the measurement of sunspot fields raises a new problem: as ordinarily used, the response of the instrument is a linear function of the longitudinal component of the field, but this is true only for weak fields. The range of linearity can be extended by choosing a line of lower magnetic sensitivity but, even then, a detailed calculation taking account of the conditions of line formation is required to define the precise range of linearity. This problem is considered in Section 5.3.5, where Severny's suggestion of using a magnetograph to measure the transverse component of the field is also briefly discussed. Section 5.3.6 describes another very sensitive photoelectric device, the white-light polarimeter developed at the Meudon Observatory, which is claimed to be capable of detecting transverse fields as small as 40 gauss.

All of the methods described suffer to a greater or lesser degree from the problems of scattered light and instrumental polarization, which are dealt with in Section 5.3.7. To complete the discussion of techniques, we summarize in Section 5.3.8 the relative merits of the existing methods and discuss some other ideas which may ultimately lead to improved observations of sunspot magnetic fields.

The available information about the magnetic fields of *individual* sunspots is set out in Section 5.4. (The magnetic properties of sunspot groups are dealt with in Chapter 6.) In view of the uncertain state of our knowledge we first describe the information which falls more or less within the classical framework (Sections 5.4.2–5.4.7) and then consider the modern observations of transverse fields in sunspot umbrae (Section 5.4.8). The topics discussed in the first group include the dependence of field strength on spot area, the variation of the field strength across a spot and its connection with the intensity distribution, the direction of the lines of force, the variation of field strength with depth, and the magnetic life-history of sunspots. Although the classical picture of the configuration of sunspot fields clearly requires major revision, it is not yet clear, on the basis of the modern magnetic observations alone, what new picture will finally emerge. However, on the assumption that some physical connection exists between the detailed distributions of field strength and brightness, some insight into the probable configuration can be obtained from the high-resolution, two-dimensional photometric maps given in Section 4.2.7. On this basis the authors have proposed a new model of the field configuration in the umbra which is consistent with the modern magnetic observations; it is described in Section 5.4.9.

5.2 The Inverse Zeeman Effect in Sunspots

5.2.1 BASIC SPECTROSCOPIC PRINCIPLES: THE EMISSION ZEEMAN TRIPLET

Consider an atom radiating in a uniform magnetic field **H**. Under the influence of the field any spectral line emitted is split into a number of components which

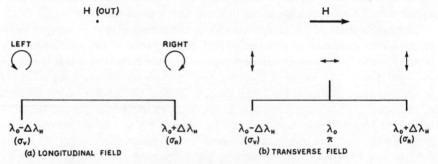

Fig. 5.1. Polarizations of the components of an emission Zeeman triplet.

(*a*) Longitudinal field directed towards the observer: only the two σ components are present. They are circularly polarized in opposite directions.

(*b*) Transverse field: both the π and σ components are present, the former being plane polarized parallel to the field and the latter perpendicular to the field.

are either linearly polarized parallel to the field or circularly polarized in planes perpendicular to the field. The former are referred to as π and the latter as σ components, σ deriving from the German word 'senkrecht'. The Zeeman components are always symmetrical about the position of the undisturbed line, both with respect to the displacements and the intensities of the components. In a conventional splitting diagram, such as Fig. 5.1 above, the π components are drawn above the horizontal line and the σ components below, the lengths of the verticals being proportional to the intensities.

The simplest possible splitting pattern is the normal triplet, derived from a line of a singlet series. If λ_0 is the wavelength of the undisturbed line expressed in cm, the wavelengths of the three components are $\lambda_0 - \Delta\lambda_H$, λ_0, and $\lambda_0 + \Delta\lambda_H$, where

$$\Delta\lambda_H = \frac{eH}{4\pi m_e c^2} \cdot \lambda_0^2, \tag{5.1}$$

H being the strength of the magnetic field in gauss, e the electron charge in e.s.u., m_e the mass of the electron in gm, and c the velocity of light in cm sec^{-1}.

The undisplaced line is a π component, and the two displaced lines are σ components. As seen by an observer looking perpendicular to the field (*transverse case*) the π component is plane polarized parallel to the field and the σ components are plane polarized perpendicular to the field. For an observer looking in the direction of the field (*longitudinal case*) the π component is absent and the σ components are circularly polarized in opposite directions; the direction of polarization of each σ component for a field directed *towards* the observer is indicated in Fig. 5.1 (*a*). The polarizations of the σ components are reversed for a field directed *away from* the observer. The polarization rules embodied in Fig. 5.1 are the same for the π and σ components of all Zeeman patterns, irrespective of the number of components.

In conventional spectroscopic notation the expression (5.1) for the displacement of the σ components of a normal Zeeman triplet is written in terms of the wavenumber $\bar{\nu}$ ($= 1/\lambda$) as

$$\Delta\bar{\nu} = \frac{eH}{4\pi m_e c^2}. \tag{5.1A}$$

$\Delta\bar{\nu}$ varies according to the strength of the magnetic field and is termed the Lorentz unit (designated by the symbol o). The displacements of the various π and σ components of more complicated Zeeman patterns are expressed in terms of this unit. For example, the displacements of the π and σ components of the sodium D_1 line are respectively $\pm\frac{2}{3}$ and $\pm\frac{4}{3}$ Lorentz units; the splitting pattern for this line is expressed in the form $\dfrac{(2)\,4}{3}$, the figure relating to the π components being placed in brackets.

Provided Russell–Saunders coupling can be assumed, the number, intensity, and displacement of the components of the Zeeman pattern of a line depend solely on the values of the S, L, and J quantum numbers of the initial and final states.[1] To obtain the Zeeman pattern one has first to calculate the Landé factors g' and g'' of the initial and final states from the formula

$$g = 1 + \frac{J(J+1) - L(L+1) + S(S+1)}{2J(J+1)}. \tag{5.2}$$

A convenient table of the g factors for Russell–Saunders terms is given by Condon and Shortley (1953: p. 382). For each state one can write down the values of the product gM corresponding to the $2J+1$ values of the magnetic quantum number M in the range $-J \leqslant M \leqslant J$. Finally, the displacements of the allowed Zeeman components may be calculated from the formula

$$\Delta\bar{\nu} = o[g'M' - g''M''], \tag{5.3}$$

[1] As the strength of the magnetic field increases the Russell–Saunders coupling eventually breaks down and the Zeeman effect passes over into the more complex Paschen–Back effect. The field strengths necessary, however, are far greater than those found in sunspots (cf. Candler, 1937: p. 106).

or, putting $\Delta M = M'' - M'$,

$$\Delta \bar{\nu} = o[(g' - g'')M' - g''\Delta M], \tag{5.3A}$$

subject to the selection rule $\Delta M = o$ for the π components and $\Delta M = \pm 1$ for the σ components. For a line of a singlet series ($S = o$) we have $L = J$, so that $g' = g'' = 1$; hence $\Delta \bar{\nu} = o$ for the π component and $\Delta \bar{\nu} = o$ for the two σ components, in agreement with (5.1A).

As we shall see in Section 5.2.5, the most suitable lines for the measurement of sunspot magnetic fields are those showing simple triplet splittings. It can be shown that triplet splittings are obtained (a) if either g' or g'' is zero, or (b) if $g' = g''$. In both cases the π component is undisplaced and the displacement of each σ component is $\Delta \bar{\nu} = g . o$, or, reverting to wavelength notation and inserting numerical values,

$$\Delta \lambda_H = 4 \cdot 67 \times 10^{-5} g . H . \lambda_0^2. \tag{5.4}$$

Equation (5.4) provides the necessary quantitative basis for the measurement of the strengths of sunspot magnetic fields.

Theory also provides rules for determining the relative intensities of the various components of a Zeeman pattern (see, for example, van den Bosch, 1957: p. 318). The relative intensities of the π and σ components of a triplet pattern are easily derived. In the transverse case the intensity of the π component is twice that of each σ component, while in the longitudinal case the intensity of each σ component is twice that in the transverse case. The relative intensities of the components of a triplet pattern in these two cases are shown diagrammatically in Figs. 5.1 (a) and (b). Generalizing to the case where the magnetic field is inclined at an arbitrary angle γ to the line-of-sight, it may easily be shown (cf. Babcock, 1949) that the relative intensities of the three components are given by

$$I_{\sigma_V} : I_\pi : I_{\sigma_R} = \tfrac{1}{4}(1 + \cos^2 \gamma) : \tfrac{1}{2} \sin^2 \gamma : \tfrac{1}{4}(1 + \cos^2 \gamma), \tag{5.5}$$

the subscripts V and R indicating the σ components at $\lambda_0 - \Delta\lambda_H$ and $\lambda_0 + \Delta\lambda_H$ respectively (cf. Fig. 5.1).

5.2.2 THEORY OF FORMATION OF AN ABSORPTION ZEEMAN TRIPLET

The results outlined in the previous section describe the Zeeman splitting of atomic lines formed in emission. We turn now to the study of the analogous magnetic splitting of atomic lines formed in *absorption* (the inverse Zeeman effect). The process of formation of a Fraunhofer line in the presence of a sunspot field constitutes a very difficult problem in the theory of radiative transfer—one which has so far not been completely solved. Nevertheless, in order to determine the strengths, directions, and polarities of sunspot fields one must know the states of polarization and the relative intensities of the Zeeman components of the lines employed. In addition, a proper understanding of the effect

of quarter- and half-wave plate analysers on the line profile is an essential pre-
requisite to any detailed discussion of the methods used in measuring sunspot
fields. However, before turning to the detailed theory of line formation in a
magnetic field, it is instructive to obtain some preliminary insight into the nature
of the inverse Zeeman effect by considering what happens when a beam of
white light is passed through an absorbing vapour pervaded by a uniform
magnetic field **H**.

Consider first the case of a longitudinal magnetic field directed *towards* the
observer. Let λ_0 be the wavelength corresponding to a certain natural frequency

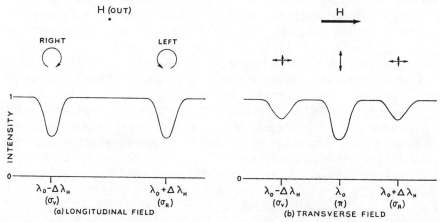

Fig. 5.2. Polarizations of the components of an *absorption* Zeeman triplet (Jenkins and
White, 1937).

(*a*) Longitudinal field directed towards the observer: only the two σ components are
present. They are circularly polarized in opposite directions.

(*b*) Transverse field: both the π and σ components are present, the former being
plane polarized perpendicular to the field and the latter partially plane polarized parallel
to the field.

of the vapour which, under the influence of the field, splits into three resonance
frequencies corresponding to the wavelengths $\lambda_0 - \Delta\lambda_H$, λ_0, and $\lambda_0 + \Delta\lambda_H$,
where $\Delta\lambda_H$ is given by (5.4). Following Jenkins and White (1937: cf. p. 425),
we may regard a beam of unpolarized light of any particular wavelength as
composed of two independent beams of right- and left-hand circularly polarized
light of equal intensity. Referring to Fig. 5.1 (*a*) we see that, in the presence of
the field, the left-hand circularly polarized beam of wavelength $\lambda_0 - \Delta\lambda_H$ will
be absorbed, whereas the right-hand circularly polarized beam will be transmitted. The result is that, at wavelength $\lambda_0 - \Delta\lambda_H$, right-hand circularly
polarized light is transmitted with an intensity equal to half that of the continuous background. Similarly, at wavelength $\lambda_0 + \Delta\lambda_H$, a beam of left-hand
circularly polarized light of the same intensity is transmitted. The resulting
absorption pattern is shown in Fig. 5.2 (*a*). Comparison of Figs. 5.1 (*a*) and

5.2 (a) shows that the σ components observed in absorption are circularly polarized in *opposite* directions to the corresponding components seen in emission.

In the case of a transverse field it is convenient to regard the incident un-polarized light as split up into components vibrating parallel and at right angles to the field (cf. Jenkins and White, 1937: p. 314). At wavelength λ_0 the parallel components are all absorbed and the perpendicular components transmitted. At $\lambda_0 - \Delta\lambda_H$ the parallel components are all transmitted; the perpendicular components, vibrating at right angles to the field, may be regarded as broken up into right- and left-hand circularly polarized beams, one of which is absorbed and the other is seen edge-on as plane polarized (Jenkins and White, 1937: p. 426). The net result is light partially polarized parallel to the magnetic field. The same is true of the other σ component at $\lambda_0 + \Delta\lambda_H$. The resulting Zeeman absorption pattern is shown in Fig. 5.2 (b). Comparison of Figs. 5.1 (b) and 5.2 (b) shows that in the transverse case the π and σ components observed in absorption are partially plane polarized at *right angles* to the corresponding emission components.

Turning now to the inverse Zeeman effect observed in sunspot Fraunhofer lines, we have to consider the far more complicated case of a gas which is both emitting and absorbing radiation. To describe the absorptive properties of the gas in the presence of a magnetic field it is necessary to introduce three selective absorption coefficients: κ_p, the volume absorption coefficient for light plane polarized in the direction of the field; κ_l, the volume absorption coefficient for light circularly polarized to the left in planes normal to the field; and κ_r, the volume absorption coefficient for light circularly polarized to the right in planes normal to the field. The values of the corresponding atomic absorption coefficients in the case of a triplet line have been given by Stepanov (1958b). However, in this section we shall not need the explicit values but only the important relations

$$\kappa_p(v) = \kappa_l(v \mp v_H) = \kappa_r(v \pm v_H) = \kappa_\lambda, \tag{5.6}$$

which express the fact that κ_p, κ_l, and κ_r are identical to the absorption coefficient κ_λ in the absence of a field, except that κ_l and κ_r are displaced in wavelength from κ_λ by amounts $\pm\Delta\lambda_H$.[2] In (5.6) the symbols v and v_H represent the wavelength and Zeeman displacement respectively, both expressed in units of the Doppler half-width $\Delta\lambda_D$. In the case of light propagating in a direction making an angle γ with that of the magnetic field, it can be shown that the effective absorption coefficients are

$$\kappa_p \sin^2\gamma, \quad \kappa_l(1 + \cos^2\gamma)/2, \quad \text{and} \quad \kappa_r(1 + \cos^2\gamma)/2. \tag{5.7}$$

Finally, it is convenient at this point to introduce the symbols

$$\eta_p = \kappa_p/\kappa, \quad \eta_l = \kappa_l/\kappa, \quad \eta_r = \kappa_r/\kappa, \quad \text{and} \quad \eta_\lambda = \kappa_\lambda/\kappa \tag{5.8}$$

[2] It should be noted that the direction of rotation of an absorbing electron changes from left-hand to right-hand (or vice versa) according as one looks along or against the direction of the field. Unless otherwise stated we shall assume that $0 \leqslant \gamma \leqslant 90°$; this involves no loss of generality since, when $\gamma > 90°$, the results for $180° - \gamma$ can be obtained merely by interchanging κ_l and κ_r.

to represent the ratios of the selective absorption coefficients defined above to the continuous absorption coefficient κ.

In order to formulate the equations of transfer governing the formation of an absorption line in the presence of a magnetic field, some suitable parametric representation of polarized light must first be introduced. In the following discussion we shall follow the treatment of Unno (1956), which is based on Stokes' theory of polarized light (cf. Chandrasekhar, 1950: p. 24; Shurcliff,

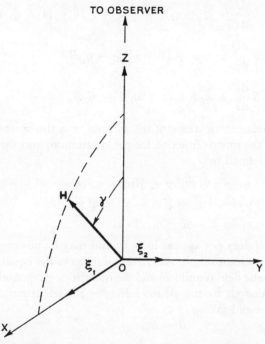

Fig. 5.3. Cartesian reference system. ξ_1 and ξ_2 are the amplitudes of the X and Y components of a light ray propagating along OZ, the line-of-sight to the observer. The magnetic field **H** lies in the plane XOZ and makes an angle γ with OZ.

1962: p. 19). Let OX, OY, OZ be a cartesian reference system such that the magnetic field **H** lies in the plane XOZ and makes an angle γ with the direction of propagation OZ (cf. Fig. 5.3). Also, let

$$\xi_x = \xi_1 \cos (\omega t - \varepsilon_1)$$
$$\xi_y = \xi_2 \cos (\omega t - \varepsilon_2)$$

be the X and Y components of a ray of polarized light traveling in the direction OZ. According to Stokes' theory, the intensity and polarization of this ray are completely determined by the values of the four parameters

$$I = I_1 + I_2$$
$$Q = I_1 - I_2$$

$$U = 2\overline{\xi_1\xi_2} \cos (\varepsilon_1 - \varepsilon_2)$$
$$V = 2\overline{\xi_1\xi_2} \sin (\varepsilon_1 - \varepsilon_2),$$

where $I_1 = \overline{\xi_1{}^2}$ and $I_2 = \overline{\xi_2{}^2}$; the bars indicate average values. Starting from this representation and assuming that the line is formed by *true absorption*, Unno derives with the aid of (5.7) and (5.8) the equations of transfer in the form

$$\cos \theta \, \frac{dI}{d\tau} = (1 + \eta_I)I + \eta_Q Q + \eta_V V - (1 + \eta_I)B$$

$$\cos \theta \, \frac{dQ}{d\tau} = \eta_Q I + (1 + \eta_I)Q - \eta_Q B \qquad (5.9)$$

$$\cos \theta \, \frac{dV}{d\tau} = \eta_V I + (1 + \eta_I)V - \eta_V B,$$

where θ is the heliocentric angle of the sunspot, τ is the optical depth in the continuum, B is the source function for the continuum, and the quantities η_I, η_Q, and η_V are defined by

$$\eta_I = \tfrac{1}{2}\eta_p \sin^2 \gamma + \tfrac{1}{4}(\eta_l + \eta_r)(1 + \cos^2 \gamma)$$
$$\eta_Q = [\tfrac{1}{2}\eta_p - \tfrac{1}{4}(\eta_l + \eta_r)] \sin^2 \gamma \qquad (5.10)$$
$$\eta_V = \tfrac{1}{2}(-\eta_l + \eta_r) \cos \gamma.$$

The parameter U does not appear in (5.9) and may be put equal to zero.[3]

It is possible to obtain simple analytical solutions of the equations of transfer (5.9) if the magnetic field is uniform and the quantities η_p, η_l, and η_r do not vary with depth (analogous to the *Milne–Eddington* model atmosphere). Making these assumptions and taking

$$B = B_0(1 + \beta_0\tau),$$

Unno obtains expressions for the emergent values $I(0, \theta)$, $Q(0, \theta)$, and $V(0, \theta)$. Using these expressions the Stokes parameters specifying the intensity and polarization of the emergent light may then be written in units of the continuous background, $I_c(0, \theta)$, as

$$r_I(\theta) \equiv \frac{I(0, \theta)}{I_c(0, \theta)} = \frac{1}{1 + \beta_0 \cos \theta}\left\{1 + \beta_0 \cos \theta \, \frac{1 + \eta_I}{(1 + \eta_I)^2 - \eta_Q{}^2 - \eta_V{}^2}\right\}$$

$$r_Q(\theta) \equiv \frac{Q(0, \theta)}{I_c(0, \theta)} = \frac{-\beta_0 \cos \theta}{1 + \beta_0 \cos \theta} \cdot \frac{\eta_Q}{(1 + \eta_I)^2 - \eta_Q{}^2 - \eta_V{}^2} \qquad (5.11)$$

$$r_V(\theta) \equiv \frac{V(0, \theta)}{I_c(0, \theta)} = \frac{-\beta_0 \cos \theta}{1 + \beta_0 \cos \theta} \cdot \frac{\eta_V}{(1 + \eta_I)^2 - \eta_Q{}^2 - \eta_V{}^2}.$$

[3] It is worth mentioning that equations (5.9) govern the formation of any line by true absorption, irrespective of the nature of its Zeeman splitting. The specialization to a Zeeman triplet is made when the relations (5.6) are used to specify the functional dependence of the parameters η_p, η_l, and η_r.

The solutions (5.11) are of fundamental importance in the study of sunspot magnetic fields. In the next section we shall use them to determine the polarizations and relative intensities of the components of an absorption Zeeman triplet, while in Section 5.2.4 we shall use them to determine the effect of quarter- and half-wave plate analysers on the profile of the triplet. Finally, in Sections 5.3.5 and 5.3.6 they are used to provide a basis for the quantitative interpretation of magnetic records obtained with the solar magnetograph and solar polarimeter.

Michard (1961) has compared these solutions with those obtained on the assumption of a Schuster–Schwarzschild model atmosphere, and has thus shown that the results depend in a rather sensitive manner on the model adopted. A more practical approach has been adopted by Stepanov (1958b). Using Michard's sunspot model (cf. Section 4.2.4), Stepanov has determined the dependence of the η-values for the much-used triplet line $\lambda6173\cdot3$ on τ_{5000} and has computed the expected profile for various values of γ, assuming a uniform magnetic field of 2500 gauss (cf. Stepanov's Fig. 9). It is noteworthy that the profiles clearly reveal the presence of the central π component, even when γ is only 20°.

The profiles of Fe $\lambda6173\cdot3$ calculated by Stepanov depend on the assumption of a uniform magnetic field. The influence of a *depth-dependent* field on the profile of this line has been considered by Hubenet (1954) for the case of a purely longitudinal field. He finds that the existence of a gradient of a few gauss per kilometre not only causes the line to be broadened but, in addition, produces an asymmetry in each of the two σ components. The asymmetry of the σ components of another important magnetic line, Fe $\lambda6302\cdot5$, has been considered similarly by Mattig (1958). As we shall see in Section 5.4.6, measurement of the asymmetry provides a possible method of estimating the height gradient of sunspot fields.

In addition to Unno's work, the theory of the formation of absorption lines in a magnetic field inclined at an arbitrary angle to the line-of-sight has been extensively developed by Stepanov (1958a, b; 1960c). His approach is to regard the light propagated through the medium as consisting of two mutually orthogonal beams of elliptically polarized light (cf. Chandrasekhar, 1950: pp. 33–34) and to compute the effective absorption coefficient for each state of polarization as a function of γ. Using this formalism Stepanov (1960d) has derived the equations of transfer in the more general case in which the process of line formation is due to both coherent scattering and true absorption. The equivalence, in the case of true absorption, between Unno's equations of transfer (5.9) and those formulated by Stepanov has been demonstrated by Stepanov (1960d) and by Rachkovsky (1961a). On the other hand, Rachkovsky (1961b) has shown that Unno's method of formulation is actually the more general of the two. Stepanov's (1960d) paper provides a convenient summary of his work.

In all the calculations of sunspot line profiles described in this section it has been assumed that the lines are formed by true absorption but, in practice, this may not always be the case. Other processes such as coherent or noncoherent scattering may play an important, or even dominant, role. In the presence of a magnetic field an additional complication appears: as Stepanov (1958b) has pointed out, a quantum absorbed at one frequency in a certain state of polarization may be re-emitted at a neighbouring frequency in a different state of polarization. This can happen even in the case of a line with a simple triplet splitting, such as Fe $\lambda6173\cdot3$. This line arises from the transition $a^5P_1 - y^5D_0$: the upper level $(J = 0)$ is not split but the lower level $(J = 1)$ splits into three sublevels with magnetic quantum numbers $M = -1, 0, +1$; hence a quantum absorbed in a particular state of polarization may be re-emitted in a different state of polarization. The necessary refinements of the theory have been considered by Stepanov (1962). Moreover, the possible effects of other factors should be considered, such as deviations from local thermodynamic equilibrium, spatial non-uniformity of the spot, and so on. The possible influence of magneto-optical effects, such as Faraday rotation, has recently been discussed by Rachkovsky (1962). Finally, it must be emphasized that the achievement of greater precision in the measurement of sunspot magnetic fields will also depend on obtaining more accurate information about the physical conditions in sunspots discussed in Chapter 4.

However, the theory has not yet developed to the stage at which all the complications described above can be taken into account in the interpretation of magnetic observations, and in the sections that follow we shall make use of Unno's solutions (5.11) for a triplet line formed by true absorption in the presence of a uniform magnetic field.

5.2.3 POLARIZATIONS AND RELATIVE INTENSITIES OF THE COMPONENTS OF AN ABSORPTION ZEEMAN TRIPLET

In this section we shall determine the states of polarization and the relative intensities of the components of a Zeeman triplet on the basis of the solutions (5.11) of the equations of transfer given above, assuming that the magnetic field is strong enough to *completely split the line* into its three components. Under these conditions the expressions (5.11) simplify considerably. We shall outline the procedure for the σ_V component, assuming as usual that γ does not exceed $90°$; the corresponding results for the other components and for the case $\gamma > 90°$ are easily obtained in a similar way.

The values of the Stokes parameters for the σ_V component are obtained by putting $\eta_p = \eta_r = 0$ in (5.10) and (5.11), yielding

$$r_I'(\theta) = \frac{1}{1 + \beta_0 \cos\theta}\left\{1 + \beta_0 \cos\theta \cdot \frac{1 + \frac{1}{4}\eta_l(1 + \cos^2\gamma)}{1 + \frac{1}{2}\eta_l(1 + \cos^2\gamma)}\right\}$$

$$r_Q'(\theta) = \frac{\beta_0 \cos\theta}{1 + \beta_0 \cos\theta}\left\{\frac{\frac{1}{4}\eta_l \sin^2\gamma}{1 + \frac{1}{2}\eta_l(1 + \cos^2\gamma)}\right\} \tag{5.12}$$

$$r_V'(\theta) = \frac{\beta_0 \cos\theta}{1 + \beta_0 \cos\theta}\left\{\frac{\frac{1}{2}\eta_l \cos\gamma}{1 + \frac{1}{2}\eta_l(1 + \cos^2\gamma)}\right\}.$$

According to a theorem in the theory of polarized light (cf. Chandrasekhar, 1950: p. 33) a beam of light specified by the four Stokes parameters $(r'_I; r'_Q; 0; r'_V)$[4] may be resolved into a beam of unpolarized light

$$(r'_I - \sqrt{\{(r'_Q)^2 + (r'_V)^2\}}; 0; 0; 0) \tag{5.13}$$

and a beam of elliptically polarized light

$$(\sqrt{\{(r'_Q)^2 + (r'_V)^2\}}; r'_Q; 0; r'_V). \tag{5.14}$$

It may also be shown that the ellipticity of the polarized beam specified by (5.14) is determined by the equation

$$\sin 2\Psi = \frac{r'_V}{\sqrt{\{(r'_Q)^2 + (r'_V)^2\}}} \quad (-\tfrac{1}{2}\pi \leqslant \Psi \leqslant \tfrac{1}{2}\pi), \tag{5.15}$$

where $\tan \Psi$ is the ratio of the axes of the ellipse parallel to the OY and OX axes, respectively (cf. Fig. 5.3). The polarization of the beam is right-handed or left-handed according as Ψ is positive or negative.

Substituting from (5.12) into (5.15) one obtains after some reduction the simple relation

$$\tan \Psi = \cos \gamma \tag{5.16}$$

between the ellipticity of the polarized beam and the inclination of the magnetic field to the line-of-sight. As we shall see in Section 5.3.3, (5.16) provides part of the theoretical basis for Treanor's 'Babinet compensator' method of determining the directions of sunspot fields. The corresponding polarization ellipse is depicted in Fig. 5.4, where the ratio of the minor and major semi-axes is OB/OA = $\cos \gamma$ in accordance with (5.16). When $\gamma = 0$ (i.e., for a *longitudinal* field directed towards the observer), equation (5.16) yields $\Psi = \tfrac{1}{4}\pi$, showing that the beam is circularly polarized to the right. When $\gamma = 90°$ (*transverse* field), $\Psi = 0$ and hence the beam is plane polarized parallel to the field. Combining the two beams specified by (5.13) and (5.14) we thus conclude that, in the case of a longitudinal field directed towards the observer, the σ_V component of the Zeeman triplet is partially circularly polarized to the right and, in the transverse case, is partially plane polarized parallel to the field, in qualitative agreement with the results of the simple analysis given at the beginning of Section 5.2.2 (cf. Fig. 5.2).

The polarizations of all three components of a well-separated Zeeman triplet are given in Table 5.1 for both the longitudinal and transverse cases. The polarization rules embodied in this table provide the basis for establishing the *polarity* of sunspot magnetic fields. By convention, the polarity is described as 'north' if the field is directed *towards* the observer, and 'south' if the field is directed *away from* the observer. In this connection the terms R (red) and V (violet) introduced by the early Mt Wilson observers are synonymous with

---

[4] The third Stokes parameter is zero since $U = 0$ throughout the analysis.

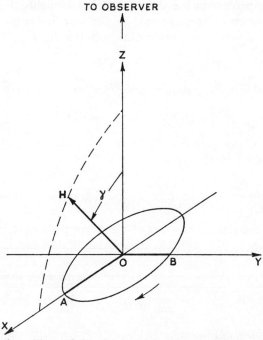

Fig. 5.4. Polarization ellipse of the short-wavelength σ component of an absorption Zeeman triplet. The ratio of the minor and major semi-axes is OB/OA = cos γ, where γ is the inclination of the magnetic field **H** to the line-of-sight OZ.

'north' and 'south' respectively (von Klüber, 1948).[5] In any actual determination of sunspot polarities care must be exercised to ensure that the results are not vitiated by spurious instrumental polarization (cf. Section 5.3.7).

Another question of great relevance to the measurement of sunspot magnetic

TABLE 5.1

Polarization Rules for an Absorption Zeeman Triplet

	COMPONENTS		
γ	σ_V	π	σ_R
0°[1]	Partially circularly polarized to the right	—	Partially circularly polarized to the left
90°	Partially plane polarized parallel to the field	Partially plane polarized perpendicular to the field	Partially plane polarized parallel to the field
180°	Partially circularly polarized to the left	—	Partially circularly polarized to the right

[1] Field directed *towards* observer.

[5] To save confusion we must point out that the polarization rules given by von Klüber (see his Fig. 1) for this purpose are actually those for the components of a Zeeman triplet seen in *emission*, whereas the ones to be used must of course be those for an *absorption* triplet.

fields is that of the relative intensities of the three Zeeman components. In this case, instead of considering residual intensities as above, it is more convenient to deal with the *depths* of the components. Again assuming that the magnetic field is strong enough to produce a complete splitting, and using equation (36) of Unno's (1956) paper[6] and also equation (5.6) above, one can show that the ratios of the depths of the σ_V, π, and σ_R components are given by

$$d_{\sigma_V} : d_\pi : d_{\sigma_R} = \frac{\frac{1}{4}\eta_0(1 + \cos^2 \gamma)}{1 + \frac{1}{2}\eta_0(1 + \cos^2 \gamma)} : \frac{\frac{1}{2}\eta_0 \sin^2 \gamma}{1 + \eta_0 \sin^2 \gamma} : \frac{\frac{1}{4}\eta_0(1 + \cos^2 \gamma)}{1 + \frac{1}{2}\eta_0(1 + \cos^2 \gamma)}$$

$$(5.17)$$

over the whole range of γ, where η_0 is the value of η_λ at the centre of the undisturbed line.

In the case of *very weak* absorption lines, for which

$$\eta_0 \ll 1, \tag{5.18}$$

these ratios simplify to

$$d_{\sigma_V} : d_\pi : d_{\sigma_R} = \tfrac{1}{4}(1 + \cos^2 \gamma) : \tfrac{1}{2} \sin^2 \gamma : \tfrac{1}{4}(1 + \cos^2 \gamma). \tag{5.19}$$

The relations (5.19) are identical to those for the relative intensities of the corresponding Zeeman components seen in *emission*, which are given by (5.5). However, as Babcock (1948) has emphasized, this equivalence applies only in the case of very weak absorption lines, whereas the lines usually used in measuring sunspot magnetic fields do not satisfy (5.18).

5.2.4   EFFECT OF QUARTER- AND HALF-WAVE PLATE ANALYSERS ON THE PROFILE OF AN ABSORPTION ZEEMAN TRIPLET

Most methods of measuring sunspot magnetic fields depend on the use of some form of polarizing analyser which, by taking advantage of the different states of polarization of the Zeeman components, increases the sensitivity of the measurements. A $\lambda/4$ analyser is used when the field is predominantly longitudinal and a $\lambda/2$ analyser when the field is largely transverse. The effect on the line profile can readily be calculated in both cases by making use of the solutions of the equations of transfer (5.11) given in Section 5.2.2. A knowledge of the results is essential for understanding the principles underlying most of the methods used to measure sunspot magnetic fields (cf. Section 5.3).

Retaining the same general notation as in the previous section, let us first consider the effect of a $\lambda/4$ plate analyser, which we may take to consist simply of a $\lambda/4$ plate followed by a Nicol prism. In Fig. 5.5 let the X and Y axes correspond to the OX and OY axes used previously (cf. Fig. 5.3). The fast and slow axes of the $\lambda/4$ plate are distinguished by the symbols f and s respectively, the angle between the f and X axes being denoted by ϕ. The letters 'o' and 'e'

---

[6] Note that Unno uses the symbol r_l to refer to the depth rather than to the residual intensity of the line.

indicate the ordinary and extraordinary axes of the Nicol prism, which is positioned so that the 'o' axis is inclined at 45° to the f axis of the $\lambda/4$ plate.

If $I_{o,\lambda/4}$ and $I_{e,\lambda/4}$ denote the emergent intensities of the ordinary and extraordinary rays, then it is convenient for the purpose of discussing the relative

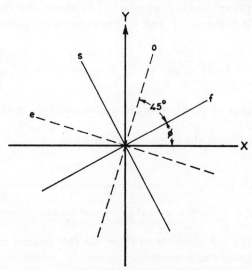

Fig. 5.5. Directions of the axes of the components of a quarter- or half-wave plate polarizing analyser (for the meanings of the symbols, see text).

intensities of the components of the Zeeman triplet to consider the normalized depressions

$$d_{o,\lambda/4} = \frac{\tfrac{1}{2}I_c - I_{o,\lambda/4}}{\tfrac{1}{2}I_c}$$

and

$$d_{e,\lambda/4} = \frac{\tfrac{1}{2}I_c - I_{e,\lambda/4}}{\tfrac{1}{2}I_c},$$

remembering that the intensity of the unpolarized continuum in each beam is $\tfrac{1}{2}I_c$. Unno (1956) has calculated $I_{o,\lambda/4}$ and $I_{e,\lambda/4}$ in terms of the Stokes parameters of the ingoing beam, given by (5.11), and from his results we obtain

$$d_{o,\lambda/4} = \frac{\beta_0 \cos\theta}{1 + \beta_0 \cos\theta}\left\{1 - \frac{1 + \eta_I - \eta_V}{(1 + \eta_I)^2 - \eta_Q{}^2 - \eta_V{}^2}\right\}$$

and $\qquad\qquad\qquad\qquad\qquad\qquad\qquad\qquad\qquad\qquad\qquad\qquad$ (5.20)

$$d_{e,\lambda/4} = \frac{\beta_0 \cos\theta}{1 + \beta_0 \cos\theta}\left\{1 - \frac{1 + \eta_I + \eta_V}{(1 + \eta_I)^2 - \eta_Q{}^2 - \eta_V{}^2}\right\},$$

where the quantities η_I, η_Q, and η_V are defined by (5.10). In the particular case of a purely longitudinal field it follows from (5.20) that

$$d_{o,\lambda/4} = \frac{\beta_0 \cos\theta}{1 + \beta_0 \cos\theta} \cdot \frac{\eta_r}{1 + \eta_r}$$

and

$$d_{e,\lambda/4} = \frac{\beta_0 \cos \theta}{1 + \beta_0 \cos \theta} \cdot \frac{\eta_l}{1 + \eta_l},$$

showing that only the σ_R component appears in the ordinary beam, while only the σ_V component appears in the extraordinary beam. Since rotating the $\lambda/4$ plate through 90° is equivalent to interchanging the ordinary and extraordinary beams, this means that each of the two σ components can be suppressed in turn.[7] As we shall see later (Sections 5.3.2, 5.3.4), practical use is made of this effect to facilitate the measurement of the separation of the two σ components in the presence of a predominantly longitudinal field.

When the magnetic field is sufficiently strong to produce an effective splitting of the line, it follows from (5.6), (5.10), and (5.20) that the depths of the σ_V, π, and σ_R components are given by

$$d_{\sigma_V,\lambda/4} = \frac{\beta_0 \cos \theta}{1 + \beta_0 \cos \theta} \cdot \frac{\tfrac{1}{4}\eta_0(1 \mp \cos \gamma)^2}{1 + \tfrac{1}{2}\eta_0(1 + \cos^2 \gamma)}$$

$$d_{\pi,\lambda/4} = \frac{\beta_0 \cos \theta}{1 + \beta_0 \cos \theta} \cdot \frac{\tfrac{1}{2}\eta_0 \sin^2 \gamma}{1 + \eta_0 \sin^2 \gamma} \qquad (5.21)$$

$$d_{\sigma_R,\lambda/4} = \frac{\beta_0 \cos \theta}{1 + \beta_0 \cos \theta} \cdot \frac{\tfrac{1}{4}\eta_0(1 \pm \cos \gamma)^2}{1 + \tfrac{1}{2}\eta_0(1 + \cos^2 \gamma)},$$

where η_0 is the value of η_λ at the centre of the undisturbed line. The expresssions (5.21) are valid for all γ between 0 and 180°, the choice of sign depending on whether the ordinary or extraordinary ray is selected. Once again we see that, provided the field is longitudinal ($\gamma = 0$), the two σ components can be suppressed alternately by rotating the $\lambda/4$ plate through 90°. Fig. 5.6 shows the profiles of the triplet line Fe λ6173·3 calculated by Stepanov (1958b) for various values of γ, assuming a uniform magnetic field of 2500 gauss. It is clear that the presence of the π component makes the separation of the σ components increasingly difficult to measure for values of γ greater than about 70°. The situation becomes rapidly worse as the strength of the field decreases and the components start to merge together even more, so that the ultimate limit for the measurement of weak fields with a $\lambda/4$ analyser is set by the blending of the σ and π components (Bumba, 1960). The practical use of the $\lambda/4$ analyser is discussed in Section 5.3.2.

In the case of *very weak* absorption lines ($\eta_0 \ll 1$) it is easily seen from (5.21) that the relative depths of the components are given approximately by

$$d_{\sigma_V,\lambda/4} : d_{\pi,\lambda/4} : d_{\sigma_R,\lambda/4} = \tfrac{1}{4}(1 \mp \cos \gamma)^2 : \tfrac{1}{2}\sin^2 \gamma : \tfrac{1}{4}(1 \pm \cos \gamma)^2.$$

[7] In the case of an absorption line the σ component suppressed is actually the one whose circular polarization is in the same sense as that transmitted by the analyser.

These simplified relations are referred to as the 'Seares formulae' after Seares (1913). They have a certain historical importance although their practical utility is very limited by the condition $\eta_0 \ll 1$.

Equations (5.21) provide a means of determining the *inclination* γ of the

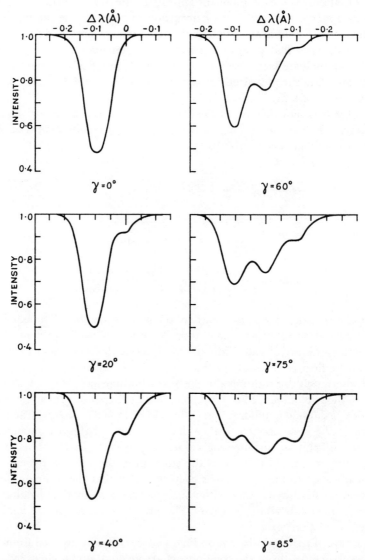

Fig. 5.6. Profiles of the Zeeman triplet line Fe I $\lambda 6173\cdot3$ viewed through a quarter-wave plate analyser for various values of the inclination γ of the magnetic field to the line-of-sight, assuming a uniform field of 2500 gauss (Stepanov, 1958b). At $\gamma = 0°$ only one of the σ components is present, but at greater angles the π component, and ultimately the other σ component, becomes prominent.

magnetic field $\mathbf{H}$ to the line-of-sight. This can be done in several different ways, e.g. by comparing the relative intensities of the σ_V and σ_R components. However, to obtain a complete knowledge of the field direction, it is also necessary to determine the azimuthal angle between the projection of $\mathbf{H}$ on the plane of the solar image and some fixed direction on the Sun. This can be done by determining the angle ϕ between the projection of $\mathbf{H}$ on the plane of the solar image and the fast axis of the wave-plate (cf. Fig. 5.5). However, to obtain ϕ it is necessary to replace the $\lambda/4$ plate by a $\lambda/2$ plate orientated in the same way. Using an analysis similar to that employed for the $\lambda/4$ analyser it can be shown that the normalized depressions are given by

$$d_{0,\lambda/2} = \frac{\beta_0 \cos\theta}{1 + \beta_0 \cos\theta} \cdot \frac{\eta_I(\eta_I + 1) - \eta_Q^2 - \eta_V^2 + \eta_Q \sin 2\phi}{(1 + \eta_I)^2 - \eta_Q^2 - \eta_V^2}$$

$$d_{e,\lambda/2} = \frac{\beta_0 \cos\theta}{1 + \beta_0 \cos\theta} \cdot \frac{\eta_I(\eta_I + 1) - \eta_Q^2 - \eta_V^2 - \eta_Q \sin 2\phi}{(1 + \eta_I)^2 - \eta_Q^2 - \eta_V^2}.$$

(5.22)

Moreover, it can be shown that rotating the $\lambda/2$ plate through 45°—so that its axes then coincide with those of the Nicol prism—is equivalent to interchanging the ordinary and extraordinary beams. Equations (5.22) provide the basis for determining the field strengths and azimuthal angles of transverse fields in sunspots.

In the special case of a purely transverse field it follows from (5.22) that

$$d_{0,\lambda/2} = \frac{\beta_0 \cos\theta}{1 + \beta_0 \cos\theta} \cdot \frac{\eta_p}{1 + \eta_p}$$

$$d_{e,\lambda/2} = \frac{\beta_0 \cos\theta}{1 + \beta_0 \cos\theta} \cdot \frac{\tfrac{1}{2}(\eta_l + \eta_r)}{1 + \tfrac{1}{2}(\eta_l + \eta_r)},$$

provided that the Nicol is orientated so that either the ordinary or extraordinary axis is parallel to the direction of the magnetic field (i.e., $\phi = 45°$). Hence only the π component appears in the ordinary beam and only the two σ components in the extraordinary beam. Since rotating the $\lambda/2$ plate through 45° is equivalent to interchanging the ordinary and extraordinary beams, this means that either the two σ components or the π component can be suppressed at will.

When the field is sufficiently strong to produce an effective splitting of the line, it follows from (5.22) that the depths of the σ_V, π, and σ_R components are given by

$$d_{\sigma_V,\lambda/2} = d_{\sigma_R,\lambda/2} = \frac{\beta_0 \cos\theta}{1 + \beta_0 \cos\theta} \cdot \tfrac{1}{4}\eta_0 \cdot \frac{1 + \cos^2\gamma \mp \sin^2\gamma \cdot \sin 2\phi}{1 + \tfrac{1}{2}\eta_0(1 + \cos^2\gamma)}$$

$$d_{\pi,\lambda/2} = \frac{\beta_0 \cos\theta}{1 + \beta_0 \cos\theta} \cdot \tfrac{1}{2}\eta_0 \sin^2\gamma \cdot \frac{1 \pm \sin 2\phi}{1 + \eta_0 \sin^2\gamma},$$

(5.23)

for all values of γ between 0 and 180°, the choice of sign depending on whether the ordinary or extraordinary ray is selected.

Finally, it must be emphasized that the simplified expressions (5.21) and (5.23) are valid only when the magnetic field is strong enough to produce an effective splitting of the line. If the field is weaker the Zeeman components blend together and one must then use the exact formulae (5.20) and (5.22). In the case of the important line Fe $\lambda6173\cdot3$, for example, Stepanov (1958b: cf. Figs. 13 and 14) has shown that blending of the π and σ components becomes increasingly significant for fields less than about 2000 gauss.

5.2.5 FRAUNHOFER LINES SUITABLE FOR MAGNETIC FIELD MEASUREMENTS

The wide diffuse character of solar absorption lines makes it difficult to resolve the individual components of complicated Zeeman patterns even in the presence of strong magnetic fields (von Klüber, 1955). The most suitable lines for making magnetic field measurements are therefore Zeeman triplets with large g values. For this reason, the theory of line formation in the presence of a magnetic field has so far been worked out in any detail only for the case of a Zeeman triplet (Section 5.2.2). According to von Klüber (1948) the best triplet splittings that can be expected in practice are $\dfrac{(0)\ 5}{2}$, $\dfrac{(0)\ 3}{1}$, $\dfrac{(0)\ 10}{3}$, and, less frequently, $\dfrac{(0)\ 3}{2}$. However, in the actual selection of a line for use in a particular case various other considerations may have to be taken into account. Following von Klüber (1948) these may be listed as follows:

(1) according to (5.4) the magnetic splitting $\Delta\lambda_H$ of a triplet line is proportional to the square of its wavelength, so to obtain the largest separations one should choose lines in the red or near infra-red region of the spectrum;[8]

(2) the contour of the Fraunhofer line must be sufficiently sharp for precision measurements to be possible;

(3) to obviate possible sources of systematic error the line must be free from blends or very close neighbours;

(4) to enable magnetic fields to be measured both inside a spot and in the activity centre outside, one should choose a line whose intensity does not change appreciably from spot to photosphere (cf. Section 4.3.1).

Taking these requirements into account von Klüber (1948) compiled a list of the most suitable lines available in the wavelength region 5000–9000Å, after carefully examining all the data then available. This list is reproduced in Table 5.2. It contains only eight lines, but this is hardly surprising in view of the number of conditions which the lines have to satisfy. The final column of the table gives for each line the value of the product $g\lambda^2$ as a relative measure of the magnitude of its Zeeman splitting.

[8] Some suitable lines of silicon are in fact available in the far infra-red, but these are hard to exploit owing to the technical difficulties of working in this spectral region (von Klüber, 1948, 1955).

TABLE 5.2

Fraunhofer Lines Most Suitable for Magnetic Field Measurements in the Spectral Region 5000–9000Å (von Klüber, 1948)

WAVE-LENGTH (Å)	ELEMENT	ROWLAND INTENSITY		EXCITATION POTENTIAL		TRANSITION	ZEEMAN PATTERN	$g\lambda^2$ (cm²)
		Spot	Disk	Lower	Upper			
5131·478	Fe	3	2	2·213	4·618	$a^5P_1 - y^5P_1$	$\dfrac{(o)\ 5}{2}$	66×10^{-10}
5247·576	Cr	5	3	0·957	3·309	$a^5D_0 - z^5P_1$	$\dfrac{(o)\ 5}{2}$	69
5250·218	Fe	3	2	0·121	2·471	$a^5D_0 - z^7D_1$	$\dfrac{(o)\ 3}{1}$	83
6173·348	Fe	5	5	2·213	4·212	$a^5P_1 - y^5D_0$	$\dfrac{(o)\ 5}{2}$	95
6258·578	V	2	−1	0·261	2·233	$a^6D_{\frac{1}{2}} - z^6D_{\frac{1}{2}}$	$\dfrac{(o)\ 10}{3}$	132
6302·508	Fe	4	5	3·671	5·629	$z^5P_1 - e^5D_0$	$\dfrac{(o)\ 5}{2}$	100
6733·162	Fe	0	1	4·618	6·451	$y^5P_1 - f^5D_0$	$\dfrac{(o)\ 5}{2}$	114
8468·417	Fe	4	2	2·213	[3·67]	$a^5P_1 - a^5P_1^0$	$\dfrac{(o)\ 5}{2}$	180

Three of the lines listed in Table 5.2 are of very great practical and historical importance: these are the iron lines $\lambda\lambda 5250\cdot2$, $6173\cdot3$, and $6302\cdot5$. The outstanding suitability of the two red lines was recognized by the early Mt Wilson observers, but that of the green line was pointed out much later by J. Evershed in 1939 (cf. von Klüber, 1948). Since then $\lambda 5250\cdot2$ has been widely used for observations with the solar magnetograph of weak photospheric fields although, as we shall see later (Section 5.3.5), the comparatively large splitting of this line actually makes it unsuited for magnetograph observations of the stronger fields in and around sunspots. On the other hand, the two red lines $\lambda\lambda 6173\cdot3$ and $6302\cdot5$ have played an extraordinarily important role in the observation of sunspot fields: the vast majority of all measurements of sunspot fields have, in fact, been made with the aid of these two lines. The $\lambda 6302\cdot5$ line has the particular advantage of lying between two well-defined telluric lines of almost constant intensity 2, namely $\lambda\lambda 6302\cdot005$ and $6302\cdot771$ of molecular oxygen, which are of great importance for wavelength measurement as well as for control purposes of all kinds (von Klüber, 1948; Bumba, 1960). Despite the large $g\lambda^2$ values of the lines $\lambda\lambda 6173\cdot3$ and $6302\cdot5$, the actual magnetic separations to be expected are very small: for example, the splitting $\Delta\lambda_H$ (cf. equation 5.4) of $\lambda 6173\cdot3$ in a field of 1000 gauss is only $0\cdot045$Å as compared with a natural half-width of approximately $0\cdot1$Å (Stepanov, 1958b).

A much longer list of Fraunhofer lines with triplet patterns but rather smaller g values than those of the lines in Table 5.2 has also been given by von Klüber (1948: cf. Table 2). In addition, von Klüber has published a list of lines showing *no* magnetic splitting ($g = 0$); these are often useful for control purposes as well as being very suitable for the measurement of Evershed velocities in sunspots (cf. Section 4.4.2). This list is reproduced in Table 5.3.

TABLE 5.3

Fraunhofer Lines with no Magnetic Splitting Suitable for Control Measurements (von Klüber, 1948)

WAVELENGTH (Å)	ELEMENT	ROWLAND INTENSITY Spot	Disk	EXCITATION POTENTIAL Lower	Upper	TRANSITION
5123·732	Fe	4	3	1·007	3·415	$a^5F_1 - z^5F_1$
5434·536	Fe	8	5	1·007	3·278	$a^5F_1 - z^5D_0$
5576·101	Fe	4	4	3·415	5·629	$z^5F_1 - e^5D_0$
5691·508	Fe (Ni)	2	2	4·283	6·451	$y^5F_1 - f^5D_0$
6613·817	Fe	1	-1	1·007	2·873	$a^5F_1 - z^7F_0$
7090·404	Fe	2	2	4·212	5·953	$y^5D_0 - e^5F_1$
7389·391	Fe	1	2	4·283	5·953	$y^5F_1 - e^5F_1$

In conclusion we may remark that use is sometimes made of Zeeman lines other than simple triplets by measuring the displacement of the 'centre of gravity' of the various σ components, weighted according to their relative intensities (cf. von Klüber, 1955; Babcock, 1962: p. 111). For example, Bumba (1960) has done this in an attempt to estimate the height gradients of the magnetic fields in sunspot penumbrae (cf. Section 5.4.6) but, in the absence of a quantitative theory of the formation of complex Zeeman lines in a magnetic field, the results must be viewed with caution.

5.3 Methods of Measuring Sunspot Magnetic Fields

5.3.1 INTRODUCTION

The simplest and most reliable method of determining the *strength* of a sunspot magnetic field is to measure, either photographically or visually, the separation of the two σ components of a suitable Zeeman triplet with the aid of a high-resolution spectrograph. Provided the field is large enough to eliminate any significant blending of the Zeeman components, its strength may be deduced at once from the basic equation

$$\Delta\lambda_H = 4\cdot67 \times 10^{-5} g \cdot H \cdot \lambda_0^2 \qquad (5.4)$$

given in Section 5.2.1. However, if information is also required about the *direction* of the field, it is necessary to measure the relative intensities of the individual components, taking advantage of their different states of polarization to obtain increased sensitivity. The classical method of measuring sunspot fields, introduced by Hale and adopted by many subsequent workers, is to place in front of the spectrograph slit a polarizing analyser, consisting basically of a $\lambda/4$ or $\lambda/2$ plate and a linear polarizer. The effects of such analysers on the profile of

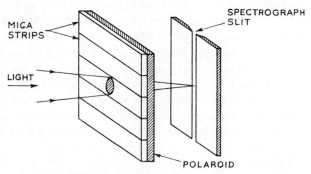

(a) COMPOUND QUARTER, OR HALF-WAVE PLATE ANALYSER

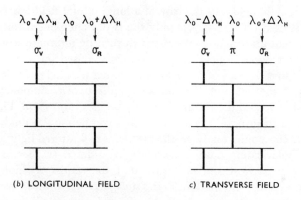

(b) LONGITUDINAL FIELD   c) TRANSVERSE FIELD

Fig. 5.7. Principles of operation of compound quarter- and half-wave plate analysers.

(a) Method of construction: for the relative orientations of the axes of the wave plate strips and the polaroid, see text.

(b) Longitudinal field: alternate suppression of the two σ components of an absorption Zeeman triplet by a $\lambda/4$ analyser.

(c) Transverse field: alternate suppression of the two σ components and the π component by a $\lambda/2$ analyser.

the line have already been discussed theoretically in Section 5.2.4. In the following section we shall discuss their practical application to the measurement of sunspot fields.

5.3.2   USE OF QUARTER- AND HALF-WAVE PLATE ANALYSERS

The polarizing analyser usually used in practice is a compound form of the simple device discussed in Section 5.2.4. The $\lambda/4$ analyser consists of a grid of mica strips cut so that the principal axes of adjoining strips are at right angles, followed by a polaroid sheet with its axis at $45°$ to the axes of the mica strips (cf. Fig. 5.7a). The $\lambda/2$ analyser is constructed similarly, except that the principal axes of the mica strips make angles of o and $45°$ alternately with the axis of the polaroid sheet. In both cases the analyser is placed as close as possible to the spectrograph slit and is made long enough to cover its entire length; in view of the small thickness of material possible with the use of mica and polaroid (von Klüber, 1948), it has a negligible effect on the quality of the solar image. The height of each mica strip must exceed a minimum value set by the convergence of the light beam and the distance of the analyser from the slit (cf. Fig. 5.7a), but otherwise should be as small as possible in order to obtain reasonable spatial resolution. The compound analyser can be used for both visual and photographic observations; further technical details can be found in the papers of Hale and Nicholson (1938), von Klüber (1948), Severny and Stepanov (1956), and Stepanov and Petrova (1958).

Recalling the results derived in Section 5.2.4, one can easily understand how the compound $\lambda/4$ and $\lambda/2$ analysers can be used to facilitate the measurement of sunspot magnetic fields. In the case of a longitudinal field, for example, the $\lambda/4$ analyser is used (cf. Fig. 5.7b): only the two σ components of the Zeeman triplet are present and each is suppressed in turn by alternate segments of the analyser. For a transverse field, the $\lambda/2$ analyser is used (cf. Fig. 5.7c): in this case, provided that the axis of the polaroid is either parallel or at right angles to the direction of the field (cf. Section 5.2.4), the two σ components are suppressed by one strip and the π component by the next and so on. Plate 5.2 gives an example of a spectrogram taken through a compound $\lambda/4$ analyser.

In using the compound analyser, the problems of scattered light and spurious instrumental polarization must be carefully considered (cf. Section 5.3.7). In addition, care must be taken to ensure that the results are not vitiated by Doppler shifts arising from motions in the sunspot, such as those associated with the Evershed effect (cf. Section 4.4.2). This is not hard to do if the inclination of the field is sufficient to give rise to a measurable π component, as is frequently the case. However, if the field is predominantly longitudinal, the technique may have to be modified somewhat. One method is to make two successive exposures between which the polaroid is rotated by $90°$. Unfortunately this method suffers from the disadvantage that two photographs taken at different times are liable to be affected by seeing in different ways (von Klüber, 1948). Various other methods, requiring only a single photograph, have been devised at the Potsdam Observatory. In one of these a birefringent prism is used in place of the polaroid to produce two identical images of a sunspot on the spectrograph slit, polarized at right angles to each other. Consequently both σ components can be observed at any required point in the spot, and the influence of any local Doppler shifts can be eliminated. In an alternative arrangement, the birefringent prism is placed behind the focal plane of the spectrograph (von Klüber, 1955: cf. Figs. 5 and 6). An even better method involves the use of a quartz plate, a few millimetres thick, which is cut parallel to its optic axis in such a way that it acts simultaneously as a $\lambda/4$ plate on one suitably chosen Fraunhofer line and as a $3\lambda/4$ plate on another suitable line nearby. In the case of a longitudinal field a polaroid can then be used to suppress the long-wavelength component of one line and the short-wavelength component of the other (or vice versa). Measurement of the difference in wavelength of the two lines then yields twice the amount of the Zeeman splitting $\Delta\lambda_H$, and the result is not affected by any local Doppler shifts. This arrangement can be modified by the addition of a grid of $\lambda/2$ strips to produce the usual 'zig-zag' type of pattern (von Klüber, 1955: cf. Figs. 7 and 8).[9]

While the use of the compound analyser has stood the test of time as a convenient technique for the *routine* determination of the polarities and maximum strengths of sunspot magnetic fields (cf. Section 5.3.9), it suffers from at least two major disadvantages as far as more refined studies are concerned. The first is, of course, that it is limited to the measurement of strong fields, the ultimate limit for weaker fields being set by the blending of the σ and π components of the Zeeman triplet employed (cf. Section 5.2.4). In fact, Severny and Stepanov (1956) have concluded that if the field is less than 1200 gauss, corrections as high as 400–500 gauss may have to be applied. The second limitation is the lack of adequate spatial resolution for the study of the fine structure of sunspot fields which, as Severny (1959) has shown, may have a scale of only 1–2″ of arc. Table 5.4 shows the resolution theoretically possible with the compound analysers used at the three observatories which have made the greatest contributions to the routine measurement of sunspot fields: Mt Wilson, Potsdam, and the Crimean Astrophysical Observatory. It is clear that in most cases the resolution

TABLE 5.4

Spatial Resolution of Compound Analysers used at Various Observatories

OBSERVATORY	TELESCOPE	IMAGE SIZE (mm)	WIDTH OF STRIPS (mm)	EFFECTIVE RESOLUTION (*sec of arc*)	REFERENCE
Mt Wilson......	150-ft tower	420	2·0	9·2	Hale and Nicholson (1938)
Potsdam........	Einstein tower	130	1·0	14·8	von Klüber
		260		7·4	(1948)
Crimean Astrophysical Observatory	Solar tower	210	0·33	3·0	Severny (1955);
		350		1·8	Stepanov and Petrova (1958)

[9] Unno (1956) has pointed out that a quartz plate cut so as to act as a $\lambda/4$ plate for one line and as a $\lambda/2$ plate for another would enable a complete determination of the strength and direction of a sunspot magnetic field to be made from a single photograph. Suitable lines for this purpose are Fe $\lambda\lambda 6173\cdot3$ and $6302\cdot5$, which originate at nearly the same optical depth.

achieved is quite inadequate for the study of the fine structure. Consequently various workers, notably A. B. Severny and his colleagues at the Crimean Astrophysical Observatory, have pioneered the application of the Babcock-type magnetograph (Section 5.3.4) to the detailed study of the magnetic fields in and around sunspots. As we shall see, the magnetograph not only has the advantage of greater sensitivity for the measurement of weaker fields but has, on occasion, enabled magnetic observations to be obtained with a resolution of about 2″ (cf. Section 5.4.4).

5.3.3 INTERFERENCE TECHNIQUES: SAVART PLATE AND BABINET COMPENSATOR

The first application of polarization interferometry to the problem of the detection and measurement of sunspot magnetic fields was made by Öhman (1950), who carried out experimental observations with a Savart polariscope at the Mt Wilson Observatory. The technique is to place the polariscope behind the eyepiece of a spectrograph (or at some other suitable place in the optical system), preferably with a $\lambda/4$ plate in front of the spectrograph slit, and to rotate the polariscope so that it gives interference fringes perpendicular to the spectral lines.[10] The Zeeman effect then shows up as a waviness in a line which in the absence of a magnetic field would be straight (cf. von Klüber, 1955: Fig. 9). The resulting pattern rather resembles the familiar zig-zag appearance produced by a conventional compound analyser. The feasibility of the Savart plate method was later independently tested during the course of experiments undertaken with the horizontal solar telescope at Simeis in 1952 (Severny and Stepanov, 1956). However, for photographic observations it was found that, due to the relatively large losses of light in the polariscope, longer exposures were required than with a compound analyser. For this reason, the latter method was adopted for the routine observation of sunspot fields at the Crimean Astrophysical Observatory.

The second application of polarization interferometry came in 1960 when P. J. Treanor at the Oxford Observatory described a new method for the spatial analysis of sunspot magnetic fields employing a Babinet compensator. The principle of the method is quite simple: as we have seen in Section 5.2.3, in the case of a strong field inclined at an arbitrary angle γ to the line-of-sight, the residual light in a σ component of a Zeeman triplet may be regarded as being composed of two independent beams, one of which is elliptically polarized. The other is unpolarized and may be disregarded in the discussion of Treanor's method, since it does not contribute to the formation of interference fringes. The polarization diagram of the elliptically polarized component is shown in Fig. 5.8, which has been drawn to conform as closely as possible to Fig. 1 of Treanor's paper. In Fig. 5.8, OE is the line-of-sight, OH the direction of the magnetic field, $\angle EOH = \gamma$, and the plane FGN is normal to OE. Then, according to Section 5.2.3, the electric vector of the elliptically polarized beam

[10] An elementary description of the technique of Savart plate interferometry is given by van Heel (1958).

may be represented by the two orthogonal components $OG = A \sin \omega t$ and $OF = A \cos \gamma \cdot \sin (\omega t + \pi/2)$, where the ratio $\tan \Psi'$ of the minor and major axes of the ellipse is given by

$$\tan \Psi' = \cos \gamma,$$

in accordance with (5.16).[11] A knowledge of Ψ' thus yields the *inclination* of the field to the line-of-sight while, if the azimuthal angle $GON = \alpha$ between the direction of the major axis OG and some fixed direction ON is also known, *the direction of the field is then completely determined.*

Treanor's experimental technique consists essentially of setting up before the slit of a large spectrograph a Babinet compensator followed by a linear polarizer

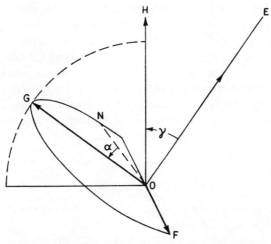

Fig. 5.8. Polarization ellipse of a σ component of an absorption Zeeman triplet. The magnetic field direction OH makes an angle γ with the line-of-sight OE. α is the angle between the major axis of the ellipse OG and some fixed direction ON.

whose axis is inclined at $45°$ to the axes of the Babinet. The resulting fringe system in a suitable Zeeman triplet is first photographed; then the Babinet and polarizer are together rotated by $45°$ and the new fringe system is photographed. The phase differences between corresponding fringes in the two σ components are measured on each photograph and the values of α and γ are then easily obtained by simple calculation (cf. Treanor, 1960). A good illustration of the fringe pattern seen in the components of the line $\lambda6173\cdot3$ is given in Plate 9 of Treanor's paper. Since the method is clearly a differential one, Treanor claims that it has the important advantage of being less affected by instrumental polarization than other methods. However, like the methods described above,

[11] The amplitude A, though taken as constant by Treanor, is actually a function of γ and, in the notation of Section 5.2.3, is given by

$$A^2 = \frac{\beta_0 \cos \theta}{1 + \beta_0 \cos \theta} \cdot \frac{\frac{1}{2}\eta_0}{1 + \frac{1}{2}\eta_0(1 + \cos^2 \gamma)}.$$

it can be used only for strong fields. A final decision on its practical utility must await the test of time.[12]

5.3.4 THE SOLAR MAGNETOGRAPH: PRINCIPLES OF OPERATION

The methods hitherto described, both visual and photographic, share the disadvantage of being applicable only to the measurement of fairly strong magnetic fields. From the time of Hale onwards various workers have experimented with the use of *photoelectric* detectors for the measurement of weak solar magnetic fields (for references see Babcock, 1953; Mogilevskii, Veller, and Val'd-Perlov, 1954). These efforts reached their culmination when, in 1952, H. W. and H. D. Babcock constructed the first solar magnetograph for recording and measuring weak longitudinal fields on the solar surface. The Babcock magnetograph was designed specifically for the large-scale mapping of weak fields (Babcock, 1953) and as such it has been an outstanding success. Quite a number of similar magnetographs have since been built (Table 5.5), but it has been left to A. B. Severny and his colleagues at the Crimean Astrophysical Observatory to exploit the potentialities of the new instrument for the detailed study of the stronger fields in and around sunspots themselves. For this reason, after first outlining the basic features of the Babcock design, we shall describe in some detail the magnetograph of the Crimean Astrophysical Observatory.

TABLE 5.5

Existing Solar Magnetographs

OBSERVATORY	REFERENCE
Hale Solar Laboratory, Pasadena	Babcock and Babcock (1952); Babcock (1953)
Mt Wilson	Babcock (1953); Howard (1959)
Cambridge Observatories	Beggs and von Klüber (1956)
Crimean Astrophysical Observatory ..	Nikulin, Severny, and Stepanov (1958); Nikulin (1960)
Pulkovo	Kotlyar (1960a)
Fraunhofer Institut	Deubner, Kiepenheuer, and Liedler (1961)

Note: another magnetograph is at present under construction at the Dominion Observatory, Ottawa (cf. *Quart. J. R.A.S.* **2**, 270: 1961).

The principle of the photoelectric method of detecting weak longitudinal magnetic fields by means of the Zeeman effect is very simple: a solar image is focused on the slit of a powerful spectrograph, in front of which is placed an

[12] It must be pointed out that there is an error in Treanor's argument which arises from his assumption that the form of the polarization ellipse of the residual light in an absorption σ component is identical to that of the corresponding *emission* component. As a consequence Treanor takes the major axis of the ellipse to lie along OF and the minor axis along OG, whereas the reverse is true (see Fig. 5.8). This in turn implies an error of $90°$ in the measurement of the azimuthal angle α of the field. However, no measured values of α are given in Treanor's paper and in other respects the validity of his procedure is not affected by the mistake.

analyser for circularly polarized light. In the focal plane of the spectrograph an exit slit is positioned on one of the wings of a Fraunhofer line with a simple triplet pattern and a high magnetic sensitivity (large Landé factor g). The light passing through the second slit falls on a photocell connected to an amplifier, whose output is displayed in a suitable manner. In the presence of a *weak* longitudinal field the spectral line is split into two overlapping, oppositely circularly polarized components whose intrinsic widths greatly exceed the magnitude of the Zeeman splitting. A change of the analyser from the right-handed to the left-handed condition will thus shift the line by a small amount

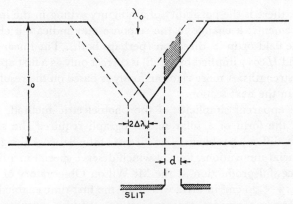

Fig. 5.9. Photoelectric detection of weak longitudinal fields by means of the Zeeman effect: the shaded area represents the change in light flux passing through the slit when the analysing device is changed from the right-hand to the left-hand circularly polarized condition.

equal to twice the Zeeman splitting given by (5.4) and hence, because of the sloping profile of the wing, the result will be a change in the amplified output signal.

In the case of a weak absorption line with a *triangular profile* (cf. Fig. 5.9) it can easily be shown that the change in the light flux δF obtained in this way is given by

$$\delta F = 2d \cdot I_c \cdot \frac{\partial r_\lambda}{\partial \lambda} \cdot \Delta\lambda_H, \qquad (5.24)$$

where d is the width of the slit, I_c is the intensity of the neighbouring continuum, and r_λ is the residual intensity in the line (Nikulin, Severny, and Stepanov, 1958). (The assumption of a triangular profile means that $\frac{\partial r_\lambda}{\partial \lambda}$ is taken to be constant.) With the aid of (5.4) this result may be written in the form

$$\delta F = 9{\cdot}34 \times 10^{-5} \cdot I_c d \cdot \frac{\partial r_\lambda}{\partial \lambda} g\lambda_0{}^2 \cdot H, \qquad (5.25)$$

showing that δF is proportional to H, the strength of the longitudinal magnetic field. δF is also proportional to I_c so that corrections should be made for solar limb darkening and changes in atmospheric transparency (Babcock, 1953; Kotlyar, 1960b).

When the magnetic field is inclined at some angle γ to the line-of-sight it can be shown that the change in light flux is proportional to the longitudinal component of the field $H_z = H \cos \gamma$, so that (5.25) becomes

$$\delta F = 9\cdot34 \times 10^{-5} \cdot I_c d \cdot \frac{\partial r_\lambda}{\partial \lambda} g \lambda_0^2 \cdot H \cos \gamma. \qquad (5.26)$$

Consequently, there is the possibility of ambiguity arising in the interpretation of the measurements: a change in the response may indicate a change in the strength of the field or in its direction (perhaps both). The linear relationship between δF and $H \cos \gamma$ implied by (5.26) is correct only as a first approximation for *small* field strengths. A more rigorous analysis based on the results of Section 5.2.4 is given in the next section.

Despite the apparent simplicity of the photoelectric method, its practical realization in the form of a solar magnetograph required the solution of a number of very difficult problems. The success of the Babcock design is due to a number of crucial innovations, four of which deserve special mention:

(1) the successful production at the Mt Wilson Observatory of fine diffraction gratings, $13 \times 20 \text{ cm}^2$ in area, which for the first time enabled a resolving power of 600,000 (measured photographically in the fifth order) to be achieved without the use of an interferometer with its attendant difficulties.[13] A further increase in resolving power would be of little value in view of the intrinsic widths of the Fraunhofer lines themselves (Babcock, 1953);

(2) the use of an electro-optic circular analyser in the form of a Z-cut crystal of ammonium dihydrogen phosphate (ADP) followed by a fixed Nicol prism. The application of an alternating voltage causes the retardation of the ADP plate to oscillate between $\pm\lambda/4$ at the applied frequency. In this way the Babcocks were able to avoid the false modulation that often results from using an analyser with moving optical parts;

(3) the introduction of a method of detecting Zeeman displacements employing not one but two exit slits and hence two photomultipliers.[14] The latter are connected to a difference amplifier, so that their response to a shift in the position of the spectral line is additive. Since the amount of useful light is

---

[13] However, routine observations of the strengths and polarities of sunspot magnetic fields have been carried out at the Kislovodsk Station of the Pulkovo Observatory using a low-resolution spectrograph equipped with an auxiliary attachment incorporating a Fabry–Pérot etalon (Polonskii, 1959).

[14] More recently Kotlyar (1960b) has devised a new kind of difference photometer for the solar magnetograph of the Pulkovo Observatory (cf. Table 5.5), which utilizes only a single photomultiplier and thus obviates the difficulty of obtaining two photomultipliers with identical characteristics. The use of a single photomultiplier would appear to have the additional advantage of making it easier to devise a method of compensating for large changes in intensity; this is particularly desirable in measuring sunspot fields.

doubled by this method, the signal-to-noise ratio is increased and, more important, the effect of any instrumental polarization introduced by the telescope mirrors (cf. Section 5.3.7) is minimized. A very interesting recent development has been the introduction of a *pair* of double exit slits which accept light from *two* magnetically sensitive lines, Fe I λ5250·2 and Cr I· λ5247·6 (cf. Table 5.2), and so further improve the overall sensitivity (Carnegie Institution, 1961: cf. p. 84);

(4) the introduction of a system for automatically scanning the whole solar disk, or a selected region of it, with conformal recording by means of a cathode-ray tube and a camera. The scanning technique is of course the feature distinguishing the 'magnetograph' from other possible types of photoelectric solar magnetometer. Further details about the design and operation of the Babcock magnetograph, as well as information about various subsequent refinements, can be found in the following papers: Babcock and Babcock (1952); Babcock (1953); Howard (1959, 1962); Howard and Babcock (1960). The question of the choice of slit width required to obtain the optimum signal-to-noise ratio is discussed by Deubner (1962).

In using a magnetograph for the measurement of *sunspot* fields the slit height should be made as small as possible. In addition, to ensure that the magnetic splitting remains small compared to the natural line width (cf. Section 5.3.5), the selected line should be rather insensitive to the Zeeman effect. The feasibility of using a magnetograph for the measurement of sunspot fields was first pointed out by Babcock and Babcock (1955), but to date all existing magnetic observations of sunspots obtained in this way have been made with the magnetograph of the Crimean Astrophysical Observatory. This instrument, which we shall now describe, provides a simultaneous record of magnetic field, radial velocity, and brightness (Nikulin, Severny, and Stepanov, 1958; Nikulin, 1960).

Plate 5.1 gives a general view of the Crimean magnetograph. The control console can be seen standing in front of the entrance slit of the 20 m horizontal spectrograph of the tower telescope of the Observatory. The large auxiliary image of the Sun on the left is used to facilitate the scanning procedure. Fig. 5.10 is a block diagram of the optical and electronic systems of the magnetograph. Light from the telescope passes through the modulator in a slightly convergent beam and is brought to a focus on the entrance slit S_1 of the spectrograph. The modulator is a circular analyser of the type used by Babcock, consisting of an ADP plate followed by a linear polarizer. The collimating mirror M_1 sends the light to a 15 × 15 cm² diffraction grating G ruled with 600 lines per mm. It is blazed in the fifth order and has an actual resolving power of about 450,000. The spectrum is brought to a focus on the compound exit slit S_2 by means of the imaging mirror M_2 and a plane mirror M_3. Before reaching S_2 the light passes through a plane-parallel glass plate P (the 'velocity compensator') which is rotated by a servo-mechanism to keep the selected spectral line accurately centred on S_2.

The exit slit S_2 actually consists of two slits a and b centred with high precision

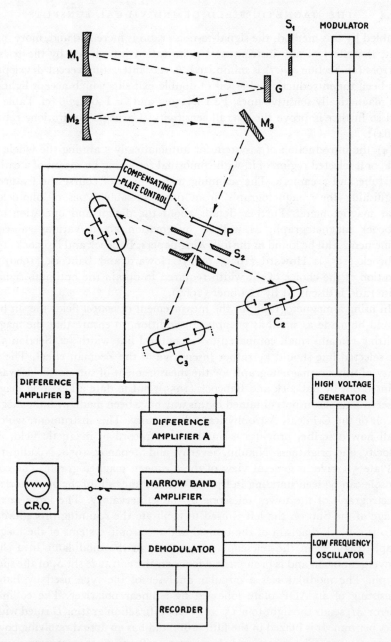

Fig. 5.10. Block diagram of the optical and electronic systems of the magnetograph of the Crimean Astrophysical Observatory (Nikulin, Severny, and Stepanov, 1958). S_1, entrance slit; M_1, collimating mirror; G, diffraction grating; M_2, imaging mirror; M_3, plane mirror; P, plane-parallel glass plate; S_2, compound exit slit; C_1, C_2, C_3, photomultipliers.

on each other (cf. Fig. 5.11). The jaws of the first slit cut off the wings of the line, while light from the core passes between the two prisms attached to the second slit onto a photomultiplier C_3, which is used for recording the brightness at the centre of the line. The beams of light from the steepest parts of the wings of the line are reflected by the prisms to the photomultipliers C_1 and C_2. The slit widths can be varied as follows: slit a from 0 to 4 mm and slit b from 0·5 to 4 mm. With a spectrograph dispersion of 10 mm/Å (fifth order), these widths correspond to wavelength intervals of 0–0·4Å and 0·05–0·4Å.

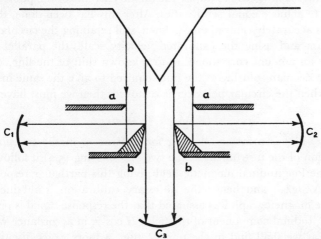

Fig. 5.11. Compound exit slit of the magnetograph of the Crimean Astrophysical Observatory (Nikulin, 1960). Light from the steepest parts of the wings of the spectral line is reflected by the prisms attached to the jaws of slit b to the photomultipliers C_1 and C_2. Light from the core of the line is received by the photomultiplier C_3.

The fluctuating components of the anode outputs of the photomultipliers C_1 and C_2 are fed to a difference amplifier A, whose output is subjected to further narrow-band amplification (Fig. 5.10). The difference signal is then demodulated and fed to the chart recorder illustrated in Plate 5.1 (c). The recorder is fitted with a simple digitising attachment, an exposed view of which is given in Plate 5.1 (b). This unit types out the field strength whenever the chart deflection attains any one of a number of pre-selected values; the numbers are typed in one colour for fields of N polarity and in another colour for fields of S polarity. Each horizontal row of numbers corresponds to a single scan across the region of the solar image under study. The radial velocity compensator is more sophisticated than that employed by Babcock (1953). The steady components of the anode outputs of C_1 and C_2 are fed to a second difference amplifier B; any unbalance due to the displacement of the spectral line gives rise to a signal which is made to actuate a mechanism for rotating the plane-parallel glass plate P, until the line is again correctly centred on S_2. A record of radial velocity is obtained simply by measuring the angle through which P

has to be rotated to compensate for the Doppler shift. The velocity scale is easily calibrated by reference to the known velocity of solar rotation.

The magnetograph is calibrated by placing in the beam before the modulator a circular polarizer consisting of a polaroid sheet followed by a $\lambda/4$ plate with its axes set at $45°$ to the axis of the polaroid (cf. Babcock, 1953). The modulation of the transmitted light is then found to exceed 98 per cent (Nikulin, Severny, and Stepanov, 1958), thus making the detector highly sensitive to any inequality of illumination. Working in the continuous spectrum near the selected spectral line it is now easy to adjust the anode currents of the photomultipliers C_1 and C_2 to achieve equal sensitivities. After this has been done, the spectral line itself is accurately centred on S_2. Then, still retaining the circular polarizer in the beam and using the calibrated velocity scale, the parallel plate P is rotated by an amount corresponding to a known shift of the line, $\Delta\lambda_c$ say. If $\Delta\lambda_H$ is the Zeeman splitting of the line required to give the same instrumental response when the circular polarizer is removed, then we must have

$$f . \Delta\lambda_c = 2 . \Delta\lambda_H,$$

where f ($= 0.2$) is a factor allowing for differences in the transmission and amplification of the instrument in the two cases. Using (5.4) it follows that the value of the longitudinal magnetic field giving this particular response is $1.07 \times 10^4 f . \Delta\lambda_c/g\lambda_0{}^2$, and hence the necessary calibration is obtained. In calibrating the magnetograph it is assumed that the response signal is proportional to the longitudinal component of the field, $H \cos \gamma$, in accordance with (5.26). However, as we shall find in the next section, a more exact discussion shows that this linear relationship holds only for *weak* fields whose strength, in the case of the line $\lambda5250.2$ often used, does not exceed a few hundred gauss. In addition, it is important to remember that the design of the associated electronic circuits also imposes a limit on the range of linearity of the response signal (cf. Severny, 1960).

An illustration of simultaneous tracings of radial velocity, longitudinal magnetic field, and brightness obtained with the Crimean magnetograph is given by Nikulin (1960: cf. Fig. 5).

5.3.5 THE SOLAR MAGNETOGRAPH: INTERPRETATION OF MAGNETIC RECORDS

In the previous section we have derived an approximate expression for the response of a magnetograph for the case of a longitudinal field, valid only for small field strengths. However, to obtain a proper theoretical basis for the interpretation of magnetograph records it is necessary once again to appeal to the theory of the formation of absorption lines in a magnetic field. In the case of a Zeeman triplet an expression for the response of the instrument can be obtained at once from the results derived in Section 5.2.4 for the profile of the line viewed through a $\lambda/4$ plate analyser. If $I_{o,\lambda/4}$ and $I_{e,\lambda/4}$ denote the emergent intensities of the ordinary and extraordinary rays of the Nicol prism, then

(b)

(c)

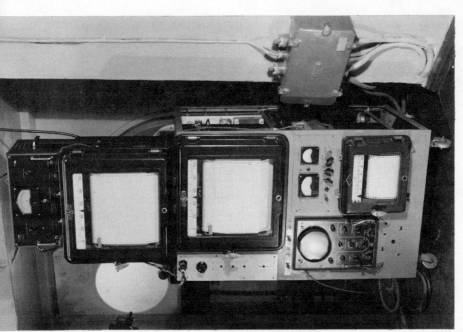

(a)

PLATE 5.1. Magnetograph of the Crimean Astrophysical Observatory.
(a) Control console; the large auxiliary image of the Sun on the left is used for guiding purposes. (b) Simple digitizer attached to chart recorder, for typing out the measured field strengths. (c) General view of chart recorder.

changing the retardation of the circular analyser from $+\lambda/4$ to $-\lambda/4$ causes a change in the intensity at a given point of the profile equal to $I_{o,\lambda/4} - I_{e,\lambda/4}$. Let us now introduce for convenience the normalized residual intensities

$$r_{o,\lambda/4} = I_{o,\lambda/4}/\tfrac{1}{2}I_c$$

and

$$r_{e,\lambda/4} = I_{e,\lambda/4}/\tfrac{1}{2}I_c$$

in place of the normalized depressions $d_{o,\lambda/4}$ and $d_{e,\lambda/4}$ defined in Section 5.2.4. Then the response signal δi from two slits S_1 and S_2 of negligible width placed

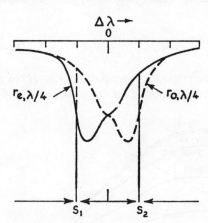

Fig. 5.12. Calculation of the response signal of a magnetograph (for details, see text). S_1 and S_2 are slits of negligible width symmetrically positioned on the wings of a Zeeman triplet.

at the same distance on either side of the centre of the undisturbed line (cf. Fig. 5.12) is given by

$$\delta i = c \cdot I_c \cdot \Delta r, \tag{5.27}$$

where c is a constant, I_c is the intensity of the neighbouring continuum, and $\Delta r \equiv r_{o,\lambda/4} - r_{e,\lambda/4}$.[15] The quantity Δr is easily derived from (5.20), yielding

$$\delta i = 2c \cdot I_c \, \frac{\beta_0 \cos\theta}{1 + \beta_0 \cos\theta} \cdot \frac{\eta_V}{(1 + \eta_I)^2 - \eta_Q{}^2 - \eta_V{}^2}. \tag{5.28}$$

Unlike the approximate relation (5.26) given in the preceding section, (5.28) is valid for all values of H and γ although, of course, it is restricted by the assumptions embodied in the theory (cf. Section 5.2.2).

With the aid of the relations (5.6) and (5.8) we can derive an approximate form of (5.28), valid for weak fields. If we write $\eta_p(v) = \eta(v)$ then, for $\gamma \leqslant 90°$, these relations give

$$\eta_l = \eta(v - v_H)$$

and

$$\eta_r = \eta(v + v_H),$$

[15] When the finite width of the slits, $d = \Delta\lambda_2 - \Delta\lambda_1$ say, is taken into account (5.27) becomes

$$\delta i = c \cdot I_c \int_{\Delta\lambda_1}^{\Delta\lambda_2} \Delta r \cdot d\lambda. \tag{5.27A}$$

where v is the wavelength and v_H the Zeeman displacement, both expressed in units of the Doppler half-width $\Delta\lambda_D$. For a weak field we can therefore write

$$\eta_l = \eta(v) - v_H \cdot \frac{d\eta}{dv}$$

and (5.29)

$$\eta_r = \eta(v) + v_H \cdot \frac{d\eta}{dv},$$

correct to the first order in v_H. Substituting from (5.29) into (5.28) we obtain with the aid of (5.10) the result

$$\delta i = 2c \cdot I_c \frac{\beta_0 \cos\theta}{1 + \beta_0 \cos\theta} \cdot v_H \cos\gamma \cdot \frac{d\eta/dv}{[1 + \eta(v)]^2}, \qquad (5.30)$$

showing that for *sufficiently weak* fields the response is proportional to $H \cos\gamma$. This provides a formal justification for the inclusion of the $\cos\gamma$ factor in (5.26).

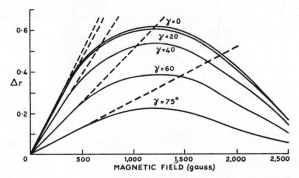

Fig. 5.13. Response signal of a magnetograph for the *photospheric* line Fe I $\lambda5250\cdot2$ (Stepanov, 1960b). It is evident that the region of linear response is limited to fields less than a few hundred gauss. γ is the inclination of the field to the line-of-sight.

Surprising as it may, seem only one attempt has so far been made to investigate the extent to which the weak-field approximation expressed by (5.26) or (5.30) may validly be used in the interpretation of magnetograph observations. Stepanov (1960b) has considered the case of observations of *photospheric* magnetic fields made with the aid of the Fe I line $\lambda5250\cdot2$. This has a large triplet splitting $\frac{(0)\ 3}{1}$ and has been widely used for observations of this kind (cf. Section 5.2.5). It has a central intensity of $0\cdot40$ and a half-width of about $0\cdot1$Å (Babcock, 1953). Stepanov's analysis is based on his own formulation of the theory of formation of absorption lines in a magnetic field and incorporates the photospheric model of V. S. Berdichevskaya. Like Unno (Section 5.2.2), Stepanov assumes the η-value of the line to be independent of depth. He then calculates the dependence of Δr on H and γ for the case of two infinitely narrow slits placed symmetrically with respect to the centre of the $\lambda5250\cdot2$ line at a distance from the centre of 0.05Å. The results are given in Fig. 5.13, which shows the variation of Δr with H for various values of γ. It is at once evident that *the region of linear response is limited to fields not exceeding a few hundred gauss.* The value of Δr increases with H up to about 1200 gauss, but thereafter *decreases*.

It is of great importance to determine the limiting value of H below which Δr (and hence δi) is proportional to $H \cos \gamma$. To do this Stepanov writes

$$\Delta r = C \cdot H \cos \gamma \cdot \Phi(H, \gamma; \beta_0, \theta) \qquad (C = \text{constant})$$

and plots $\Phi(H, \gamma)$ as a function of H for various values of γ (Stepanov, 1960b: cf. Fig. 4). The resulting curves show that $\delta i \propto H \cos \gamma$ only for $H < 300\text{--}400$ gauss. Moreover, by evaluating the integral (5.27A) he showed that when slits of finite width ($d = 0.035$Å) are used instead of infinitely narrow ones, the limiting value of H is reduced to 200 gauss. For lines with smaller Zeeman splitting the region of linear response does of course extend to higher values of H.

Despite their obvious importance no calculations of this kind have yet been attempted in respect of observations of *sunspot* magnetic fields. One difficulty is, of course, that unlike the case of the photosphere no really reliable sunspot model is at present available to serve as an adequate basis for calculations of this kind. Nevertheless, the preceding discussion makes it clear that the choice of a suitable Zeeman triplet really depends on the magnitude of the field to be measured: failure to ensure that the magnetic splitting is small compared to the natural width of the line (or to the separation of the exit slits) may result in a very serious underestimate of the field (cf. Fig. 5.13).[16]

The use of the magnetograph has by no means been restricted to observations made in photospheric lines with simple triplet splittings. Various workers at the Crimean Astrophysical Observatory have obtained magnetic records in the Na D_1 resonance line, the Hβ line, and the H and K lines of Ca II, all of which display more complicated Zeeman patterns (see, for example, Severny and Bumba, 1958; Bumba, 1960; Severny, 1960; Stepanov, 1960a, 1961). Unfortunately, however, as we have seen in Section 5.2.2, the theory of formation of Fraunhofer lines in a magnetic field has so far been worked out only for lines with simple triplet patterns formed by 'true absorption'. This means that no adequate theoretical basis really exists for the quantitative interpretation of records obtained in the strong photospheric and chromospheric lines listed above. Consequently, for the moment, it seems desirable to regard the results obtained in these lines as *qualitative* indications of the presence of magnetic fields in the corresponding levels of the solar atmosphere rather than as reliable measurements of actual field strengths.

---

[16] For the measurement of strong magnetic fields in the neighbourhood of sunspot groups Severny and his co-workers (see, for example, Bumba, 1960) have extensively used the Fe I line $\lambda4886.3$ in place of the more sensitive line $\lambda5250.2$ normally employed. Their observations were originally calibrated on the assumption that the $\lambda4886.3$ line was a simple triplet with a g factor of $\frac{1}{2}$ (Severny, 1958), but further investigation has shown that it has a more complicated Zeeman structure with an 'effective' g value close to 1.3 (Severny, 1960).

Very recently Severny and Stepanov have extended the use of the magneto-graph to the measurement of *transverse* fields in sunspots (Severny, 1962). This is done by replacing the $-\lambda/4$ to $+\lambda/4$ modulator by a o to $\lambda/2$ modulator and by adding, instead of subtracting, the fluctuating photomultiplier outputs. Neglecting the finite width of the exit slits, the response signal is then $\delta i = c \cdot I_c$. Δr, where c is a constant, I_c the continuum intensity, and $\Delta r \equiv r_{o,\lambda/2} - r_{e,\lambda/2}$ in the notation of Section 5.2.4. Using an analysis similar to that for the longi-tudinal case, it can easily be shown that

$$\delta i = 2c \cdot I_c \cdot \frac{\beta_0 \cos \theta}{1 + \beta_0 \cos \theta} \cdot \frac{\eta_Q}{(1 + \eta_I)^2 - \eta_Q^2 - \eta_V^2} \cdot \sin 2\phi, \quad (5.31)$$

where ϕ represents the azimuthal angle of the projection of the magnetic field on the plane of the solar image (cf. Section 5.2.4: Fig. 5.5). Like the corres-ponding result for a longitudinal field (equation 5.28) this relation is true for all values of H and γ, subject to the assumptions in the theory. Proceeding as in the longitudinal case, we can now derive for sufficiently weak fields the approxi-mate expression

$$\delta i = \tfrac{1}{2}c \cdot I_c \cdot \frac{\beta_0 \cos \theta}{1 + \beta_0 \cos \theta} \cdot (v_H \sin \gamma)^2 \cdot \frac{\mathrm{d}^2\eta/\mathrm{d}v^2}{[1 + \eta(v)]^2} \cdot \sin 2\phi, \quad (5.32)$$

showing that for weak fields the response is proportional to $H^2 \sin^2 \gamma$.

It is evident from (5.32) that the measurement of the transverse component of the field differs from that of the longitudinal component in two respects: in the first place the technique is inherently less sensitive, the response for weak fields being proportional to $H^2 \sin^2 \gamma$ whereas in the longitudinal case it is proportional to $H \cos \gamma$. Secondly, the azimuthal angle ϕ of the field enters into the equation as well as the inclination γ to the line-of-sight. Hence each scan must be made twice, the $\lambda/2$ modulator being rotated as a unit by a known amount between each scan; in this way both $H \sin \gamma$ and ϕ can be determined. Combining the values of these quantities with those of $H \cos \gamma$ (obtained in the normal way) it is thus possible to determine both the strength and direction of solar magnetic fields by means of magnetograph observations alone, although three times as many scans are necessary as for the longitudinal component.

5.3.6 THE SOLAR POLARIMETER

In 1958 A. Dollfus, working at the Meudon Observatory along lines originally suggested by Lyot, perfected a very sensitive white-light polarimeter which he and Leroy have subsequently applied to the measurement of *transverse* magnetic fields in and around sunspots. The polarimeter is carried on an equatorial telescope which is directed at the part of the Sun to be studied and can be rotated as a unit around its optical axis. The design is basically simple (Dollfus, 1958a, b; Dollfus and Leroy, 1962; Leroy, 1962: cf. Fig. 1): a 10 cm objective forms a small primary image which is magnified by a second lens to give a final image 10 cm in diameter. A small circular diaphragm, 5″ of arc in diameter, is placed in the final image plane concentric with the optical axis, thus isolating the region of the Sun to be studied. Near the focal plane is placed a thin glass or celluloid plate suitably inclined to compensate for the instrumental polarization introduced by the telescope optics. Behind this is placed a second such device whose function is to produce a small, known degree of polarization for the purpose of calibrating the instrument. The light then passes through an optical modulator designed to rotate the plane of polarization of the incident light 30 times a second in discrete steps, each of exactly 90°. The modulator consists of a rotating $\lambda/2$ plate device followed by a fixed birefringent prism, which divides the emergent light into two beams. These are directed to two vacuum photocells whose outputs are connected to a *difference* amplifier. The net result is that any plane polarized component of the incident sunlight is directed in turn to each of the photocells, thus producing an alternating output signal.

The Dollfus polarimeter is extremely sensitive: it will detect a degree of polarization as low as 1 in 10^5. The limiting sensitivity is set by instrumental polarization and this has been reduced to a very small amount by careful design of the optical system. In the first place a cemented doublet is used as the objective, thereby eliminating the partially polarized parasitic image that would otherwise be formed by reflection at the inner surfaces. The polarization acquired by even the most inclined rays passing through the lens does not exceed 4×10^{-5} and this is compensated for in the manner described above. A second possible source of parasitic polarization arises from the asymmetry of the ring of diffracted light around the edge of the objective lens when the telescope is directed towards the limb. The diffracted light is eliminated by placing a suitably-sized diaphragm in the plane of an image of the objective formed in front of the modulator by means of a field lens (Dollfus, 1958b). Finally, since the $\lambda/2$ modulator contains a moving optical component care must be taken to ensure that no resulting false modulation is imposed on the unpolarized component of the incident light (in the case of the magnetograph an electro-optic analyser is used to obviate this difficulty (cf. Section 5.3.4)).

Using this device Dollfus and Leroy have found that the integrated radiation from sunspots and the surrounding active regions is frequently partially plane polarized (cf. Section 5.4.8). They attribute the origin of the polarization to the presence of transverse magnetic fields, which cause the light of the magnetically sensitive Fraunhofer lines in the integrated radiation to show a weak, net polarization in the direction of the field as a result of the transverse Zeeman effect. For simplicity let us first suppose that the spectral region selected contains just *one* Fraunhofer line and that this line has a simple triplet splitting. Then following Leroy (1960, 1962) the origin of the residual polarization in the line can be explained qualitatively as follows: in the presence of a transverse field the line is split into its three components, the two σ components being partially plane polarized parallel to the field and the π component perpendicular to the field (Section 5.2.3). Moreover, provided the strength of the line is such that it lies on the transitional region of the curve of growth, all of the components will have an intensity nearly equal to that of the original line (cf. Section 4.3.3). Hence the total radiation in the line will be partially plane polarized parallel to the field.

For the case of a single triplet line discussed above a quantitative discussion of the response of the polarimeter can be given on the basis of the results derived in Section 5.2.4. Using the same notation, let $r_{o,\lambda/2}$ and $r_{e,\lambda/2}$ be the residual intensities of the line in the emergent ordinary and extraordinary beams of the birefringent prism. Then the response signal is

$$\delta i = c \cdot I_c \cdot \int_{\lambda_1}^{\lambda_2} (r_{o,\lambda/2} - r_{e,\lambda/2}) \mathrm{d}\lambda, \qquad (5.33)$$

where c is a constant, I_c is the intensity in the continuum, and λ_1 and λ_2 are two arbitrarily chosen wavelengths in the continuum on either side of the

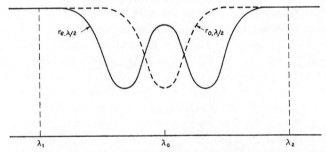

Fig. 5.14. Calculation of the response signal of a Dollfus white-light polarimeter (for details, see text). λ_0 is the central wavelength of an absorption Zeeman triplet; λ_1 and λ_2 are arbitrarily chosen wavelengths in the continuum on either side of the line.

undisturbed line (cf. Fig. 5.14). Using (5.31) and taking $\sin 2\phi = 1$, so that the polarimeter gives a maximum response, (5.33) becomes

$$\delta i = 2c \cdot I_c \cdot \frac{\beta_0 \cos \theta}{1 + \beta_0 \cos \theta} \int_{\lambda_1}^{\lambda_2} \frac{\eta_Q}{(1 + \eta_I)^2 - \eta_Q{}^2 - \eta_V{}^2} \mathrm{d}\lambda. \qquad (5.34)$$

By analogy with (5.31) and (5.32) of Section 5.3.5 it follows from (5.34) that, for sufficiently weak fields, δi is proportional to $H^2 \sin^2 \gamma$, where γ is the angle between the magnetic field and the line-of-sight. According to calculations made by Leroy (1962) this simple relationship holds so long as H is less than about 600 gauss. For larger fields the dependence of δi on H and γ is considerably more complicated.

In order to calibrate the actual polarization measurements in terms of magnetic field strength one has to sum the contributions arising from *all* the magnetically sensitive Fraunhofer lines in the spectral region employed. However, this is a matter of considerable difficulty since most of the lines involved have more complicated Zeeman patterns than the triplet case discussed above. The calibration curve actually given by Leroy (1962) has been obtained on the assumption that all of the lines have simple triplet splittings and is therefore at best only a first approximation. According to this curve the degree of polarization reaches a maximum value of about 10^{-2} for fields greater than 3000 gauss; the smallest degree detectable, 10^{-5}, corresponds to a field of 40 gauss. Although the use of the polarimeter appears to offer an attractively simple new method of measuring transverse fields in and around sunspots, some doubt must remain as to the reliability of the measurements until the calibration problem has been fully solved.

Some further uncertainty arises from the assumption that the observed polarization is produced solely as a consequence of the transverse Zeeman effect. Several authors have predicted theoretically that the continuous radiation from the normal photosphere should be slightly polarized as a result of scattering by both free and bound electrons, the effect being greatest at the limb but absent at the centre of the disk (for references see Minnaert, 1953: p. 130). Detailed measurements of the polarization of the photospheric continuum have been made by Dollfus and Leroy (1960) as a function of both effective wavelength and distance from the limb: these show that the degree of polarization over regions 15″ of arc in diameter does not exceed 2×10^{-5} at distances 4′ or more from the limb, whereas the values observed in sunspots can exceed 20×10^{-4} (cf. Section 5.4.8). This suggests that measurements of transverse magnetic fields in the central region of the solar disk are unlikely to be vitiated by polarization of the continuous radiation. Finally, it should be mentioned that the light of some Fraunhofer lines formed by coherent scattering may also be expected to show some degree of polarization; in fact, in the case of a favourable line like the Ca resonance line $\lambda4227$, the polarization actually observed near the limb is of the order of a few per cent (see, for example, Jäger, 1957). This provides an additional reason for confining measurements of sunspot fields with the Dollfus polarimeter to the central region of the solar disk.

5.3.7   INFLUENCE OF SCATTERED LIGHT AND INSTRUMENTAL POLARIZATION

The general problem of the possible adverse influence of scattered light and instrumental polarization is common to all the methods of measuring sunspot magnetic fields described above. If the field in a spot were predominantly longitudinal, the presence of parasitic light in the umbra would be revealed by the appearance in a triplet line of an *unpolarized* central 'π component' due to light scattered from the surrounding photosphere. In such a case the amount of scattered light could be estimated from the intensity of the 'π component' and, as Unno (1956) has indicated, the magnetic observations could then be corrected by Wanders' method (cf. Section 4.2.6). However, there is now strong evidence for the existence of appreciable *transverse* magnetic fields in sunspots; hence the mere appearance of a π component certainly cannot be taken as a priori evidence of scattered light unless it can be shown to be entirely unpolarized. (The occurrence of transverse magnetic fields in sunspots is discussed in Section

5.4.8.) In fact, Bumba (1962a) has pointed out that there is sometimes very strong evidence to the contrary: in some spots there are many lines which, although faint in the photosphere, show π components several times *more intense* in the umbra. We should like to emphasize once more that the best answer to the problem of scattered light is to be found in proper telescope design and choice of site (cf. Section 2.4) rather than in recourse to tedious and uncertain correction procedures.

In all types of magnetic field measurements where we are directly concerned with the polarization of the incident light, the presence of instrumental polarization is liable to cause serious error. Moreover, if the observational technique requires the use of a spectrograph, two distinct aspects of the problem have to be considered: (*a*) spurious polarization introduced by the optical system of the telescope; and (*b*) the polarizing action of the spectrograph itself. We shall discuss each aspect in turn.

As a specific example of the first aspect we may take the case of the 150-foot solar tower at Mt Wilson, whose polarizing properties were carefully considered by Hale (1913) and Seares (1913). The spurious polarization is produced mainly by the two coelostat mirrors at the top of the tower, the amount depending on the position of the Sun and on the condition of the mirror coatings. Seares concluded that the effect should be small for moderate hour angles.[17] It could be compensated by placing a tilted glass plate at some appropriate point in the optical system. However, since the required tilt of the plate would depend on the hour angle, Babcock (1953) has rejected this idea and relies instead on the use of a difference amplifier (cf. Section 5.3.4). As a general rule it is clearly desirable to keep the number of reflecting surfaces in the telescope to an absolute minimum. Moreover, the use of auxiliary devices introducing extra reflections, such as the three-mirror image rotator for the Potsdam tower telescope described by von Klüber (1948), is certainly not to be recommended in magnetic observations, as Evershed's experience clearly indicates (*Observatory* **64**, 155). Finally, it should be emphasized that the problem of instrumental polarization can be overcome more effectively with a *refracting* telescope than with a reflector, as we have already seen in the case of the Dollfus white-light polarimeter (Section 5.3.6).

The second aspect of the polarization problem concerns the possible polarizing action of the spectrograph, which in practice can sometimes be very considerable (see, for example, Hale and Nicholson, 1938). For this reason measurements of the relative intensities of the Zeeman components of a line should always be made with the final linear polarizer of the analysing device kept in a fixed position in front of the spectrograph slit, so that the light of each Zeeman component is polarized in the same direction on entering the slit. Difficulties

---

[17] Bumba and Topolova-Ruzickova (1962) find that the degree of polarization introduced by the aluminized mirrors of the coelostat feeding light to the solar spectrograph of the Ondrejov Observatory is of the order of 1% around noon but increases to nearly 6% at large hour angles. In addition, the direction of polarization changes continually throughout the course of a day.

arise when this conventional procedure is not followed, as the experience gained in routine observations of sunspot fields at the Crimean Astrophysical Observatory indicates: here a compound analyser was used consisting of a single $\lambda/4$ plate followed by a grid of narrow polaroid strips with their axes at right angles to one another and at 45° to the axis of the $\lambda/4$ plate (Severny and Stepanov, 1956). This meant that light from the two σ components of a triplet line entered the spectrograph in alternate beams polarized at right angles to one another. As a consequence of spectrograph polarization, the zig-zag spectra so obtained showed an annoying non-uniformity of brightness, which decreased the accuracy of the measurements (Stepanov and Petrova, 1958).[18]

Another problem sometimes encountered, although a minor one, is a reduction in the intensity of a beam of plane polarized light during its passage through a spectrograph when its direction of polarization happens not to coincide with the preferred direction of the spectrograph (Hale and Nicholson, 1938; Treanor, 1960). This difficulty can easily be overcome by placing a $\lambda/2$ plate after the final linear polarizer of the analysing device, and using it to rotate the plane of polarization to the most favourable angle, the effective rotation being twice the angle through which the $\lambda/2$ plate is turned (cf. Ditchburn, 1952: p. 371).

The actual change in the polarization of sunlight during its passage through a telescope and spectrograph can be determined by following the procedure devised by von Klüber (1948) for the Einstein tower telescope at Potsdam. A left-hand circular polarizer (see Section 5.3.4) is inserted in the path of the incident light before the first mirror of the telescope, the width of the transmitted beam being limited to the diameter of the polarizer. An analysis of the light in the focal plane of the spectrograph then reveals any change in polarization suffered by the light during its passage.

5.3.8   RELATIVE MERITS OF EXISTING METHODS; POSSIBLE FUTURE TECHNIQUES

Despite the attention that has been given for more than half a century to the development of techniques for the measurement of sunspot magnetic fields, an urgent need remains for the development of new methods more suited to meet the requirements of modern sunspot research. The most pressing requirement at the present time is to obtain observations with higher *spatial* and *temporal* resolution than hitherto achieved. Although recent improvements in the solar magnetograph have led to some encouraging progress in these directions, the final answer is not in sight. Consequently, in this section we shall merely discuss some of the more important advantages and disadvantages of the methods described above and shall mention some current ideas which may ultimately lead to the development of improved techniques.

The techniques described in the preceding sections fall naturally into two

[18] The difficulty was ultimately overcome by using a compound analyser made in the form of a sector, the individual polaroid strips being cut parallel to the circumference: the analyser was rotated until brightness uniformity was achieved.

classes whose usefulness in any particular situation depends on the ratio of the magnetic splitting $\Delta\lambda_H$ to the natural line width $\Delta\lambda_L$. In the first class we have the $\lambda/4$ and $\lambda/2$ analysers, the Savart plate, and the Babinet compensator technique, which all depend on having $2\Delta\lambda_H/\Delta\lambda_L \gtrsim 1$; they are therefore suitable only for the measurement of *strong* magnetic fields. In the second class we have the magnetograph and the Dollfus polarimeter, which have linear response curves only when $\Delta\lambda_H/\Delta\lambda_L \ll 1$; these are therefore most suited to the measurement of *weak* fields. The use of the magnetograph can be extended to the case of stronger fields by choosing lines of lower magnetic sensitivity, but the precise range of linearity for any particular line can be determined only by detailed calculation (cf. Section 5.3.5).

The methods of the first group have the disadvantage that one observation gives information about the field only along a single line through the spot. Using these methods it would be extremely tedious to determine the complete field distribution over a spot or spot group. This disadvantage is to some extent overcome by the magnetograph, which uses a scanning technique with conformal recording, and presumably a similar technique could also be employed with the solar polarimeter. Unfortunately, even with a scanning aperture $5''$ of arc in diameter, it takes a considerable time to obtain a magnetogram. As an example we may cite the attempt of Howard and Babcock (1960) to detect a possible change in the photospheric magnetic field distribution in an active region during the course of a solar flare. Using an effective aperture $5''$ wide these authors obtained 14 magnetograms of a region $4\cdot5' \times 4\cdot5'$ in extent at the rate of one every 15 minutes. Only the longitudinal component of the field was measured, but to obtain both the longitudinal and transverse components according to the scheme proposed by Severny would have required three times the number of magnetograms (Section 5.3.5). To obtain 'fine-scan' magnetograms in a time as short as 15 minutes represents a considerable achievement, but the time resolution of such observations is still very low compared with the time-scale, for example, of the physical changes occurring during solar flares.

Seeing changes and instrumental vibrations become increasingly important as the size of the magnetograph aperture is decreased. The effect in both cases is to cause the image to oscillate about the aperture in a random, two-dimensional manner. As a result significant differences may be apparent in the *fine detail* recorded on successive magnetograms even though no actual change in the magnetic field has occurred (Howard and Babcock, 1960: cf. Fig. 3). In view of the short duration (no more than a few seconds) of most moments of good solar seeing (cf. Section 2.2.4) there seems little prospect of ultimately obtaining *systematic* observations with a resolving power of $1''$ of arc or better with a scanning technique of the type used in the magnetograph. For this reason considerable interest must be attached to the ingenious method devised by Leighton (1959) for 'photographing' longitudinal magnetic fields with the aid of an ordinary spectroheliograph, a technique which also offers the possibility of achieving much better time resolution.

Leighton's method is based on the comparison, by means of a photographic subtraction technique, of two spectroheliograms which are identical except for small local differences brought about by the presence of a magnetic field. The sunlight coming to the entrance slit S of the spectroheliograph (cf. Fig. 5.15) is passed through a beam splitter B so that two images I,I' of exactly equal size and approximately equal brightness are formed on S. A mica $\lambda/4$ plate Q is

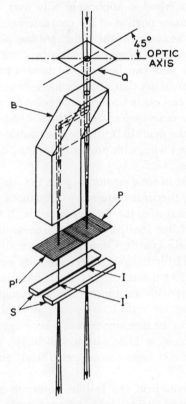

Fig. 5.15. Use of a spectroheliograph to measure solar magnetic fields (Leighton, 1959). Q, mica quarter-wave plate; B, beam splitter; P, P', polaroid sheets; I, I', solar images; S, entrance slit of spectroheliograph.

placed directly in front of the beam splitter, while behind it a polaroid sheet P,P' is inserted in each beam, so orientated that in one case right-hand, and the other left-hand, circularly polarized light is suppressed. The light from the two images I,I' then proceeds through the spectroheliograph in the usual manner. The second slit of the spectroheliograph is set on one wing of a suitable magnetically sensitive line, and a spectroheliogram is taken in the usual way. The two images so obtained will show the same variations in brightness at all points where these represent true intensity changes, but will show *opposite* variations in brightness at points where there is a sufficiently strong longitudinal magnetic field (cf. Section 5.3.4: Fig. 5.9). In order to eliminate all brightness variations other than those due to the field, a contact-printed transparency of one of the images is developed to gamma unity. If this positive transparency is placed in register with its negative a uniform grey image results but if it is superimposed instead on the negative of the second image, a uniform grey results only at points where there is no magnetic field, whereas differences due to a field are *enhanced* twofold. The method is applicable only over the range of intensity values covered by the *linear* portion of the characteristic curve, and great care must be taken during the course of the rather tedious photographic manipulations to ensure that spurious brightness variations are not introduced. The smallest magnetic field that can be detected is about 20 gauss. The spatial resolution attained approaches that of a normal spectroheliogram, so that the 'magnetic' photograph can easily be compared point for point with an ordinary spectroheliogram of the same region. So far the method has been applied only to the measurement of the fields in facular regions outside spots (cf. Section 7.4) but there seems no reason why, with proper technique, it should not be applied to sunspots themselves.

Although a large gain in time resolution over the magnetograph is afforded by Leighton's method, the time required to obtain each spectroheliogram is still considerable: for example, the time required with the exit slit positioned on one wing of the Ca I line $\lambda 6102 \cdot 8$ is about three minutes. To obtain shorter exposure times Giovanelli (1958), Öhman (1960), and others have suggested the use of narrow band birefringent filters specially designed to isolate lines suitable for Zeeman effect measurements. Such filters could be made to give twin transmission bands with an appropriate wavelength separation, one for left-hand and the other for right-hand polarization. However, no practical application of this idea to the measurement of solar magnetic fields has yet been made, although a birefringent filter with a wavelength resolution comparable to that required has in fact been constructed (Steel, Smartt, and Giovanelli, 1961).

With the single exception of the Dollfus polarimeter, present methods of measuring sunspot magnetic fields depend on the use of a large spectrograph, in practice often one associated with a tower-type telescope. However, as we have seen in Chapter 2, there are strong reasons for avoiding the use of tower telescopes and fixed spectrographs if high spatial resolution is required. One interesting alternative to the large spectrograph, which in the years to come could play an important role in the measurement of sunspot fields, is the *optical resonance tube* (Scrimger and Hunten, 1957; Blamont and Roddier, 1961). Such a device has the great advantage that it would be very easy to accommodate on the mounting of a high-resolution telescope of the type suggested in Section 2.4.4.

The resonance tube employed by Blamont and Roddier utilizes the optical resonance of strontium vapour passed through the tube in the form of an *atomic beam*, which is generated by heating the metal to a temperature of

600°C. The direction of the atomic beam is at right angles to that of the incident sunlight, while the light re-emitted, which is proportional to the incident intensity, is observed in the direction at right angles to both. A magnetic field is applied to the strontium vapour in the direction of observation: by varying the field the frequency of the scattered light is varied and thus the instrument is made to scan across the profile of the incident light in the manner of a high-resolution spectrometer. The line excited is the resonance transition $^1P_1 - {}^1S_0$ at $\lambda 4607\cdot3$; it has a normal triplet Zeeman splitting ($g = 1$) and no hyperfine structure. The effective half-width of the absorption profile is only 1–2×10^{-3}Å. It is interesting to note that Blamont and Roddier have already used their strontium resonance tube to observe the Zeeman splitting of $\lambda 4607\cdot3$ in the neighbourhood of a sunspot group. A somewhat similar resonance technique has been discussed by Öhman (1960). Nevertheless, the actual potentiality of the resonance tube technique for the measurement of sunspot fields, like that of the other suggestions discussed in this section, remains to be explored.

5.3.9   ROUTINE OBSERVATIONS

As a result of the stimulus provided by the International Geophysical Year (July 1, 1957 to December 31, 1958) the effort devoted to the routine measurement of sunspot magnetic fields has increased markedly in recent years. In addition to the long-established programmes at Mt Wilson and Potsdam (cf. Section 5.1), routine observations are now also undertaken at a number of stations in the U.S.S.R., namely, at the Pulkovo, Kislovodsk, and Crimean Astrophysical Observatories, and at the Scientific Institute of Earth Magnetism and Propagation of Radio Waves (NIZMIR) near Moscow. In fact, at the present time the Soviet Union carries the major burden of this important work. Details of the equipment and measuring technique used at each of the six institutions mentioned above are summarized in Table 2 of an important publication by Stepanov, Shaposhnikova, and Petrova (1962), which also gives the collected results of routine measurements of sunspot magnetic fields made during the I.G.Y. Table 3 of the same volume gives the systematic deviations from the overall mean of the results of each observatory, as a function of field strength. According to this table the Mt Wilson observers underestimate a field of 1000 gauss by no less than 700 gauss, on the average. (Some of the possible sources of error in magnetic measurements of this kind have already been discussed in Sections 5.3.2 and 5.3.7.)

Summaries of the Mt Wilson sunspot magnetic field observations are published regularly in the *Publications of the Astronomical Society of the Pacific*, while the Potsdam results appear in the *Mitteilungen des Astrophysikalischen Observatoriums Potsdam*.[19] The Soviet magnetic measurements are incorporated in the daily solar maps given in the bulletin, *Solar Data* ('Solnechnye Dannye'), and

[19] Prior to April, 1957, the Potsdam results were published in the *Astronomischen Nachrichten*: see Steen and Maltby (1959).

are also given separately in a supplement to the bulletin entitled *Sunspot Magnetic Fields*. Since the beginning of 1962 the maps given in *Solar Data* have also incorporated Potsdam results.

5.4 The Magnetic Field of Individual Sunspots

5.4.1 INTRODUCTION

As a result of the extensive pioneering observations of Hale and his collaborators at Mt Wilson, a classical picture of the magnetic field configuration inside single, regular sunspots emerged which, following Hale and Nicholson (1938), can be described as follows:

(1) the field is symmetrical around the axis of the spot;

(2) it has its maximum value at the centre of the umbra, the lines of force at this point being perpendicular to the solar surface (*longitudinal* field);

(3) away from the centre of the umbra the field becomes smaller and inclined to the vertical, an inclination of about 70° being attained at the outer border of the penumbra. This picture was elaborated by a number of subsequent workers and its validity remained unchallenged for several decades. In recent times, however, observations have been obtained which indicate the existence of significant *transverse* fields in the umbrae of sunspots. These observations contradict the classical picture and indicate that some radical revision of our ideas is now necessary. On the other hand, it must be admitted that at the moment it is not clear precisely what form this revision will take. Consequently, in the following pages we shall first describe the information obtained about sunspot fields which falls more or less within the classical framework and shall then consider the observations which indicate the need for a revision of the classical picture. We shall end the chapter by suggesting a new picture which is consistent not only with the presence of transverse fields but also with our knowledge of the complex light intensity distribution in the umbra.

5.4.2 FIELD STRENGTHS IN SUNSPOTS; DEPENDENCE ON AREA

All sunspots have detectable magnetic fields and the strength of the field increases with spot area. Hale and Nicholson (1938) found that the observed strengths range from a maximum of about 4000 gauss for the largest spots down to values of the order of 100 gauss for the smallest spots measurable with their equipment. Although fields exceeding 4000 gauss are sometimes observed (for example, the field of 4300 gauss recorded by von Klüber, 1948), the magnetic fields of most spots do not exceed 2000 gauss. In fact, according to some statistics compiled by Bell (see Ringnes and Jensen, 1960: Table 4), only about 20% of spot groups show fields greater than 2000 gauss, while only 5% have fields exceeding 3000 gauss.

The relationship between the maximum field strength of a spot and its area has been investigated statistically by a number of workers: Nicholson (1933);

Houtgast and van Sluiters (1948); Mattig (1953); and Ringnes and Jensen (1960). Their results are broadly concordant with the empirical formula

$$H_m = 3700 \cdot \frac{A}{A + 66} \qquad (5.35)$$

given by Houtgast and van Sluiters, H_m being the maximum field strength in gauss and A the area of the spot in millionths of the visible hemisphere.[20] According to (5.35) the limiting value of the field for a very large spot is 3700 gauss although, as remarked above, somewhat larger fields do sometimes occur. In any event too close a relationship between H_m and A is not to be expected since, as we shall see below, the maximum field strength remains almost the same during much of the lifetime of a spot whereas the area rises to a maximum and thereafter steadily declines (cf. Fig. 5.18).

Taken in conjunction with the result that the radiation intensity decreases with increasing spot area (cf. Section 4.2.2), the relation (5.35) also implies a correlation between the darkness of a spot and its magnetic field strength. This accords with the observation of Nicholson (1933) that a spot darker than normal usually has a field strength above the average for its area.

5.4.3 DEPENDENCE OF FIELD STRENGTH ON HELIOCENTRIC ANGLE

The relation (5.35) is strictly applicable only to spots in the central region of the solar disk. This is due to the existence of a centre-limb effect (Cowling, 1946; Houtgast and van Sluiters, 1948), which takes the form of an apparent systematic decrease in the maximum field strength as a spot approaches the limb. Two possible explanations have been proposed:

(1) an increase in the relative importance of parasitic light as the spot nears the limb, causing an apparent decrease in the observed field strength;

(2) the existence of a vertical field gradient, leading to a lower field strength in the higher levels of the spot seen near the limb.

Houtgast and van Sluiters accepted the existence of a vertical gradient as the cause of the centre-limb effect observed in large sunspots (diameter 50″ of arc or more), and so obtained a value for the gradient (Section 5.4.6). On the other hand, Mattig (1958: cf. Table 6) has since shown that the influence of parasitic light alone is quite sufficient to explain the observed effect, even in large spots. Moreover, some recent observations of Treanor (1960), although few in number, have failed to reveal any statistically significant effect. At the moment we cannot conclude that the existence of a *true* centre-limb effect (i.e., one corrected for the effect of parasitic light) has been reliably established.

5.4.4 VARIATION OF FIELD STRENGTH ACROSS A SUNSPOT

On the basis of the extensive observations made at Mt Wilson, Hale and Nicholson (1938) concluded that, on the average, the magnetic field strength

[20] Ringnes and Jensen found pronounced secular variations in the values of the numerical coefficients in (5.35) over the period 1917 to 1956, but the variations did not appear to be correlated with the solar cycle.

decreases across a spot from a maximum at the centre of the umbra to a very small value at points not far outside the outer boundary of the penumbra. As a general rule the field strength decreases only slowly across the umbra and then drops very sharply across the penumbra (von Klüber, 1948; Bumba, 1960). Although all authors agree that the field strength is small at the outer border of the penumbra, the field is not confined solely within the visible outline of the spot but extends for some distance into the surrounding photosphere (Bumba, 1960; Leroy, 1962), where it appears to merge with the magnetic field of the faculae surrounding the spot or spot group (cf. Howard, 1959). Bumba finds the strength of the field at the outer border of the penumbra to be about 300 gauss, irrespective of the size of the spot.

As an example of the empirical formulae that have been proposed by various authors to represent the observed variation of field strength across individual regular sunspots we may quote the one given by Broxon (1942), which is based on unpublished observations of Nicholson. Broxon's relation is

$$H = H_m(1 - r^2/b^2),$$ (5.36)

where H is the strength of the magnetic field at a radial distance r from the centre of the spot, H_m is the maximum field strength, and b is the radius of the spot. This relation and a somewhat more complicated one proposed by Mattig (1953) have recently been criticized by Bumba (1960) on the ground that across the penumbra they predict a rate of decrease of H smaller than that actually observed. Bumba attributes the cause of this discrepancy to the lower spatial resolution of the earlier observations. All the results concerned are based on measurements of the separation of the σ components of a Zeeman triplet with the aid of a polarizing analyser (cf. Section 5.3.2). However, since the accuracy of measurement by this technique drops considerably as the strength of the field decreases, not too much reliance should be placed on the measured field strengths in the penumbra, particularly in the outer parts.

Recently, Leroy (1962) has investigated the variation of the *transverse* component of the field across single, regular sunspots, using the solar polarimeter described in Section 5.3.6. He finds that the strength of the transverse component increases from a zero value at the centre of the umbra to a maximum of nearly 1000 gauss at the umbra–penumbra border and thereafter slowly decreases to a few hundred gauss at the outer penumbral border.

So far the only systematic attempt to investigate the relationship between variations in field strength and *brightness* across spots has been that made by von Klüber (1948). His paper gives cross-sections through a number of spots, which clearly reveal a significant correlation between *large-scale* variations in field strength and brightness: points of maximum field strength roughly coincide with those of minimum intensity. A typical example is shown in Fig. 5.16, where the arrows indicate the approximate limits of the spot umbra.

All the results based on observations of relatively low spatial resolution,

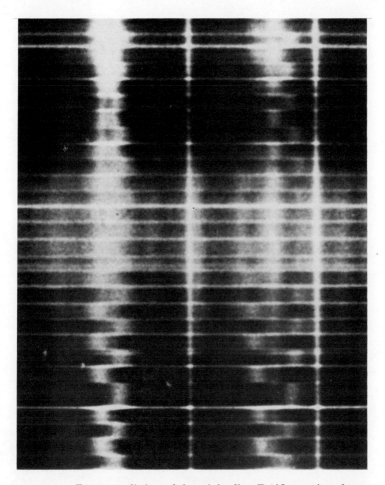

PLATE 5.2. Zeeman splitting of the triplet line Fe $\lambda6302\cdot5$ in a large sunspot near the centre of the disk (Severny, 1959). The spectrogram was taken through a 'polaroid grid' quarter-wave plate analyser (cf. Section 5.3.7). Over most of the penumbra only the two σ components of the line appear (*longitudinal field*), but both the π and σ components are present in the umbra (*transverse field*).

which include Broxon's relation, von Klüber's cross-sections, and a two-dimensional magnetic map of a sunspot published by Schröter (1953), imply a smooth decrease in field strength from the centre to the edge of the spot. However, observations made with greater resolving power reveal the existence of conspicuous irregularities (Severny, 1959). This author has plotted the longitudinal component of the magnetic field in spot groups with the aid of the magnetograph of the Crimean Astrophysical Observatory, using a scanning aperture of 1″ of arc and the line $\lambda 4886 \cdot 3$ (cf. Section 5.3.5). The magnetic traces show marked irregularities (see Severny's Fig. 1): in one instance a single

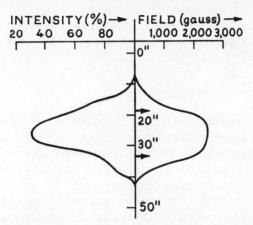

Fig. 5.16. Variation of light intensity and magnetic field strength along a line through a sunspot (von Klüber, 1948). Note that the point of minimum intensity roughly coincides with that of maximum field strength. The arrows indicate the approximate limits of the umbra.

spot appears to contain no less than three points of maximum field strength. Severny's observations also suggest the existence of *small-scale* magnetic irregularities, of the order of 50–100 gauss, both inside and outside spots.[21] Since the magnetograph measures the longitudinal component of the field $(H \cos \gamma)$, both the large- and small-scale fluctuations may correspond to variations in either the strength or *direction* of the field. It is to be hoped that Severny's work will stimulate others to try to obtain high-resolution observations of sunspot magnetic fields and, in particular, to compare these with photometric measurements made with comparable resolution.

5.4.5   DIRECTION OF LINES OF FORCE

The classical picture of the way in which the direction of the magnetic field varies across a sunspot emerged as a result of the work of Nicholson and Joy at

---

[21] The scale of the magnetic fine structure appears to be of the order of 2″ of arc; this is comparable to the scale of the umbral granulation (cf. Section 3.6.4), but it is not known whether the two phenomena are related in any way.

the Mt Wilson Observatory (Hale and Nicholson, 1938). Two different methods were employed. The first method involved a comparison of the intensities of the two σ components of a Zeeman triplet when viewed through a $\lambda/4$ plate analyser (cf. Section 5.2.4), while the second involved a determination of the direction of polarization of the π component in spots near the limb (cf. Section 5.2.3). In applying the first method, use was made of the fact, easily verified from (5.21), that the direction of the field is normal to the line-of-sight at any point in the spot where the σ components are of equal intensity. By determining the positions of such points and repeating the observations at different helio-centric angles as each spot was carried around the disk by solar rotation, Joy and Nicholson derived a law of direction based on 105 observations of 61 different spots. Their results (cf. Hale and Nicholson, 1938: Fig. 5) are represented to a fair approximation by the empirical relation

$$\delta = \frac{r}{b} \cdot 70^\circ, \qquad (5.37)$$

where δ is the inclination of the field to the vertical at a point distant r from the centre of the spot and b is the radius of the spot. According to (5.37) the field is vertical at the centre of the spot but, as one moves away from the centre, it becomes more and more inclined to the vertical, reaching an angle of 70° at the outer boundary of the penumbra (see Fig. 5.17).

The results derived by the first method refer to the lines of force lying in the plane containing the line-of-sight and the solar radius to the spot. To investigate the inclination of the field in the plane at right angles to this one, Nicholson studied the direction of polarization of the π component of a Zeeman triplet in three spots near the limb, measurements being made in each case at a number of points along a line running parallel to the limb and passing through the centre of the spot. Then, assuming that the lines of force lay in the meridional plane through the centre of the spot and were thus approximately normal to the line-of-sight (transverse Zeeman effect), he identified the direction of vibration of the π component at each point with the direction of the magnetic field and obtained excellent agreement with the results of the first method. However, as we have seen in Section 5.2.3, in the case of an *absorption* triplet the π component is actually partially plane polarized *perpendicular*, rather than parallel, to the field (cf. Table 5.1). This means that the apparent agreement found by Nicholson between the results of the two methods is illusory: in fact, the results of the second method would appear to show that the field is *horizontal* at the centre of the spot and becomes nearly vertical at the outer edge of the penumbra, in complete contradiction to (5.37). However, too much reliance should not be placed on these results in view of the obvious difficulties arising in limb observations of this kind; nevertheless, as we shall see in Section 5.4.8, modern observations have also demonstrated the existence of transverse fields in the umbra.

In more recent times the first method has been employed by Bumba (1960) to determine the inclination of the magnetic field in the outer parts of a number of isolated sunspots. He used the magnetograph of the Crimean Astrophysical Observatory to obtain maps of the longitudinal component of the field and assumed that the field was normal to the line-of-sight at points where the

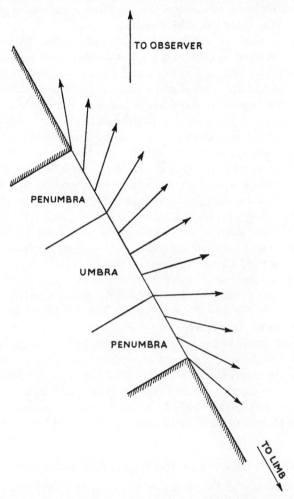

Fig. 5.17. Variation in the inclinations of the lines of force across a sunspot according to the classical picture. When the spot is near the limb, the lines of force are directed towards the observer on the side of the spot remote from the limb and away from the observer on the side near the limb. This results in an apparent difference in polarity on the two sides of the spot.

longitudinal component vanished. The results are in satisfactory agreement with those obtained by Joy and Nicholson (Bumba, 1960: cf. Fig. 9). Moreover, the greater sensitivity of the magnetograph enabled the observations to be extended considerably beyond the outer boundaries of the spots. More recently,

Leroy (1962) has attempted to investigate the variation of the inclination of the lines of force across the penumbrae of single, regular sunspots on the basis of observations made with the solar polarimeter (cf. Section 5.3.6). His results also accord with the classical picture, but imply somewhat greater angles of inclination than those found by Bumba.

The recent work of Treanor (1960), based on the 'Babinet compensator' technique described in Section 5.3.3, has yielded some information about the inclination of the field in the umbra and inner penumbra. Although Treanor's method is potentially capable of yielding a complete specification of the field direction with respect to axes fixed in the spot, the results actually presented (Treanor, 1960: cf. Table 2) give only the inclination of the field to the line-of-sight at a number of points across two spots. One isolated spot showed a high degree of circular symmetry, but in both cases the change in the direction of the lines of force is significantly less than that predicted by (5.37). However, Treanor's work was only in the nature of a preliminary investigation and further studies using the same technique are now in progress (Oxford Observatory, 1961).

The particular distribution of inclinations of the lines of force implied by (5.37) leads to an immediate explanation of why opposite polarities are observed on the two sides of a large sunspot near the limb (Hale and Nicholson, 1938). The situation to be considered is illustrated in Fig. 5.17, where the inclinations of the field lines at various distances from the spot centre have been calculated from (5.37). On the side of the spot away from the limb, the lines of force are directed towards the observer and the polarity determined with a $\lambda/4$ plate analyser will clearly be the same as that shown by the umbral field at the centre of the disk. On the side of the spot towards the limb, where the field is inclined away from the observer, the $\lambda/4$ analyser will indicate the presence of a weaker field of opposite polarity. Consequently, as Hale and Nicholson have emphasized, to determine the true polarity of a sunspot from observations made near the limb one should observe the field in the penumbra on the side facing the centre of the disk.

5.4.6 MAGNITUDE OF THE VERTICAL FIELD GRADIENT

It was pointed out by Houtgast and van Sluiters (1948) that a rough estimate of the rate at which the field strength in a sunspot varies with depth can be obtained by a simple geometrical argument from the law of direction of the field lines established by Joy and Nicholson (equation 5.37). According to this relation the inclination of the lines of force to the vertical increases linearly outwards along a radius of a spot at the rate of $7°$ ($=0.12$ radians) for each tenth of the radius. Taking a spot radius of 10,000 km, the lines of force which at one level pass through an area of the spot equal to, say, $\pi \times (1000)^2$ km^2 will at a second level 100 km above the first pass through an area equal to $\pi \times (1000 + 100 \times 0.12)^2$ km^2. If the field strength over the area considered at the first level is taken to be 2000 gauss, then the field strength at the second level is reduced to $\left(\dfrac{1000}{1012}\right)^2 \times 2000 = 1950$ gauss, implying that the field strength increases with depth at the rate of about 0·5 gauss/km. However, this result should be viewed with great caution since it is based on the classical picture of the field configuration which, as we shall see in Sections 5.4.8 and 5.4.9, is inconsistent with modern observations.

A second potential method of estimating the height gradient is provided by the recent detection of magnetic fields in the chromosphere above sunspots (cf. Section 4.3.5). For example, Bumba (1960) has measured the Zeeman splitting of the H, K, and Hα lines above a number of spots with the aid of a $\lambda/4$ plate analyser and has derived maximum field strengths which are sometimes only a few hundred gauss less than those measured in the umbrae of the spots at the photospheric level. Unfortunately, a precise determination of the mean field gradient cannot be made from the results since the effective heights of formation of these lines above sunspots are not accurately known. Moreover, the interpretation of magnetic measurements made with the aid of lines other than simple Zeeman triplets introduces an additional uncertainty. However, assuming a height of formation of about 2000 km, Bumba concludes that the mean gradient above the umbra does not exceed about 0·5 gauss/km.

A third possible method of determining the gradient of sunspot magnetic fields arises from the fact that the existence of a gradient produces an *asymmetry* in the individual σ components of a magnetically sensitive line. This was first pointed out by Hubenet (1954), who calculated the effect for the Fe I line $\lambda6173\cdot3$ (cf. Section 5.2.2). An actual attempt to determine the field gradient from observations of line asymmetry was made by Baranovsky and Stepanov (1959), who preferred to use in place of the comparatively weak line $\lambda6173\cdot3$ the strong Na D_1 line, which has a Zeeman splitting pattern $\dfrac{(2)\,4}{3}$. Spectra of four sunspots were taken through a $\lambda/4$ plate analyser and the observed profiles of the σ components were compared with profiles calculated theoretically on the basis of Michard's sunspot model. Values of the field gradient ranging from 1·0 to 1·8 gauss/km were obtained for the four spots. Unfortunately, these results are subject to considerable uncertainty on both observational and theoretical grounds. On the observational side, accurate *photospheric* profiles of the D_1 line obtained by one of the present authors (Bray, 1956) show unmistakable asymmetry in both the core and wings of the line due to the presence of numerous disturbing lines, mostly of atmospheric origin. On the theoretical side, the numerical value of the gradient is greatly affected by the actual choice of sunspot model. For example, on the basis of Mattig's model the values would be many times greater since the height scale of Mattig's model is much less than Michard's (see Section 4.2.4).

The same uncertainty regarding the particular choice of a sunspot model affects the results obtained by those authors who have attempted to derive values of the field gradient from magnetic measurements made in lines of different strengths (Tanaka and Takagi, 1939; Houtgast and van Sluiters, 1948; Bumba, 1960; Treanor, 1960). Of these, the recent observations of Treanor and Bumba deserve special mention. The former made careful measurements of the central field strengths in a number of spots, using the Fe I line $\lambda 6173\cdot3$ and the Fe II line $\lambda 6149\cdot3$. The results suggested a systematic negative gradient, imply-ing a decrease of field strength with depth. However, taking account of possible errors and uncertainties Treanor concluded that his results were really to be taken not as evidence for the existence of a negative gradient but rather as evidence against the existence of a large positive gradient. Bumba, on the other hand, measured the fields in the penumbrae of two sunspots. Both photospheric and chromospheric lines were used and the observations were interpreted on the basis of Minnaert's model of the photosphere and van der Hulst's model of the chromosphere, respectively. The results suggest that the field gradient in the penumbra is only of the order of a few tenths of a gauss per kilometre and may, at times, take zero or even negative values (Bumba, 1960: cf. Fig. 21).

As remarked in Section 5.4.3, Houtgast and van Sluiters also attempted to derive a value of the field gradient in large sunspots from a statistical study of the apparent centre-limb variation of sunspot magnetic fields. Interpreted in terms of Michard's sunspot model their data lead to a value of $1\cdot5$ gauss/km for the rate of increase of field strength with depth (Michard, 1953). However, little weight can be attached to this result in view of the fact that the existence of a true centre-limb effect has not yet been reliably established (see Section 5.4.3).

Still another method of estimating the height gradient of sunspot fields has recently been proposed by Leroy (1962) for the case of single, regular spots. It is based on measurements of the transverse component of the field made with the solar polarimeter described in Section 5.3.6, together with the equation div $\mathbf{H} = 0$ (see Section 8.2.2). Leroy finds that the vertical component of the umbral field decreases with height at the rate of $0\cdot2$–$0\cdot4$ gauss/km; the rate of decrease is smaller in the penumbra.

Summarizing, we may remark that the results of the various authors, though all subject to considerable uncertainties, do indicate small values for the field gradient and thus confirm the conclusion that sunspot magnetic fields penetrate into the overlying atmosphere with relatively little attenuation. This may explain why sunspots are observed to persist as coherent structures for at least several thousand kilometres into the chromosphere (cf. Section 4.3.5).

5.4.7 MAGNETIC LIFE-HISTORY OF SUNSPOTS

The growth and decay of the magnetic field in long-lived sunspots were studied by Cowling (1946), using the Mt Wilson magnetic observations for the years 1917–1924. The analysis was carried out for two groups of spots having lifetimes of about 55 and 30 days, and in each case the measured field strengths were corrected for the apparent centre-limb effect discussed in Section 5.4.3. The results are represented by the solid curves in Fig. 5.18 (a) (55-day spots) and

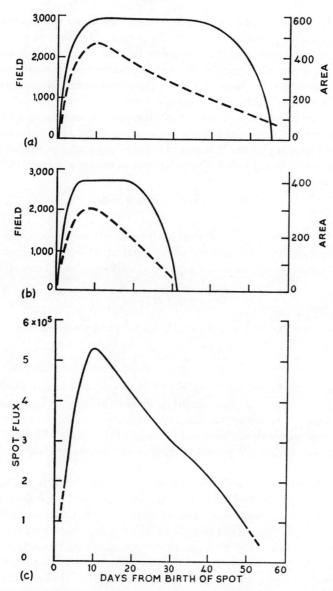

Fig. 5.18. Magnetic life-history of sunspots (Cowling, 1946). The magnetic field curves (gauss) are represented by solid lines and the area curves (millionths of the visible hemisphere) by broken lines. (a) and (b) give the results for 55- and 30-day spots respectively. (c) gives the corresponding flux curve for 55-day spots, calculated with the aid of equation (5.38).

Fig. 5.18 (*b*) (30-day spots), while the broken lines show corresponding area curves for 60- and 30-day leader spots obtained by the Greenwich observers. In both cases the curves show that the magnetic field increases very rapidly to a maximum value and then remains practically constant for much of the lifetime of the spot. On the other hand, although the area increases almost as quickly to a maximum value, it thereafter steadily declines.

Using these results we can now study the growth and decay of the total *magnetic flux* through a regular sunspot, assuming that the field distribution conforms to the classical picture. It is easy to show (Chapman, 1943; Mattig, 1953) that the magnetic flux through a circular spot of radius *b* is given by

$$\Phi = 2\pi \int_0^b H(r) \cdot \cos \delta \cdot r \, dr,$$

where $H(r)$ is the intensity of the field at distance *r* from the centre of the spot and δ is the inclination of the field to the vertical. If we now substitute from Broxon's relation (5.36) and from Joy and Nicholson's law of direction (5.37) we obtain

$$\Phi = 0.383 \, H_m \cdot \pi b^2$$

which, in terms of the spot area $A = \pi b^2$, can be written with sufficient accuracy as

$$\Phi = 0.4 \, H_m \cdot A. \qquad (5.38)^{[22]}$$

The flux curve obtained with the aid of (5.38) for the longer-lived spots is shown in Fig. 5.18 (*c*): the flux attains its maximum value at the same time as the spot area and thereafter steadily declines, even though the maximum field strength remains practically constant for most of the spot's lifetime. It may be significant that during the period of decline the flux curve follows an almost linear law.

Despite the general stability of sunspot magnetic fields it is nevertheless possible that they may show fluctuations of a few hundred gauss over periods of the order of a day (Brunnckow and Grotrian, 1949; Grotrian and Künzel, 1949). However, in view of the many uncertainties associated with the measurement of sunspot fields, the evidence in general is not particularly conclusive. Vyal'shin (1961) has observed an interesting case in which the field of a spot steadily decreased from 2900 to 2100 gauss in a period of $4\frac{1}{2}$ hours, then rapidly increased again to about 2500 gauss, and thereafter remained constant. Two direct photographs, one taken just before the commencement of observation and the other shortly after the field returned to a steady value, showed significant changes in the visible structure of the spot. Systematic observations of the short time scale behaviour of sunspot fields, particularly during the periods of formation and decay, would be very valuable.

---

[22] The presence of the factor 0.383 in place of the 0.315 found by Chapman is due to a slight difference in the adopted form of Joy and Nicholson's law.

5.4.8   OBSERVATIONS OF TRANSVERSE FIELDS IN SUNSPOT UMBRAE

In contradiction to what one would expect on the basis of the classical picture (cf. Fig. 5.17), many observers have found that the π component of a triplet line is very often clearly visible in the umbrae of sunspots near the centre of the disk. To quote the words of Hale and Nicholson (1938): '. . . but in many spots the p [π]-component is strong, even at the centre of the umbra, when the spot is near the centre of the Sun (see plate 6 i, j)'. However, while the Mt Wilson observers readily accepted the appearance of the π component in spots near the *limb* as evidence of transverse fields (cf. Section 5.4.5), they did not draw the same conclusion from the presence of the π component in spots near the centre of the disk. It is true that some caution was justified in view of the possibility of scattering from the penumbra and the surrounding photosphere (Section 5.3.7); on the other hand, owing to foreshortening, the effect of scattered light is actually far more serious in the case of spots near the limb (cf. Section 5.4.3)! However, in 1941 Evershed (*Observatory* **64,** 154) drew attention to a vital fact which had hitherto passed unnoticed. Evershed studied the Zeeman effect in two large sunspots in the particularly sensitive triplet lines $\lambda\lambda 5250\cdot 2$ and $6173\cdot 3$ and found that while both the σ and π components were present in the umbra, only the σ components appeared in the penumbra. *This indicates at once that the presence of the π component in the umbra cannot be due merely to scattering.*

Evershed's observations have since been confirmed by the more extensive work of Severny (1959) and Bumba (1962a, b). Plate 5.2 shows the appearance of the triplet line $\lambda 6302\cdot 5$ in the spectrum of a large sunspot near the centre of the disk, photographed by Severny through the 'polaroid grid' $\lambda/4$ analyser described in Section 5.3.7. An examination of this spectrogram shows that only the two σ components of the line are present over most of the penumbra, where the typical zig-zag appearance indicates the presence of a predominantly *longitudinal* field. On the other hand, both the π and σ components are present in the umbra; the intensity of the π component is actually greater than that of the two σ components which, according to Severny, are of approximately equal intensity in each step of the grid. In fact, the relative strengths of the π and σ components accord qualitatively with those expected on the basis of the intensity formulae (5.21) of Section 5.2.4 in the case of a predominantly *transverse* field. According to Severny, the same situation is found in the umbrae of all large sunspots and, to a lesser degree, in the umbrae of smaller spots. The observations of Evershed and Severny thus suggest a field configuration very different from the classical picture.[23] On the new view, a predominantly trans-

---

[23] Neither Evershed nor Severny (1959) actually attributed the appearance of the π component in sunspot umbrae to the presence of transverse magnetic fields. In fact, the latter proposed an explanation in terms of *depolarization* resulting, for example, from collisions between hydrogen and iron atoms (see also Stepanov, 1960b). However, in a later report Severny (1962) presents maps showing the direction of the *transverse* component of umbral magnetic fields which he and Stepanov have recently measured with the aid of a magnetograph equipped with a $\lambda/2$ modulator (cf. Section 5.3.5). Bumba (1962a, b) attributes the appearance of the π component in sunspot umbrae directly to the presence of transverse fields.

verse field in the umbra is surrounded by a largely vertical field in the penumbra. This interpretation receives strong support from some magnetograph observations made by Severny (1959: cf. Fig. 2), which actually show a region of zero longitudinal field at the very centre of the umbra of a sunspot.

Further confirmation is provided by the polarization measurements made by Dollfus and Leroy. These authors have found that the integrated radiation from sunspots is partially plane polarized: on the average the degree of polarization is about 5×10^{-4} (Dollfus, 1958b), but it may rise to 20×10^{-4} and sometimes even higher (Leroy, 1959).[24] As explained in Section 5.3.6, the existence of the polarization is due to the effect of transverse fields on the magnetically sensitive Fraunhofer lines in the integrated radiation. In order to obtain maximum sensitivity, the observations are usually made in the green region of the spectrum, but those made in the ultra-violet, blue, or yellow show no essential differences. The diameter of the scanning aperture used by Leroy (1959) was 15″ of arc, but this has since been decreased to 5″ (Dollfus and Leroy, 1962; Leroy, 1962). The polarization measurements have not yet given much information about the actual configuration of the umbral field. Nevertheless, it is interesting to note that in *large* spots the greatest polarization is found in small regions situated indifferently in the umbra or penumbra, while large regions of progressively smaller polarization cover the greater part of the spot. According to Leroy the lifetime of the strongly polarized regions is a few days, but more rapid changes often occur. The greatest transverse fields do not always coincide with the centre of the spot,[25] while often transverse fields of 200 gauss or more penetrate into the surrounding photosphere (cf. Section 5.4.4).

Transverse magnetic fields in sunspots have also been detected by a number of other workers, using conventional spectroscopic techniques (von Klüber, 1948; Semel, 1960; Michard, Mouradian, and Semel, 1961). Of particular interest is the observation by von Klüber of a strong transverse field between two regions of opposite polarity in the umbra of a large spot. On a normal photograph (see von Klüber's Fig. 11) the umbra of the spot appears as a single structure with only an incomplete light-bridge separating the regions of opposite polarity. A similar case was recorded by the Mt Wilson observers, in which a spot umbra was divided into two parts of opposite polarity by a well-marked light-bridge; the published spectrogram clearly indicates the presence of a transverse field in the neighbourhood of the bridge (Hale and Nicholson, 1938: Plates 6m, n). The occurrence of large complicated sunspots containing several, separate umbrae of different polarities within a single penumbra is fairly well known (cf. Hale and Nicholson, 1938), but von Klüber, on the basis of his own

[24] The failure of the attempt made by Babcock (1960) to detect plane polarization in the integrated radiation of sunspots can presumably be attributed to the much lower sensitivity of his equipment.
[25] In the case of small, regular spots Leroy (1962) finds that there is *no* transverse component at the centre of the umbra. For such spots the polarimetric results are in good general accord with the classical picture of the field configuration.

observations, has suggested that regions of opposite polarities may frequently occur even within single, isolated sunspots.

5.4.9   A POSSIBLE MODEL OF THE FIELD CONFIGURATION IN SUN-SPOTS

We cannot pretend at the moment to have a reliable knowledge of the exact morphology of the magnetic field of the individual sunspot. As we have seen in the previous section, the classical picture requires major revision, but it is not yet clear, on the basis of magnetic observations alone, what new picture will

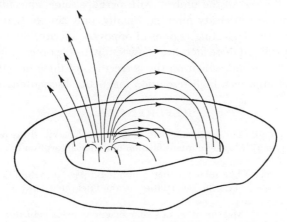

Fig. 5.19. A proposed model of the magnetic field configuration in the umbra.

finally emerge to replace the old. However, it is possible that some insight into this question can be obtained by appealing to high-resolution observations of the *intensity* distribution inside sunspots. The work of von Klüber described in Section 5.4.4 has established a significant correlation between large-scale variations in field strength and brightness, in the sense that the greater the field, the smaller the intensity. It would be most surprising if some relationship did not also exist between the *small-scale* variations in field strength and intensity. The two-dimensional photometric maps given in Section 4.2.7, which were made with an effective resolution of 2″ of arc, illustrate the striking complexity of the intensity distribution inside sunspots. This complexity appears to be characteristic of all sunspots, from mere pores to the largest and most irregular spots. If we accept the existence of a physical connection between magnetic field and intensity, then we must expect a similar complexity in the field configuration in both large and small sunspots.

Perhaps the most important question posed by the photometric maps concerns the significance of the small, extra-dark umbral regions which we have called 'cores'. They are generally only a few seconds of arc in diameter and are situated not at the centre of the umbra but near one end. Do the umbral cores represent the regions of greatest magnetic field strength?

In a recent paper (Bray and Loughhead, 1962) the authors have suggested that this is indeed the case and have outlined a new picture of the field configuration in the umbra, whose main features are illustrated in Fig. 5.19. Assuming that the magnetic field is greatest in the core, located near one end of the umbra, we may suppose that some of the field lines emanating from it return to other parts of the umbra rather than to neighbouring spots or the photosphere. Since the area of these parts is much greater than that of the core, the field strength there must be correspondingly weaker. It also follows that the umbra must possess *two* polarities, one concentrated in the core and the other scattered over the rest of the umbra, with perhaps some concentration in the regions of subsidiary intensity minima. Finally, one can see that over certain parts of the umbra, separating regions of opposite polarity, the magnetic field must be *transverse*.[26] All these facts are consistent with the observations described in Section 5.4.8, but judgement on the correctness of these new ideas must be postponed until improved high-resolution magnetic observations are obtained.

REFERENCES

BABCOCK, H. W. [1948] 'The magnetic field of Gamma *Equulei*', *Astrophys. J.* **108**, 191.

BABCOCK, H. W. [1949] 'Magnetic intensification of stellar absorption lines', *Astrophys. J.* **110**, 126.

BABCOCK, H. W. [1953] 'The solar magnetograph', *Astrophys. J.* **118**, 387.

BABCOCK, H. W. [1960] 'Test for polarisation in the integrated light of sunspots', *Pub. Astron. Soc. Pac.* **72**, 204.

BABCOCK, H. W. [1962] 'Measurement of stellar magnetic fields', *Astronomical Techniques*, ed. W. A. HILTNER, p. 107. Univ. Chicago Press.

BABCOCK, H. W., and BABCOCK, H. D. [1952] 'Mapping the magnetic fields of the Sun', *Pub. Astron. Soc. Pac.* **64**, 282. Reprinted in *The Sun*, ed. G. KUIPER, p. 704. Univ. Chicago Press, 1953.

BABCOCK, H. W., and BABCOCK, H. D. [1955] 'The Sun's magnetic field, 1952-1954', *Astrophys. J.* **121**, 349.

BARANOVSKY, E. A., and STEPANOV, V. E. [1959] 'The profile of the sodium D_1 line and the gradient of the magnetic field of a sunspot', *Izv. Crim. Astrophys. Obs.* **21**, 180.

BEGGS, D. W., and KLÜBER, H. VON [1956] 'A solar magnetograph', *Nature* **178**, 1412.

BLAMONT, J. E., and RODDIER, F. [1961] 'Precise observation of the profile of the Fraunhofer strontium resonance line. Evidence for the gravitational red shift on the Sun', *Phys. Rev. Letters* **7**, 437.

BOSCH, J. C. VAN DEN [1957] 'The Zeeman effect', *Handbuch der Physik*, ed. S. FLÜGGE, vol. 28, p. 296. Berlin, Springer.

BRAY, R. J. [1956] 'The centre-limb variation of the sodium D line intensities', *Mon. Not. R.A.S.* **116**, 395.

BRAY, R. J., and LOUGHHEAD, R. E. [1962] 'Isophotal contour maps of sunspots', *Aust. J. Phys.* **15**, 482.

BROXON, J. W. [1942] 'Relation of the cosmic radiation to geomagnetic and heliophysical activities', *Phys. Rev.* **62**, 508.

---

[26] Another picture of the field configuration in sunspot umbrae has recently been suggested by Bumba (1962a). In his model the lines of force in the umbra are twisted into spirals whose pitch increases rapidly with height (cf. Bumba, 1962b: Fig. 12); consequently, the field lines in the lower levels of the umbra form an almost horizontal ring, whereas in the upper levels they are directed more or less vertically.

BRUNNCKOW, K., and GROTRIAN, W. [1949] 'Über die zeitliche Änderung der magnetischen Feldstärke von Sonnenflecken im Laufe eines Tages', Z. Astrophys. **26**, 313.

BUMBA, V. [1960] 'Results of an investigation of the magnetic field of single sunspots', Izv. Crim. Astrophys. Obs. **23**, 212.

BUMBA, V. [1962a] 'Contribution to the study of the magnetic field in sunspot umbrae', Bull. Astron. Inst. Czech. **13**, 42.

BUMBA, V. [1962b] 'Magnetic fields in sunspot umbrae', Bull. Astron. Inst. Czech. **13**, 48.

BUMBA, V., and TOPOLOVA-RUZICKOVA, B. [1962] 'Polarization of light in solar spectrograph. I. Polarization on coelostat mirrors', Bull. Astron. Inst. Czech. **13**, 95.

CANDLER, A. C. [1937] Atomic Spectra, vol. 1. Cambridge Univ. Press.

CARNEGIE INSTITUTION [1961] Year Book 60. Washington.

CHANDRASEKHAR, S. [1950] Radiative Transfer. Oxford, Clarendon Press.

CHAPMAN, S. [1943] 'Magnetism in the Sun's atmosphere', Mon. Not. R.A.S. **103**, 117.

CONDON, E. U., and SHORTLEY, G. H. [1953] The Theory of Atomic Spectra. Cambridge Univ. Press.

COWLING, T. G. [1946] 'The growth and decay of the sunspot magnetic field', Mon. Not. R.A.S. **106**, 218.

DEUBNER, F.-L. [1962] 'Über die Anwendung weiter Spektrographenspalte bei lichtelektrischen Polarisationsmessungen an Fraunhoferlinien', Z. Astrophys. **56**, 1.

DEUBNER, F.-L., KIEPENHEUER, K. O., and LIEDLER, R. [1961] 'Der Magnetograph des Fraunhofer Instituts', Z. Astrophys. **52**, 118.

DITCHBURN, R. W. [1952] Light. London, Blackie and Son.

DOLLFUS, A. [1958a] 'Un polarimètre photoélectrique très sensible pour l'étude du Soleil', C.R. Acad. Sci. **246**, 2345.

DOLLFUS, A. [1958b] 'Premières observations avec le polarimètre solaire,' C.R. Acad. Sci. **246**, 3590.

DOLLFUS, A., and LEROY, J.-L. [1960] 'Polarisation de la lumière au bord du disque solaire', C.R. Acad. Sci. **250**, 665.

DOLLFUS, A., and LEROY, J.-L. [1962] 'Un magnétomètre mesurant les champs magnétiques perpendiculaires au rayon visuel. Applications à l'étude des champs radiaux autour des taches', Trans. I.A.U. **11 B**, 438.

GIOVANELLI, R. G. [1958] Trans. I.A.U. **10**, 197.

GROTRIAN, W., and KÜNZEL, H. [1949] 'Statistische Untersuchung der täglichen Änderung der magnetischen Feldstärke von Sonnenflecken', Z. Astrophys. **26**, 325.

HALE, G. E. [1913] 'Preliminary results of an attempt to detect the general magnetic field of the Sun', Astrophys. J. **38**, 27.

HALE, G. E., and NICHOLSON, S. B. [1938] 'Magnetic observations of sunspots 1917–1924. Part I', Pub. Carnegie Inst., No. 498.

HEEL, A. C. S. VAN [1958] 'Interferometry with Savart's plate', Concepts of Classical Optics, by J. STRONG, Appendix D, p. 400. San Francisco, W. H. Freeman.

HOWARD, R. [1959] 'Observations of solar magnetic fields', Astrophys. J. **130**, 193.

HOWARD, R. [1962] 'Preliminary solar magnetograph observations with small apertures', Astrophys. J. **136**, 211.

HOWARD, R., and BABCOCK, H. W. [1960] 'Magnetic fields associated with the solar flare of July 16, 1959', Astrophys. J. **132**, 218.

HOUTGAST, J., and SLUITERS, A. VAN [1948] 'Statistical investigations concerning the magnetic fields of sunspots, I', Bull. Astron. Inst. Netherlands **10**, 325.

HUBENET, H. [1954] 'Fraunhofer-Linien im inhomogenen Magnetfeld', Z. Astrophys. **34**, 110.

JÄGER, F. W. [1957] 'Zur Theorie und Beobachtung der Polarisation in solaren Fraunhoferlinien', Z. Astrophys. **43**, 98.

JENKINS, F. A., and WHITE, H. E. [1937] Fundamentals of Physical Optics. New York, McGraw-Hill.

KLÜBER, H. VON [1948] 'Über den Nachweis und die Messung lokaler Magnetfelder auf der Sonnenoberfläche', *Z. Astrophys.* **24**, 121.

KLÜBER, H. VON [1955] 'Spectroscopic measurements of magnetic fields on the Sun', *Vistas in Astronomy*, ed. A. BEER, vol. 1, p. 751. London, Pergamon.

KOTLYAR, L. M. [1960a] 'The solar magnetograph of the Pulkovo Observatory. Part 1', *Izv. Gl. Astron. Obs. Pulkovo* **21**, No. 163, p. 73.

KOTLYAR, L. M. [1960b] 'A new design of a differential photometer for a solar magnetograph', *Astron. J. U.S.S.R.* **37**, 469 (*Sov. Astron. AJ* **4**, 445).

LEIGHTON, R. B. [1959] 'Observations of solar magnetic fields in plage regions', *Astrophys. J.* **130**, 366.

LEROY, J.-L. [1959] 'Polarisation de la lumière des taches solaires', *C.R. Acad. Sci.* **249**, 2492.

LEROY, J.-L. [1960] 'Interprétation de la polarisation de la lumière des taches solaires', *C.R. Acad. Sci.* **251**, 1720.

LEROY, J.-L. [1962] 'Contributions à l'étude de la polarisation de la lumière solaire', *Ann. Astrophys.* **25**, 127.

MATTIG, W. [1953] 'Die radiale Verteilung der magnetischen Feldstärke in normalen Sonnenflecken', *Z. Astrophys.* **31**, 273.

MATTIG, W. [1958] 'Zur Linienabsorption im inhomogenen Magnetfeld der Sonnenflecken', *Z. Astrophys.* **44**, 280.

MICHARD, R. [1953] 'Contribution à l'étude physique de la photosphère et des taches solaires', *Ann. Astrophys.* **16**, 217.

MICHARD, R. [1961] 'Formation des raies de Fraunhofer en présence d'un champ magnétique', *C.R. Acad. Sci.* **253**, 2857.

MICHARD, R., MOURADIAN, Z., and SEMEL, M. [1961] 'Champs magnétiques dans un centre d'activité solaire avant et pendant une éruption', *Ann. Astrophys.* **24**, 54.

MINNAERT, M. [1953] 'The photosphere', *The Sun*, ed. G. KUIPER, p. 88. Univ. Chicago Press.

MOGILEVSKII, M. A., VELLER, A. E., and VAL'D-PERLOV, V. M. [1954] 'Determination of local magnetic fields on the Sun using a modulated photoelectric spectrophotometer', *Dokl. Akad. Nauk U.S.S.R.* **95**, 957.

NICHOLSON, S. B. [1933] 'The area of a sunspot and the intensity of its magnetic field', *Pub. Astron. Soc. Pac.* **45**, 51.

NIKULIN, N. S., SEVERNY, A. B., and STEPANOV, V. E. [1958] 'The solar magnetograph of the Crimean Astrophysical Observatory', *Izv. Crim. Astrophys. Obs.* **19**, 3.

NIKULIN, N. S. [1960] 'Some improvements of the magnetograph of the Crimean Astrophysical Observatory', *Izv. Crim. Astrophys. Obs.* **22**, 3.

ÖHMAN, Y. [1950] 'The use of Savart fringes in the observation of Zeeman effects in sunspots', *Astrophys. J.* **111**, 362.

ÖHMAN, Y. [1960] 'Magneto-optics in solar physics', *Scientia* **54**, 1.

OXFORD OBSERVATORY [1961] *Quart. J. R.A.S.* **2**, 263.

POLONSKII, V. V. [1960] 'Determination of the strength and polarity of sunspot magnetic fields by means of an attachment with the Fabry–Pérot etalon', *Solnechnye Dannye 1959*, No. 10, p. 78.

RACHKOVSKY, D. N. [1961a] 'On the formation of absorption lines in a magnetic field. Remarks on the papers by W. Unno and V. E. Stepanov', *Izv. Crim. Astrophys. Obs.* **25**, 277.

RACHKOVSKY, D. N. [1961b] 'A system of radiative transfer equations in the presence of a magnetic field', *Izv. Crim. Astrophys. Obs.* **26**, 63.

RACHKOVSKY, D. N. [1962] 'Magneto-optical effects in spectral lines of sunspots', *Izv. Crim. Astrophys. Obs.* **27**, 148.

RINGNES, T. S., and JENSEN, E. [1960] 'On the relation between magnetic fields and areas of sunspots in the interval 1917–1956', *Astrophysica Norvegica* **7**, 99.

SCHRÖTER, E. H. [1953] 'Ein Versuch zur Bestimmung des Verlaufes der magnetischen Feldstärke über die Fläche eines Sonnenfleckes', *Z. Astrophys.* **33,** 20.

SCRIMGER, J. A., and HUNTEN, D. M. [1957] 'Absorption of sunlight by atmospheric sodium', *Canadian J. Phys.* **35,** 918.

SEARES, F. H. [1913] 'The displacement-curve of the Sun's general magnetic field', *Astrophys. J.* **38,** 99.

SEMEL, M. [1960] 'Observations simultanées du champ magnétique et du champ des vitesses dans un grand groupe de taches solaires', *C.R. Acad. Sci.* **251,** 1346.

SEVERNY, A. B. [1955] 'The solar tower of the Crimean Astrophysical Observatory of the U.S.S.R. Academy of Science', *Izv. Crim. Astrophys. Obs.* **15,** 31.

SEVERNY, A. B. [1958] 'The appearance of flares in neutral points of the solar magnetic field and the pinch effect', *Izv. Crim. Astrophys. Obs.* **20,** 22.

SEVERNY, A. B. [1959] 'Fine structure of the magnetic field and depolarization of radiation in sunspots', *Astron. J. U.S.S.R.* **36,** 208 (*Sov. Astron. AJ* **3,** 214).

SEVERNY, A. B. [1960] 'An investigation of magnetic fields connected with solar flares', *Izv. Crim. Astrophys. Obs.* **22,** 12.

SEVERNY, A. B. [1962] 'Magnetically active regions on the Sun', *Trans. I.A.U.* **11 B,** 426.

SEVERNY, A. B., and BUMBA, V. [1958] 'On the penetration of solar magnetic fields into the chromosphere', *Observatory* **78,** 33.

SEVERNY, A. B., and STEPANOV, V. E. [1956] 'First observations of the magnetic fields of sunspots made at the Crimean Astrophysical Observatory', *Izv. Crim. Astrophys. Obs.* **16,** 3.

SHURCLIFF, W. A. [1962] *Polarized Light.* Harvard Univ. Press.

STEEL, W. H., SMARTT, R. N., and GIOVANELLI, R. G. [1961] 'A $\frac{1}{8}$ Å birefringent filter for solar research', *Aust. J. Phys.* **14,** 201.

STEEN, O., and MALTBY, P. [1959] 'On the correlation between observations of magnetic fields of sunspots at Mt Wilson, Potsdam and the Crimea,' *Astrophysica Norvegica* **6,** 113.

STEPANOV, V. E. [1958a] 'The absorption coefficient of atoms in the case of reverse Zeeman effect for an arbitrary directed magnetic field', *Izv. Crim. Astrophys. Obs.* **18,** 136.

STEPANOV, V. E. [1958b] 'On the theory of the formation of absorption lines in a magnetic field and the profile of Fe $\lambda 6173$ in the solar sunspot spectrum', *Izv. Crim. Astrophys. Obs.* **19,** 20.

STEPANOV, V. E. [1960a] 'The determination of the magnetic field gradient in the solar photosphere', *Izv. Crim. Astrophys. Obs.* **22,** 42.

STEPANOV, V. E. [1960b] 'The dependence of the readings of the solar magnetograph on the strength and direction of the field', *Izv. Crim. Astrophys. Obs.* **23,** 291.

STEPANOV, V. E. [1960c] 'The absorption coefficient of atoms in the reverse Zeeman effect for arbitrary multiplicity', *Izv. Crim. Astrophys. Obs.* **24,** 293.

STEPANOV, V. E. [1960d] 'The absorption coefficient in the inverse Zeeman effect for arbitrary multiplet splitting and the transfer equation for light with mutually orthogonal polarisation', *Astron. J. U.S.S.R.* **37,** 631 (*Sov. Astron. AJ* **4,** 603).

STEPANOV, V. E. [1961] 'The determination of the mean gradient of the chromospheric magnetic field', *Izv. Crim. Astrophys. Obs.* **25,** 174.

STEPANOV, V. E. [1962] 'Radiative equilibrium equations in atmospheres of magnetic stars', *Izv. Crim. Astrophys. Obs.* **27,** 140.

STEPANOV, V. E., and PETROVA, N. N. [1958] 'The polarities and maximum strengths of the magnetic fields of sunspots in 1956', *Izv. Crim. Astrophys. Obs.* **18,** 66.

STEPANOV, V. E., SHAPOSHNIKOVA, E. F., and PETROVA, N. N. [1962] 'Catalogue of strengths and polarities of magnetic fields of sunspots for the period of the International Geophysical Year (1 July 1957 – 31 December 1958)', *Ann. I.G.Y.* **23,** 1.

TANAKA, T., and TAKAGI, Y. [1939] 'Zeeman effect in sun-spots', *Proc. Phys.-Math. Soc. Japan* **21,** 421.

TREANOR, P. J. [1960] 'The spatial analysis of magnetic fields in sunspots', *Mon. Not. R.A.S.* **120**, 412.

UNNO, W. [1956] 'Line formation of a normal Zeeman triplet', *Pub. Astron. Soc. Japan* **8**, 108.

VYAL'SHIN, G. F. [1961] 'Rapid changes in the magnetic field of sunspots', *Solnechnye Dannye 1960*, No. 10, p. 79.

CHAPTER 6

The Properties of Sunspot Groups

6.1 Introduction

In Chapters 3, 4, and 5 we have been concerned solely with the individual sunspot, dealing successively with its morphology, physical conditions, and magnetic field. In this chapter we shall turn our attention to sunspot groups. The physical significance of the sunspot group is that its individual members *belong to the same magnetic field system*, the spots themselves acting as anchorage points for loops of magnetic flux which pass from spot to spot. It follows that a knowledge of the properties of spot groups must throw some light on the behaviour of the associated magnetic flux loops.

The chapter is divided into two main sections dealing in turn with the basic properties of individual sunspot groups (Section 6.2) and the statistics of their occurrence and of their distribution over the solar surface (Section 6.3). Among the basic properties described in Section 6.2 are the mode of development, lifetimes, areas, orientation of the axis, and proper motions. Section 6.2 also contains an account of two systems of classifying sunspot groups, which are both based on fundamental properties of the groups. These are the Zürich system, a nine-step classification based on the characteristic stages which a spot group may pass through during the course of its development and decline (Section 6.2.2), and the Mt Wilson system, which is based on the magnetic polarities of the individual spots in a group and, to some extent, on the distribution relative to the group of the surrounding K faculae (Section 6.2.6).

It has long been known that the frequency of occurrence of sunspot groups follows an 11-year cycle, a property shared by other forms of solar activity. There are a number of possible indices which can be used as a measure of solar activity, but by far the most widely used index is the *sunspot number* (R). This index is based on the total numbers of both groups and spots present on the visible hemisphere, and was introduced by R. Wolf in 1848. The definition, determination, and significance of the sunspot number are discussed in Section 6.3.1, while the characteristic features of the 11-year sunspot cycle revealed by records of this index are presented in Section 6.3.2. As we shall see in Section 6.3.4, the arrangement of magnetic polarities in spot groups is intimately related to the 11-year cycle. The chapter ends with an account of the latitude distribution of sunspot groups and its variation with the 11-year cycle (Section 6.3.5), and of irregularities in the distribution of spot groups between the northern and southern hemispheres (Section 6.3.6).

References to many of the investigations of sunspot groups of a purely statistical nature have been omitted from this account. The continuous publication since 1874 of the areas and positions of sunspot groups in the *Greenwich Photoheliographic Results* has led a number of authors to undertake statistical studies into the frequency of occurrence, latitude and longitude distributions, etc., of sunspot groups and, in fact, the number of published investigations of a purely statistical nature would seem to exceed those having a more direct physical basis. Such studies have had various aims in view: for example, some authors have tried to find an additional periodicity in the sunspot cycle, while others have sought to explain the apparent asymmetry in the occurrence of spot groups between the eastern and western hemispheres. However, the majority of such investigations have led to no new insight into the physical nature of sunspots or the solar cycle.[1]

6.2 The Properties of Individual Sunspot Groups

6.2.1 DEVELOPMENT OF SUNSPOT GROUPS

Sunspot groups originate as one or more sunspot *pores* (cf. Section 3.4) in a region of hitherto undisturbed photosphere. In the initial stage the group consists either of a handful of pores concentrated within an area of 5–10 square heliographic degrees or of a single spot at the preceding (west) end of what may subsequently become a fully developed group. Most groups, however, never develop beyond this stage, and by the following day have disappeared. The subsequent development of those that survive may be described—following Waldmeier (1955: p. 165) and Kiepenheuer (see Table 7.1)—as follows:

2nd day: the group has increased in area and become elongated; the individual spots are concentrated at the preceding (west) and following (east) ends of the group. (According to Kiepenheuer, the preceding spots appear 1 day before the following spots.) One spot at each end of the group usually develops much more than the others;

3rd day: the spots continue to grow and the main preceding spot develops a penumbra;

4th day: the main following spot also develops a penumbra; between the two main spots numerous small spots have appeared and the total number of spots may now be ~ 20–50;

5th–12th days: the group attains its maximum area during this period, often on or about the 10th day;

13th–30th days: the small spots between the preceding and following components disappear; then the following component disappears, usually by

---

[1] The interested reader will find investigations of this kind described in papers by U. Becker, B. Bell, V. F. Chistyakov, W. Gleissberg, M. A. Kliakotko, M. Kopecky, P. R. Romanchuk, J. Tuominen, Y. I. Vitinsky, and J. Xanthakis, chiefly in the following journals: *Astronomical Journal of the U.S.S.R.*, *Bulletin of the Astronomical Institutes of Czechoslovakia*, *Solnechnye Dannye*, and *Zeitschrift für Astrophysik*. Gleissberg's short monograph, *Die Häufigkeit der Sonnenflecken* (Akademie-Verlag: Berlin, 1952), is a useful source of references to statistical investigations carried out prior to 1952.

breaking up into several small spots, all of which gradually diminish in size (Hale and Nicholson, 1938). Meanwhile the preceding spot assumes a roundish shape;

30th–60th days: the preceding spot gradually becomes smaller until it, too, disappears; unlike the following component, the dissolution of the preceding spot seldom takes place by breaking up into smaller spots (Hale and Nicholson).

The above description gives a good general idea of the process of development of (long-lived) sunspot groups, but an adequate knowledge of the exact details of the process is still lacking. Our understanding would be greatly improved if sequences of high-quality photographs of sunspot groups taken at appropriately spaced intervals were available. In order to eliminate gaps in the records, such observations should be carried out by several cooperating observatories (cf. Section 3.4.5).

Good photographs of spot groups are to be found in Plates 3.4, 3.5 (*a*), 3.9, 3.20, and 3.25, and also in the following references: Abetti (1929: Fig. 36; 1957: Plates 24–26, 108); Kiepenheuer (1953: Fig. 18); Newton (1955: Fig. 1); Waldmeier (1955: Fig. 57); and de Jager (1959: Fig. 38).

6.2.2   THE ZÜRICH CLASSIFICATION OF SUNSPOT GROUPS

This scheme, devised in 1938 by M. Waldmeier (1955: p. 166) is a 9-step classification based on the characteristic stages which a spot group may pass through during the course of its development and decline. The individual classes in the Zürich classification are defined as follows (cf. Fig. 6.1):

CLASS A : a single pore or group of pores showing no bipolar[2] configuration;

CLASS B : a group of pores showing a bipolar configuration;

CLASS C : a bipolar group, one spot of which possesses a penumbra;

CLASS D : a bipolar group whose main spots possess penumbrae; at least one of the spots has a simple structure. The length of the group is generally $< 10°$ of heliographic longitude;

CLASS E : a large bipolar group; the two main spots possess penumbrae, and generally have a complex structure. Numerous smaller spots lie between the main spots. The length of the group is $> 10°$;

CLASS F : a very large bipolar or complex group; length $> 15°$;

CLASS G : a large bipolar group containing *no* small spots between the main spots; length $> 10°$;

CLASS H : a unipolar[2] spot possessing a penumbra; diameter $> 2°5$;

CLASS J : a unipolar spot possessing a penumbra; diameter $< 2°5$.

Four examples of each of the nine classes are given in Fig. 6.1, which is taken from Waldmeier. Photographs illustrating the classes have been published by de Jager (1959: Fig. 38).

The Zürich classification is essentially an *evolutionary* one. Large spot groups pass through all the classes during the course of their growth and decay, but a

---

[2] Note: the words 'bipolar' and 'unipolar' are here used in a loose sense and do not necessarily correspond to the correct magnetic designations (see Section 6.2.6).

medium-sized group might pass only through the sequences A-B-C-D-C-H-J-A or A-B-C-B-A. A small group might pass only through the sequence A-B-A or perhaps never progress beyond class A. It should be noted that all sunspot groups, irrespective of size, begin *and end* as class A groups.

The evolutionary course of a large sunspot group, passing through all the Zürich classes, is shown in Fig. 7.1. During the group's rapid rise to maximum

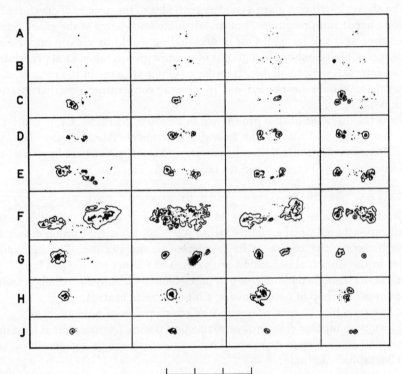

Fig. 6.1. The Zürich classification of sunspot groups. Four examples of each class are shown. The scale at the bottom indicates degrees of heliographic longitude.

area, it quickly passes through the classes A to E; it spends most of its time in classes G to J, when its area is slowly decreasing to zero. The associated flare activity reaches its maximum during classes D, E, and F (cf. Section 7.3.4). In contrast to the behaviour of large groups, small groups show more or less symmetrical development curves; with increasing maximum area and lifetime the curves become more asymmetrical.

The practical utility of the Zürich classification lies in its evolutionary basis. Once a given spot or spot group has been placed in its appropriate class, some indication is often immediately available of the stage which it has reached in its development. Such information can be of importance in planning certain types of observational programmes, particularly when it is necessary to choose a stable spot, such as, for example, in the study of the Wilson effect (cf. Section 3.8).

6.2.3 LIFETIMES AND AREAS OF SUNSPOT GROUPS

The *lifetimes* of sunspot groups were systematically investigated by M. N. Gnevishev (see Waldmeier, 1955: p. 164 and Fig. 44), using Greenwich observations during the period 1912–1934 of some 3000 groups. He found that more than half of the groups had lifetimes of less than 2 days, no less than 90% having lifetimes of less than 11 days. The frequency of occurrence rapidly decreases with increasing lifetime. Spot groups at high latitudes within the spot zones appear, on the average, to have somewhat shorter lifetimes than those nearer the equator.

According to Waldmeier, the lifetime of a spot group is related to the maximum area attained by the rule-of-thumb

$$T = 0 \cdot 1\, A_{max},\qquad(6.1)$$

where T = lifetime in days and A_{max} = maximum area in millionths of the visible hemisphere. Thus a group which attained a maximum area of 400 millionths—a large group—would have a lifetime of 40 days. Very large groups sometimes last for 100 days or more. However, equation (6.1) should be taken only as a rough guide, because wide individual variations can occur. In fact, very small groups sometimes persist for days, while very large groups sometimes appear and disappear during a single rotation (27 days).

The distribution of the *areas* of sunspot groups is given in Table 6.1, after Newton (1958: p. 86). The majority of spot groups have areas of less than 250 millionths, but the largest group ever recorded (in 1947) had a maximum area of 6132 millionths. On the average, about 8 groups having an area greater than 1500 millionths occur during each solar cycle (Newton, 1955). A compact group having an area greater than 500 millionths is visible to the naked eye; during times of high sunspot activity there may be 4 such groups simultaneously present on the disk.

TABLE 6.1

Distribution of Sunspot Groups According to Area

MEAN AREA DURING DISK PASSAGE (millionths of visible hemisphere)	PERCENTAGE OF ALL GROUPS %
1–250	85·6
250–500	9·2
500–750	3·0
750–1000	1·0
1000–1500	0·61
1500–2000	0·38
> 2000	0·23

6.2.4 ORIENTATION OF THE AXES OF SUNSPOT GROUPS

Sunspot groups occupy a roughly oval area (see Fig. 6.1) whose major axis is slightly inclined to the parallels of latitude. The sense of the inclination is such that the preceding (west) spots always lie at lower latitudes than the following (east) spots. Although the inclination α is necessarily a somewhat imprecise parameter, there is some evidence that it increases with the latitude of the group (see Waldmeier, 1955: Table 36); for a group with a latitude of 10–14°, $\alpha \simeq 5$–6°. According to Waldmeier, α also depends on the stage of development of the group, decreasing by a few degrees as the group progresses from the initial stage (class B) to its maximum development (class F). The decrease of α as the group develops is due to motions of the individual spots relative to one another (cf. Section 6.2.5).

6.2.5 PROPER MOTIONS OF SUNSPOTS IN A GROUP

As we have seen in the previous section, the preceding spot of a group lies at a lower latitude than the follower. It follows that differential solar rotation will cause a divergence in longitude between the spots, even in the absence of motions of the individual spots relative to their immediate surroundings. The *proper motion* of a spot is therefore defined as the motion remaining after subtracting, using the appropriate formula,[3] the component due to differential rotation.

When this subtraction is carried out, it is found that both the preceding and the following spots show proper motions in longitude. Somewhat smaller motions in latitude also occur. At the initial stage of formation of a group, the preceding spot moves westward and the following spot eastward. The net result is therefore an increase in the separation of the components. Two typical examples are illustrated in Fig. 6.2, taken from Waldmeier (1955: Fig. 47). Fig. 6.2 (a) shows the divergent motions in a newly-formed group. The preceding component moves nearly 5° of longitude to the west in a period of 6 days; the following component moves 3° to the east in the same period. Fig. 6.2 (b) shows a large spot which splits up into two components which then proceed to move apart, each component traveling some $2°5$ in 6 days.

According to the Greenwich observers, the westerly motion of the preceding spot ceases when the group attains its maximum area; thereafter the motion is *easterly*. The following spot moves only eastward, but it comes to rest after a few days. These facts are well illustrated in Fig. 6.3, which is also due to Waldmeier (1955: Fig. 48). The group in question was both large and long-lived; it appeared at latitude 30° north. The origin of coordinates in Fig. 6.3 is chosen as the place of birth of the preceding spot. The curves represent the subsequent motions of this and the following spot; the lines at right angles to the curves indicate 2-day

---

[3] Waldmeier (1955: p. 169) uses the formula $\omega = 14°37 - 2°60 \sin^2 B$, where ω is the daily angle of rotation at latitude B. However, this formula is subject to some uncertainty, which could lead to error in the derived proper motions.

intervals. The initial motion of the preceding spot is westerly and it continues to travel west and slightly north until, after about 18 days, it is some 7° west of its initial position. It then proceeds to travel northeast, ending up some weeks later 6° east and 5° north of its initial position. The following spot travels eastward throughout its life. It will be noticed that the velocity of both spots is greatest at the beginning of the group's development.

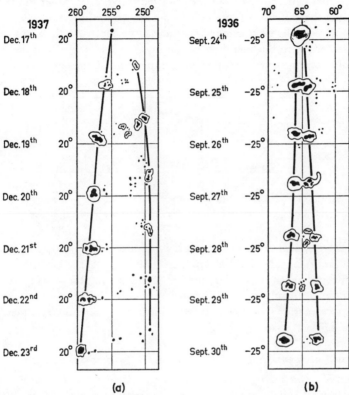

(a)   (b)

Fig. 6.2. Longitudinal proper motions in (a) a newly-formed group, and (b) a group resulting from the splitting of a large spot (Waldmeier, 1955). The scales indicate heliographic latitude and longitude.

Both spots in Fig. 6.3 show a well-marked latitudinal drift in addition to their longitudinal motions. The latitudinal drift of long-lived spots has been systematically studied by Tuominen with the aid of the Greenwich observations. He found that at latitudes greater than 16° the spots move polewards and at latitudes less than 16° equatorwards (see Waldmeier, 1955: Fig. 49), in confirmation of a result obtained earlier by Spörer. The latitudinal proper motion is always smaller than the *initial* longitudinal motion (cf. Fig. 6.3) and is therefore difficult to detect in short-lived spots.

An interesting consequence of the existence of longitudinal proper motions is that closely adjacent sunspot groups sometimes interpenetrate, apparently without mutual interference. Several excellent illustrations of this phenomenon have been published by Waldmeier (1955: see Fig. 51). Fig. 50 of Waldmeier's book illustrates the formation of a pseudo-bipolar group by an interpenetration of this kind. This apparently bipolar group had the following abnormal properties: (a) the following spot was larger than the preceding spot; (b) both spots had north magnetic polarity; (c) they showed converging rather than diverging motion; (d) the following spot rather than the preceding spot was nearer the equator. This strange behaviour was due to the fact that the spots, although close together, were not members of the same group: they were the surviving leaders of two groups whose following components had disappeared.

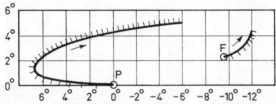

Fig. 6.3. Latitudinal and longitudinal proper motions of the preceding and following components of a long-lived group (Waldmeier, 1955). For description, see text.

From the physical point of view, it is evident that the proper motions of the spots in a group must reflect the dynamical behaviour of the magnetic flux loops anchored to them. In particular, the occasional formation of a pseudo-bipolar group would seem to suggest that two apparently independent magnetic field regions can sometimes come together and interpenetrate one another (see also Section 6.2.6).

6.2.6 THE MAGNETIC CLASSIFICATION OF SUNSPOT GROUPS

In Section 6.2.2 we have described the Zürich classification of sunspot groups— a developmental classification based on their visual appearance. In this section we shall describe the scheme of classification devised in 1919 by Hale and his colleagues at the Mt Wilson Observatory. This scheme is based on the magnetic polarities of the individual spots in a group and, to some extent, on the distribution of the surrounding K faculae (cf. Section 7.3.1). It is the product of the assiduous measurement by the Mt Wilson observers of the polarities of the individual members of several thousand spot groups.[4]

---

[4] The Mt Wilson observers found no difficulty in assigning a definite polarity (north or south) to each spot or, if need be, to each separate umbral region in large complex spots. On the other hand, as we have seen in Section 5.4.8, even apparently simple spots may contain regions of opposite polarity. However, the routine magnetic observations made at Mt Wilson were of low spatial resolution, which in general would be adequate to determine only the *dominant* polarity of each spot. In this section, therefore, and also in Sections 6.2.7 and 6.3.4, the reader must bear in mind that we are relying on a rather coarse description of the field of the individual spot.

Hale and his colleagues found that sunspot groups usually consist of two members, which may be single or multiple, of opposite magnetic polarity (Hale and Nicholson, 1938). The resulting concept of a *bipolar* spot group forms the basis of the Mt Wilson scheme of classification, all spot groups being regarded as variants of this fundamental type. Under this scheme, which is shown schematically in Fig. 6.4, spot groups are divided into three classes distinguished by the symbols α (unipolar), β (bipolar), and γ (complex). We shall now describe each of these classes in turn.

(1) *Unipolar.* Unipolar groups are composed of one or more spots of the same magnetic polarity. However, unipolar spots (and also the main components of bipolar groups) may occasionally have small companions of opposite polarity. On the other hand, these play such a minor and sporadic role that they are disregarded in classifying the group.

Unipolar groups are subdivided according to the distribution of the associated K faculae as follows:

(α): those in which the distribution of the faculae preceding and following the group is fairly symmetrical;

(αp): those which are followed by an elongated area of faculae;

(αf): those which are preceded by an elongated area of faculae.

Hale was led to introduce the K faculae into his scheme of classification by the observation of localized magnetic fields in the facular regions surrounding spot groups (Hale's so-called 'invisible spots'). The early Mt Wilson observations of facular fields lacked the sensitivity necessary to give a complete picture of the field distribution in the neighbourhood of spot groups. However, this void has since been filled by modern magnetograph observations, and it is now known that the distribution of the magnetic field throughout an activity centre is closely correlated with the distribution of the K faculae (see Section 7.4). The facular field is essentially bipolar in character (see, for example, Howard, 1959: Fig. 3). Moreover, unipolar groups are usually accompanied by extensive fields of opposite polarity, whose total flux is probably quite adequate to balance that of the spots themselves (Howard). In this way modern observations support Hale's original contention that αp and αf spots should be regarded as the preceding and following members, respectively, of incomplete *bipolar* groups.

(2) *Bipolar.* The simplest form of bipolar group comprises two spots of opposite polarity. However, most bipolar groups consist of a stream of spots, with spots of one polarity at one end and spots of opposite polarity at the other end. The change in polarity normally occurs near the centre of the stream, but it may occur some distance from the centre, on either side. In many cases the visual appearance of a spot group is by itself highly suggestive of a bipolar character but, as Hale and Nicholson (1938) have emphasized, it is not safe to regard all paired spots or binary groupings of small spots as true bipolar groups unless magnetic confirmation is available. For example, Hale and Nicholson have described pseudo-bipolar groups (see their Figs. 10 and 11) whose seemingly bipolar appearance was contradicted by the observed distribution of

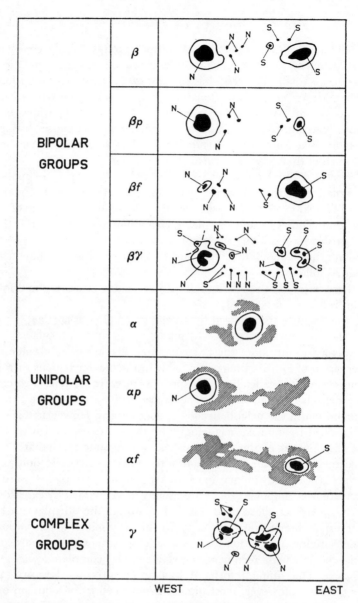

		WEST ... EAST
BIPOLAR GROUPS	β	
	βp	
	βf	
	$\beta\gamma$	
UNIPOLAR GROUPS	α	
	αp	
	αf	
COMPLEX GROUPS	γ	

Fig. 6.4. The magnetic classification of sunspot groups. 'N' and 'S' indicate north and south polarities respectively. The shaded areas represent K faculae.

polarities. An example of the formation of a pseudo-bipolar group as a consequence of the proper motions of two adjacent groups has already been described in Section 6.2.5.

Bipolar groups are subdivided according to the distribution of the field strengths and the areas of the constituent spots as follows:

(β): those in which the preceding and following members, whether single or multiple, are approximately equal in area;

(βp): those in which the preceding member is the principal component of the group;

(βf): those in which the following member is the principal component of the group;

($\beta \gamma$): those in which bipolar characteristics are present, but lacking a well-marked dividing line between regions of opposite polarity. This includes cases in which the preceding or following components are accompanied by small spots of the 'wrong' polarity.

Strictly speaking, the principal component of a bipolar group, whether single or multiple, is defined as the one with the greater magnetic flux. The extent to which the magnetic flux through the principal component of a group can exceed that through the minor component has been investigated quantitatively by Grotrian and Künzel (1950). However, when account is taken of the weak fields extending beyond the visible boundaries of the spots, an overall flux balance is maintained between regions of opposite polarity during most of the lifetime of the group (Bumba, 1961).

(3) *Complex.* Complex groups are those in which the polarities are so irregularly distributed that they cannot be classified as bipolar. Their number is small, but they include some of the largest and most active groups (cf. Section 7.3.4). In some cases, individual spots in complex groups contain sharply delineated regions of opposite polarity within a single penumbra (cf. Section 5.4.8). Grotrian and Künzel (1951) have shown that the regions of both north and south polarity in complex groups are distinguished by very high magnetic flux values.

It is not yet known what general rules govern the evolution of complex groups. The formation and subsequent development of one such group has recently been described by the Potsdam observers (Künzel, Mattig, and Schröter, 1961), but this example may represent a pathological case. The group arose as the result of the coalescence of an old unipolar spot with the following component of a newly-formed bipolar group. The unipolar spot was of south polarity, while the following component of the new bipolar group was of north polarity. As a result of the coalescence of these two spots, which took place within three days of the birth of the bipolar group, a complex group was formed in which the following spot contained within its penumbra several separate umbral regions of north polarity as well as an umbral region of south polarity. The evolution of this complex group is very clearly illustrated by the photographs published by Künzel *et al.*

The two schemes of classifying sunspot groups described in this chapter, namely the Zürich classification and the Mt Wilson classification, should be regarded as complementary rather than competitive. Both schemes, in fact, have proved to be extremely useful. For example, the flare activity of a group is closely correlated with its Zürich class (see Section 7.3.4), while the magnetic scheme of classification led Hale to the discovery of the now-famous laws of sunspot polarities (Section 6.3.4).

Magnetic classifications of sunspot groups are published regularly in the *Publications of the Astronomical Society of the Pacific*, which contains summaries of the Mt Wilson magnetic observations. It should be noted that in these summaries the letter l (preceding, following) the group designation means that on the day of observation the group (appeared, disappeared) at the (east, west) limb. The letter d (preceding, following) the group designation indicates that the group (originated, dissolved) on the day of observation. The letter x inserted in place of a group designation signifies that the group was seen or photographed at Mt Wilson but was not observed magnetically.

6.2.7 DISTRIBUTION OF SUNSPOT GROUPS ACCORDING TO MAGNETIC CLASS

The problem of determining the statistical distribution of spot groups according to magnetic class is complicated by the fact that many groups pass from one class to another during the course of their lives. According to Hale and Nicholson (1938: cf. Table 7), most groups are initially bipolar. This, however, is not true of short-lived groups lasting only a single day: these are mainly unipolar. Long-lived groups usually evolve from the β (or βp) stage fairly quickly, remaining as αp groups for the greater part of their lives. In some cases γ groups are observed to evolve out of α and β groups.

Table 6.2 gives the frequency of spot groups of each magnetic class expressed as a percentage of the total daily number of groups classified, derived from the

TABLE 6.2

Distribution of Sunspot Groups According to Magnetic Class
1917–1924

α	αp	αf	β	βp	βf	$\beta \gamma$	γ	UNCLASSIFIED
15	28	3	11	29	10	3	1	18

Note: the entry in the last column is expressed as a percentage of the total daily number of groups observed, while the entries in the other columns are expressed as percentages of the total daily number of groups classified.

Mt Wilson magnetic observations for the period 1917–1924 (Hale and Nicholson, 1938). The table shows that the frequencies of unipolar, bipolar, and complex groups are 46%, 53%, and 1% respectively. It is also evident that αp and βp groups predominate over αf and βf groups respectively, thus reflecting the well-marked tendency of preceding spots to outlive their following companions (cf. Section 6.2.1).

6.3 Statistics of the Occurrence and Distribution of Sunspot Groups

6.3.1   THE WOLF NUMBER AS A MEASURE OF SUNSPOT ACTIVITY

In 1848 the Swiss astronomer R. Wolf introduced the relative sunspot number R as a measure of the spot activity of the Sun. The Wolf sunspot number is defined as the quantity

$$R = K(\text{10}g + f), \qquad (6.2)$$

where f is the total number of spots on the visible disk, irrespective of size, and g is the number of spot groups. The factor K is a personal reduction coefficient which depends on the observer's method of counting the spots and of subdividing them into groups, on the size of his telescope and the magnification employed, and on the seeing conditions. For his own observations Wolf took $K = 1$, thus fixing the scale of the relative numbers. In 1882 Wolf's successors at the Zürich Observatory changed the counting method somewhat (see Waldmeier, 1961). However, from simultaneous observations using both methods they were able to derive a new value for K, namely 0·60, which reduced the new observations to the old scale. This value of K has been in continuous use at the Zürich Observatory from 1882 to the present day.

The sunspot numbers are usually determined from visual observations with a refractor of modest size, using a fairly low magnification. Wolf used a Fraunhofer refractor of 8 cm aperture and 110 cm focal length and a magnification of 64; this instrument, in a virtually unchanged form, is still in use at Zürich for the daily determination of the sunspot number.

In order to fill in gaps in the records Wolf himself initiated an international collaboration. At the present time more than 30 observatories and a similar number of amateur astronomers cooperate by sending their sunspot statistics to Zürich. These observations can be reduced to the Zürich scale by determining the values of the coefficient K, provided the telescopes and counting methods used do not differ too much from those in use at Zürich.

The foreign observations are sometimes received only after some delay, so in order to be able to publish at least provisional sunspot numbers promptly without gaps in the observations, Waldmeier and the other Swiss observers established additional observatories at Arosa (1939) and Locarno (1957). Provisional sunspot numbers based on observations made at the three Swiss-operated observatories are despatched at the end of each month to more than 100 interested establishments and, in addition, are broadcast by the Swiss Shortwave Service. Definitive sunspot numbers, incorporating the results from foreign observatories, are published annually in the *Astronomische Mitteilungen der Eidgenössischen Sternwarte Zürich*. They are also published in the *Quarterly Bulletin on Solar Activity*, in the *Zeitschrift für Meteorologie*, and in the *Journal of Geophysical Research*.[5]

[5] Lists of additional published sources of data on sunspot activity are given by Kiepenheuer (1953: p. 459); de Jager (1959: p. 333); and Athay and Warwick (1961: p. 74).

The definition of R is of course quite arbitrary. Moreover, a good deal of judgement has to be exercised in its measurement; even under identical instrumental and seeing conditions different observers will obtain different values for both f and g. A more objective measure of sunspot activity is provided by the *area* of the spots; however, even the measurement of this quantity is subject to some uncertainty, as an examination of some of the photographs in Chapter 3 will readily show. Nevertheless, areas of sunspot groups have been measured photographically at Greenwich since 1874. The data are published annually in the *Greenwich Photoheliographic Results*. More recently, measurement of sunspot areas has been undertaken at the Monte Mario Observatory in Rome and at the U.S. Naval Observatory, Washington, D.C. The relationship between area and sunspot number for a single day is rather loose, but the monthly averages reveal a close connection which takes the form (Waldmeier, 1955: p. 140)

$$A = 16 \cdot 7\, R, \qquad (6.3)$$

where A is the area in millionths of the visible hemisphere, corrected for foreshortening. Thus, near the peak of the solar cycle, when R might be, say, 100 (see Fig. 6.5), the total area of all the spots on the visible disk would be 1670 millionths, which would correspond to 4 fairly large groups, each about 400 millionths in area.

The determination of sunspot areas not only requires more elaborate equipment than the determination of sunspot numbers but, in addition, is far more time-consuming. It is not surprising, therefore, that the sunspot number is the most widely-used index of sunspot activity and, indeed, of solar activity as a whole.[6] In fact, the importance of the sunspot number is now far greater than when it was first introduced, largely as a result of increased interest in solar–terrestrial relationships (cf. Waldmeier, 1961); literally hundreds of such investigations have been published in which this index has been used. Moreover, it is the only index which reaches so far back in time. Thus, by searching early records Wolf was able to extend the daily sunspot numbers back to 1818; monthly means can be extended back to 1749, annual means back to 1700, and the years of maxima and minima are known even as far back as 1611, the year of the first telescopic observations of sunspots (cf. Section 1.2).

The most extensive compilation of sunspot numbers is that of Waldmeier (1961), who gives the following tables: (a) the years of minimum and maximum sunspot activity during the period 1610–1960; (b) annual mean sunspot numbers, 1700–1960; (c) monthly mean sunspot numbers, both smoothed and unsmoothed, 1749–1960; and (d) daily sunspot numbers, 1818–1960. In addition, Waldmeier gives graphs showing the annual mean sunspot numbers, 1700–1960; the smoothed and unsmoothed monthly means, 1755–1960; and the daily sunspot numbers, 1825–1960. The first of these is reproduced in Fig. 6.5.

[6] For a discussion of other possible indices of solar activity, see Athay and Warwick (1961).

6.3.2 THE 11-YEAR CYCLE OF SUNSPOT ACTIVITY

One of the most important and interesting properties of the Sun is the periodicity of the sunspot activity. This is illustrated in Fig. 6.5, which shows the variation of the annual mean sunspot number during the period 1700–1960. Several features of the sunspot periodicity are immediately evident from the figure:

(1) the periodicity is not strict, the time between successive maxima or minima showing considerable variation;

(2) there is a large range in the heights of the maxima;

(3) during any one cycle, the curve shows considerable irregularities;

(4) for the medium and high maxima, the rise to maximum is generally steeper than the fall to minimum.

Some of the numerical features of the sunspot cycle are summarized in Table 6.3. It is evident that the basic period of the sunspot cycle, defined as the

TABLE 6.3

Periodicity of the Sunspot Cycle[1]

	AVERAGE	RANGE
Period between maxima	10·9 years	7·3–17·1 years
Period between minima.......	11·1 years	9·0–13·6 years
Time of rise to maximum	4·5 years	2·9–6·9 years
Time of fall to minimum......	6·5 years	4·0–10·2 years
Maximum sunspot number[2] ...	108·2	48·7–201·3
Minimum sunspot number[2] ...	5·1	0–11·2

[1] Based on data given by Waldmeier (1961) for the period 1750–1958.

[2] Smoothed monthly means.

time between successive maxima or minima, is approximately 11 years. However, the great variations in the height of the maxima have led numerous authors to seek secondary cycles which, when superimposed on an 11-year cycle, correctly reproduce the observed curves (for references, see Waldmeier, 1955: pp. 152, 156–157). It is easy to represent the past course of the sunspot number variation by such means, but the real test lies in a prediction of the *future* variation: in this test such techniques have not been successful. It must be emphasized, therefore, that the only reliably established periodicity of the sunspot cycle is the well-known 11-year period.

H. H. Turner, H. Ludendorff, and later, Waldmeier himself, have pointed out that there is some tendency for high and low maxima to alternate. However, it would seem from an examination of Fig. 6.5 that this rule has too many exceptions for it to be accorded much significance.

According to Waldmeier (1955: p. 154), the course of the sunspot curve during any given cycle is largely determined by a single parameter, namely R_{max}, the maximum sunspot number attained. The shapes of the curves for high, medium, and low cycles are shown in Fig. 6.6, due to Waldmeier (cf. his Fig. 41); each curve is the average for a number of cycles. The most significant deduction from these curves is that for *high* and *medium* cycles the time of rise

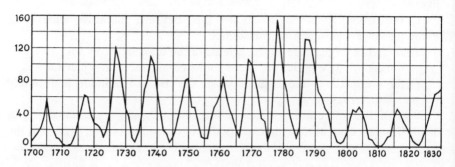

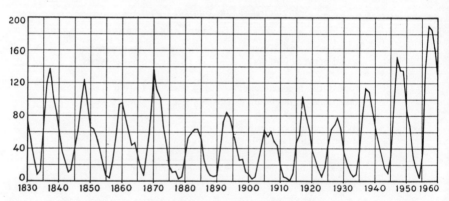

Fig. 6.5. The 11-year cycle of sunspot activity. The graph shows the variation of the annual mean sunspot number for the period 1700–1960 (Waldmeier, 1961). For a description of the characteristic features of the sunspot periodicity, see text.

T_R is less than the time of fall T_F, the asymmetry increasing with the height of the maximum. For *low* cycles T_R is approximately equal to T_F. Waldmeier gives the following empirical formulae for T_R and T_F, expressed in years:

$$(0 \cdot 17 \pm 0 \cdot 02) T_R = (2 \cdot 69 \pm 0 \cdot 09) - \log R_{max}, \qquad (6.4\text{A})$$

for even-numbered cycles;

$$(0 \cdot 10 \pm 0 \cdot 02) T_R = (2 \cdot 48 \pm 0 \cdot 10) - \log R_{max}, \qquad (6.4\text{B})$$

for odd-numbered cycles; and

$$T_F = (3 \cdot 0 \pm 0 \cdot 6) + (0 \cdot 030 \pm 0 \cdot 006) R_{max}, \qquad (6.5)$$

for both odd and even cycles.[7] In (6.4)–(6.5), R_{max} represents the largest value of the smoothed monthly sunspot number; in (6.5), T_F represents the time taken for the sunspot number to fall from R_{max} to a value of 7·5. Waldmeier also gives an empirical formula for the sunspot number 5 years after maximum:

$$R_5 = (0\cdot29 \pm 0\cdot06)R_{max} - (11\cdot4 \pm 6\cdot7). \qquad (6.6)$$

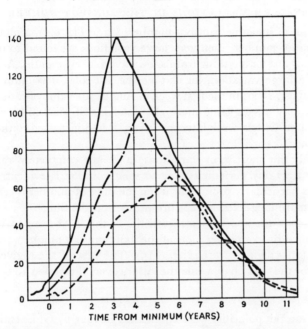

Fig. 6.6. Sunspot number curves for high, medium, and low cycles (Waldmeier, 1955). Each curve is the average for a number of cycles. Note the asymmetry of the curves for high and medium cycles.

The fact that the sunspot curve for any given cycle depends essentially on a single parameter is of some practical importance, since observations over only a short part of the cycle allow its future course to be forecast with some degree of success. Several examples of such short-term forecasts have been given by Waldmeier (1955: p. 156). On the other hand, only limited reliance can be placed on such forecasts or on the formulae given above; for example, (6.4B) fails to give even a nearly correct value of T_R for the anomalously high maximum of 1957·9.

There is no known method of making reliable *long*-term forecasts of sunspot activity.

The 11-year cycle of solar activity is one of the most important properties of the Sun. Its origin is still not fully understood, but it is clear that the ultimate explanation will involve the theory of the internal constitution of the Sun. Most

[7] The numbers of the cycles occurring between 1879 and 1944 are given in Fig. 6.7.

modern theories assume the existence of a weak *general* solar magnetic field and seek to find ways whereby portions of this field are periodically brought to the surface and amplified to give field strengths comparable to those observed in sunspots. The theory of the origin of the solar cycle is discussed in some detail in Section 8.4.

6.3.3 SHORT-TERM VARIATIONS IN THE SUNSPOT NUMBER

When, instead of smoothed sunspot numbers (as in Fig. 6.5), a plot is made of the *daily* sunspot number, then very large variations are found to occur; these variations may last for anything from a few days to a few months. As an extreme example of a large variation, let us take the last solar maximum during the month of August, 1960: in the first week of August the value of R declined from about 70 to only 25—a value more characteristic of solar minimum than of solar maximum. Only a week later, however, R had increased to the very high value of 250.

In general, the short-term variations in R are completely random, but occasionally a 27-day periodicity due to solar rotation is evident. The latter periodicity is particularly prominent when the sunspot number is relatively low and one or more long-lived groups are present on the Sun.

A good idea of the general characteristics of the short-term variations in R can be obtained by examining plots of the daily sunspot numbers published by Waldmeier (1955: Fig. 42; 1961: p. 115). The first of these gives the daily values of R for the year 1937, while the latter covers the period 1825–1960.

6.3.4 HALE'S LAWS OF SUNSPOT POLARITIES

The laws of sunspot polarities were enunciated in 1925 by G. E. Hale and S. B. Nicholson of the Mt Wilson Observatory. They apply (with few exceptions) to all spot groups which are *essentially bipolar in character*. These include not only the β groups (cf. Section 6.2.6) but also the αp and αf groups, the latter two being regarded as the preceding and following members, respectively, of incomplete bipolar groups. As can be seen from Table 6.2, no less than 84% of the total daily number of classified groups are essentially bipolar. The laws themselves can be stated as follows (Hale and Nicholson, 1938):

(1) the arrangement of magnetic polarities in essentially bipolar groups in the northern hemisphere is opposite to that in the southern hemisphere;

(2) the entire system of polarities remains unchanged during any one 11-year cycle of sunspot activity, but reverses with the beginning of the next cycle.

During the last cycle (Cycle No. 19), for example, the preceding and following spots of bipolar groups in the northern hemisphere had north and south polarities respectively; the polarities were opposite in the southern hemisphere. According to Hale and Nicholson, the solar equator forms a sharp line of demarcation between the northern and southern polarity systems. In fact, closely neighbouring groups on opposite sides of the equator usually conform strictly to the appropriate polarity patterns, although occasional exceptions have been noted. It is noteworthy that the preceding and following parts of the large-scale bipolar magnetic regions of the solar photosphere (BM regions) obey the same laws of magnetic polarity as the sunspots themselves (Babcock and Babcock, 1958).

Exceptions to the laws of sunspot polarities are rare. Out of 5814 essentially bipolar groups observed at Mt Wilson during the years 1917–1946, Richardson (1948) found that only 180 (3·1%) definitely failed to obey the polarity laws. He found that such anomalous groups differ little from normal groups although, on the average, they tend to be slightly smaller and somewhat shorter-lived.

On the basis of the laws of sunspot polarities Hale introduced the concept of the 'complete' solar cycle, corresponding to the interval between successive appearances in high latitudes of spot groups with the same arrangement of polarities and having a duration of about 22 years. This period is twice as long as the 11-year periodicity in the sunspot number discussed in Section 6.3.2.

The reversal of polarities is a sudden phenomenon, which begins with the appearance of spots of the new cycle in high latitudes before the spots of the old cycle have completely disappeared. Consequently, at sunspot minimum four separate sunspot zones are simultaneously present (cf. Section 6.3.5), each characterized by its own particular arrangement of polarities. It is now believed that the reversal of the sunspot polarities is intimately related to the reversal of the Sun's general magnetic field; this topic is discussed in Section 8.4.

6.3.5 THE LATITUDE DISTRIBUTION OF SUNSPOT GROUPS AND ITS VARIATION DURING THE 11-YEAR CYCLE

Since the days of the very early visual observers it has been known that sunspots are mainly confined to a relatively narrow belt in each hemisphere parallel to the solar equator. These belts do not possess precisely defined limits, but the vast majority of spots occur between the equator and latitudes $\pm 35°$. At latitudes greater than $40°$ sunspots are increasingly rare and, for the most part,

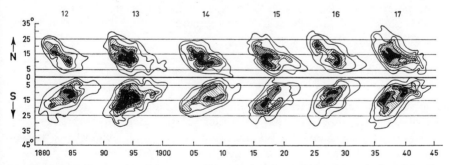

Fig. 6.7. The latitude distribution of sunspot groups and its variation during the solar cycle (Becker, 1955). The 'iso-lines' represent lines of equal spot frequency. The numbering of the cycles is shown at the top.

small and short-lived. However, on occasion spots have been observed at 50° or even 60°, and sunspot pores have been observed up to 75° (Waldmeier, 1955: p. 160).

Later observers (cf. Section 1.4) showed that the average latitude of the spots depends on the phase of the solar cycle, being greatest at the beginning of the cycle and decreasing steadily until the end. This is well illustrated in Fig. 6.7, due to Becker (1955); this figure shows the latitude distribution for the six cycles occurring between 1879 and 1944. It will be noticed that at any one epoch there is a considerable spread in latitude; nevertheless, the overall drift to the equator is well-marked for each cycle in both hemispheres. The average latitude as a function of time from minimum is given in Table 6.4, which is based on a curve given by Waldmeier (1955: cf. Fig. 41). The average latitude at any epoch is found to be independent of R_{max}, the maximum sunspot number attained during the cycle.

TABLE 6.4

Latitude Drift of the Sunspot Belts during the Solar Cycle

TIME FROM MINIMUM (*years*)	AVERAGE LATITUDE
0	±28°
2	21
4	16
6	12
8	10
11	7

Near the time of solar minimum, the first spots of the new cycle appear before the last spots of the old cycle have disappeared. In fact, the start of the new cycle is heralded by the appearance of new spots at high latitudes. According to Waldmeier, the overlapping of the two cycles lasts, on the average, for about two years. During the period of overlap there are thus *four* sunspot zones; the arrangement of magnetic polarities for groups in any given zone is fixed by the hemisphere and cycle to which the zone belongs (cf. Section 6.3.4).

The cause of the latitude drift is very obscure. Existing theoretical ideas on the subject are discussed in Section 8.4.

6.3.6 IRREGULARITIES IN THE DISTRIBUTION OF SUNSPOT GROUPS BETWEEN THE NORTHERN AND SOUTHERN HEMISPHERES

The northern and southern hemispheres often show very unequal sunspot activity, and the inequality may persist for several years. Moreover, the cycle may both commence and reach a maximum earlier in one hemisphere than in the other. Cycle No. 18 well illustrates these facts (Waldmeier, 1955: Table 37): in this cycle, which started in 1943, the activity in the southern hemisphere persistently exceeded that in the northern until 1948, after which the situation was reversed. Moreover, the southern maximum occurred in 1947, but the weaker northern maximum was delayed until 1949. According to Newton and Milsom (1955), the asymmetry in the activity of the two hemispheres does not appear to follow any definite laws.

6.4 Concluding Remarks

This concludes our account of the properties of sunspot groups seen at the *photospheric* level. However, as we have seen in Sections 4.3.5 and 5.4.6, there is evidence that sunspots and, in particular, their magnetic fields, extend upwards into the overlying chromosphere and probably beyond. The arrangement of magnetic polarities in spot groups, described in Sections 6.2.6, 6.2.7, and 6.3.4, strongly suggests that the upward extension of the magnetic field is in the form of gigantic loops of magnetic flux linking spots of opposite polarity. These flux loops may extend far beyond the visible surface of the Sun. It follows that the spot group is an integral and essential part of a complex field system; as such, it plays an important role in an even larger region of disturbance on the Sun, namely the *activity centre*. The role played by sunspots in activity centres is the subject of the following chapter.

REFERENCES

ABETTI, G. [1929] 'Solar physics', *Handbuch der Astrophysik*, ed. G. EBERHARD, A. KOHL-SCHÜTTER, and H. LUDENDORFF, vol. 4, p. 57. Berlin, Springer.

ABETTI, G. [1957] *The Sun*. London, Faber and Faber.

ATHAY, R. G., and WARWICK, C. S. [1961] 'Indices of solar activity', *Advances in Geophysics* **8**, 1.

BABCOCK, H. W., and BABCOCK, H. D. [1958] 'Photospheric magnetic fields', *Electromagnetic Phenomena in Cosmical Physics*, ed. B. LEHNERT, p. 239. Cambridge Univ. Press.

BECKER, U. [1955] 'Untersuchungen über die Herdbildung der Sonnenflecken', *Z. Astrophys.* **37**, 47.

BUMBA, V. [1961] 'Magnetic fields in sunspot groups', *Bull. Astron. Inst. Czech.* **12**, 82.

GROTRIAN, W., and KÜNZEL, H. [1950] 'Über den Induktionsfluss durch die Sonnenflecken', *Z. Astrophys.* **28**, 28.

GROTRIAN, W., and KÜNZEL, H. [1951] 'Über den Induktionsfluss durch Sonnenfleckengruppen der Klassen βγ und γ', *Z. Astrophys.* **29**, 173.

HALE, G. E., and NICHOLSON, S. B. [1938] 'Magnetic observations of sunspots 1917–1924. Part I', *Publ. Carnegie Inst.*, No. 498.

HOWARD, R. [1959] 'Observations of solar magnetic fields', *Astrophys. J.* **130**, 193.

JAGER, C. DE [1959] 'Structure and dynamics of the solar atmosphere', *Handbuch der Physik*, ed. S. FLÜGGE, vol. 52, p. 80. Berlin, Springer.

KIEPENHEUER, K. O. [1953] 'Solar activity', *The Sun*, ed. G. KUIPER, p. 322. Univ. Chicago Press.

KÜNZEL, H., MATTIG, W., and SCHRÖTER, E. H. [1961] 'Beobachtungen einer eigentümlichen Verschmelzung zweier Sonnenfleckengruppen', *Die Sterne* **37**, 198.

NEWTON, H. W. [1955] 'The lineage of the great sunspots', *Vistas in Astronomy*, ed. A. BEER, vol. 1, p. 666. London, Pergamon.

NEWTON, H. W. [1958] *The Face of the Sun*. London, Penguin Books.

NEWTON, H. W., and MILSOM, A. S. [1955] 'Note on the observed differences in spottedness of the Sun's northern and southern hemispheres', *Mon. Not. R.A.S.* **115**, 398.

RICHARDSON, R. S. [1948] 'Sunspot groups of irregular magnetic polarity', *Astrophys. J.* **107**, 78.

WALDMEIER, M. [1955] *Ergebnisse und Probleme der Sonnenforschung*, 2nd ed. Leipzig, Geest u. Portig.

WALDMEIER, M. [1961] *The Sunspot-Activity in the Years 1610–1960*. Zürich, Schulthess and Co.

CHAPTER 7

Sunspots and Activity Centres

7.1 Introduction

The transient phenomena on the Sun occur within localized regions known as activity centres, whose areas usually do not exceed about one-tenth of the visible hemisphere (cf. de Jager, 1961). Besides sunspots, the phenomena associated with activity centres include photospheric and chromospheric faculae, flares, surges, prominences, and coronal condensations. In addition, activity centres are responsible for the emission of enhanced solar radio noise and X-ray radiation and also, perhaps, solar cosmic rays. Finally, they are the site of magnetic fields, usually of a bipolar character. Several activity centres may be present simultaneously on the disk, but they originate independently and have different lifetimes. Many of them are short-lived, existing for only a few weeks; others last for as long as 100–200 days, or even longer. The number of activity centres present at any given time depends on the sunspot number, and hence on the phase of the solar cycle (cf. Section 6.3.2). The various stages of development of a long-lived activity centre are indicated in Table 7.1, which is based on data published by Kiepenheuer (1953: p. 432; 1960). This table gives a useful bird's-eye view of some of the phenomena associated with activity centres. However, the phenomena are so diverse and their inter-relationships so complex that a full discussion lies far outside the scope of this book.[1]

It is now generally accepted that the many diverse and often spectacular phenomena associated with activity centres owe their origin to the presence of magnetic fields which have risen from the solar interior and pushed their way through the photospheric layers into the chromosphere and corona above. Little is known about the three-dimensional configuration of the field in an activity centre; however, it can be pictured in an idealized way as a system of gigantic loops of magnetic flux whose ends are anchored to sunspot umbrae of opposite polarity (see, for example, Babcock and Babcock, 1958). Although the magnetic field precedes the appearance of the first sunspot of the group and survives the disappearance of the last spot (cf. Table 7.1), the spots nevertheless play a unique role in this complex field system: not only do they serve as anchorage points for the flux loops but, in addition, they are the regions of greatest field strength. Moreover, it is evident from Table 7.1 that the period of greatest activity of the activity centre coincides with the period during which spots are

---

[1] The reader is referred to the articles by Kiepenheuer and de Jager cited above, and also to articles by Dodson and Hedeman (1956); de Jager (1959: p. 183); Christiansen *et al.* (1960); Krivsky and Letfus (1960); and Athay and Warwick (1961).

present. In a certain sense it is permissible to consider the magnetic field of an activity centre as 'belonging' to the associated sunspot group although, as we shall see in Chapter 8, the spots themselves also owe their origin to magnetic fields.

It is evident, therefore, that we must expect to find some relationships between sunspots and other features of activity centres, the magnetic field providing the connecting link. In this chapter we shall examine the observational evidence for such relationships, with the particular aim of throwing light on the role played by the sunspots. We shall, however, restrict ourselves to those activity phenomena observed to show a more or less definite relationship with spots and, moreover, for each phenomenon we shall discuss only those aspects having a definite bearing on the relationship. The phenomena discussed include photospheric faculae (Section 7.2), chromospheric faculae (Section 7.3.1), the dark filamentary patterns observed in Hα in the neighbourhood of sunspots (Section 7.3.2), prominences (Section 7.3.3), and chromospheric flares (Section 7.3.4). Finally, the available information concerning the role played by magnetic fields in chromospheric faculae, prominences, and flares is briefly discussed in Section 7.4.

7.2 Sunspots and Photospheric Faculae

Photospheric ('white-light') faculae are visible only near the limb and, as a rule, are distributed over a large area in the neighbourhood of a sunspot group and its surroundings. It has long been known that there is a close *statistical* association between spots and faculae. No spots near the limb are observed without attendant faculae and, conversely, except for the polar faculae, which occur in zones where spots do not normally appear, there are few faculae which are not associated with spot groups at some stage of their lives. The details of the close association between spots and faculae have been very lucidly summarized by the Greenwich observers (see Kiepenheuer, 1953: pp. 368–70), on the basis of extensive observations of faculae carried out over a number of years.

The association found by the Greenwich observers is essentially a statistical one. On the other hand, evidence of a more direct connection, suggestive of an actual physical relationship, is provided by good-quality photographs of sunspot groups near the limb, such as the Frontispiece. On this photograph the facular material surrounding the group appears to invade the spot nearest the limb. In fact the bright material within the spot, which at the centre of the disk would be called a 'light-bridge' (Section 3.7), is indistinguishable both with regard to brightness and granular appearance from the faculae outside. The spot furthest from the limb contains a light-bridge whose brightness exceeds that of the faculae outside the spot at the same heliocentric angle, but in this case the bridge does not link up with any facular material. Another spot group near the limb is shown in Plate 7.1. In this case a massive tongue of bright material, equally as bright as the faculae outside the spot, penetrates deeply into the penumbra at the top of the main spot. The contrast between the faculae and their surroundings is less on this photograph than on the Frontispiece, the spot group being further from the limb.

More striking evidence for a possible physical relationship between spots and faculae is provided by the observation that sunspot *pores* near the limb are often bordered on one side by bright streaks of facular material (Loughhead and Bray, 1959). The facular streaks closely follow the outlines of the pores and occur almost invariably on the *limb* side. Plate 7.2 gives two illustrations of this phenomenon: Plate 7.2 (*a*) shows portion of an isolated group of small spots, while Plate 7.2 (*b*) shows some of the debris between the main components of a well-developed bipolar group. All of the pores shown appear somewhat elongated in the vertical direction owing to foreshortening (the heliocentric angles are $58°$ and $64°$ respectively). The bright borders are very conspicuous on some of the pores; their apparent widths lie between 1 and $2''$ of arc. It is noteworthy that the bright streaks, as well as bordering the pores on the limb side, in some cases also extend around the top or bottom parts of the pores. The effect is sometimes observed even in very small pores only $1–2''$ in diameter; in these cases it takes the form of a bright point of light adjacent to the pore on its limb side. However, the phenomenon is by no means universal: many pores similar to those illustrated in Plate 7.2, observed at comparable heliocentric angles, have failed to show the effect. Nevertheless, when a bright border is present, it occurs almost invariably on the limb side. Although the interpretation of the facular bordering effect is not yet clear, the observations do suggest a close physical relationship between spots and faculae in at least some cases.[2]

7.3 Sunspots and Chromospheric Phenomena

7.3.1   CHROMOSPHERIC FACULAE

As Kiepenheuer (1953: p. 371) has remarked, the transition from photospheric to chromospheric faculae is a continuous one. As far as form and frequency are concerned the two phenomena bear a close resemblance, and this also applies to their relationship with sunspots. One important difference is that whereas photospheric faculae are visible only near the limb, chromospheric faculae are equally well visible at the centre of the disk. This circumstance makes the chromospheric faculae much easier to observe.

Chromospheric faculae are much longer-lived than the associated sunspots. They often outlive the spots by several rotations and, in addition, they actually precede the formation of the first spot of the group by periods ranging from a few hours to a day or so (cf. Table 7.1). In the early stage they present the appearance of a bright, compact patch, particularly prominent in the K line of

[2] One obvious interpretation is that over the pores lie facular 'clouds' of similar dimensions which, at the heliocentric angles concerned, appear displaced towards the limb by parallax; the observed displacements imply a height of about 1200 km (Loughhead and Bray, 1959). However, direct attempts by the authors to detect facular material beyond the limb have not been successful.

TABLE 7.1

Development of an Activity Centre

DAY	SUNSPOTS[1]	PHOTOSPHERIC AND CHROMO- SPHERIC FACULAE	FLARES	MAGNETIC FIELD	FILAMENTS	CORONA
1		Appearance of a small bright speck visible in Hα, K, and, if near the limb, in white light.		Strength of longitudinal component increases to > 50 gauss.		Brightness increases in λ5303 and in white light.
2	First spot appears, at W end of facular region.	Faculae increase in brightness and area.				
3	Another spot (or spots) of opposite polarity to existing spot appears at E end of facular region.	Faculae continue to increase in brightness and area, especially the K faculae.		Area occupied by field increases.		Brightness continues to increase.
4	Main spot of the group (the W spot) forms a penumbra.	Faculae cover an increasing area around spot group.	First flares are observed.	Field is now definitely bipolar and its area is still increasing.	Small unstable filaments appear near the main spot.	Brightness still increasing and coronal emission line λ5694 now appears.
5	E spot (or spots) now forms penumbra. Between E and W spots numerous small spots have appeared; the total number of spots may now be ~ 20.		Flare activity increases.			Coronal condensations in the form of loops, visible in Hα and λ5303, are seen for the first time.

6–13	Spot group attains its maximum area.	Brightness of faculae still increasing.	Flare activity reaches a maximum.	Area of field region continues to increase.		Brightness in $\lambda5303$ and white light continues to increase; condensations and violent motions are observed in Hα, $\lambda5303$, $\lambda5694$, and white light.
14–30	All spots except W spot disappear.	K faculae are now very extensive, but Hα faculae have begun to shrink.	Flare activity decreases.	Field region shows its largest flux.	A stable filament, of length 50,000 km, points towards surviving spot.	Corona is still bright in $\lambda5303$ but has begun to fade in white light.
30–60	Surviving spot slowly disappears.	Brightness of Hα and K faculae decreases.		Field is still detectable, but now shows an irregular distribution.	Filament grows steadily in length ($\sim 10^5$ km per rotation), and divides the activity centre into two halves.	Brightness of corona in $\lambda5303$ decreases.
60–100		Hα faculae disappear; K faculae lose their compactness, resuming the form of a bright network.		Field still detectable.	Filament reaches its greatest length and is now almost parallel to the equator.	
100–250				Field region may last for 100–250 days.	Filament may last for 100–250 days. Its disintegration may occur simultaneously with that of the field.	

*[1] Note: for a more detailed description of the development of sunspot groups, see Section 6.2.1.

Ca II (cf. Dodson and Hedeman, 1956: Plate 1); later on, they spread out over the whole area of the group and become more ragged in appearance.

Chromospheric faculae are not so conspicuous in Hα as in the K line of Ca II. In the K line they not only occupy a greater area but, in addition, show a greater contrast with their surroundings. To gain an idea of the general relationship between chromospheric faculae and their associated spot groups the reader cannot do better than to examine some of the excellent Hα, K, and white-light photographs reproduced in various books and articles: see, for example, Abetti (1929: Figs. 57, 59–67, 119–121); Kiepenheuer (1953: Figs. 27, 35, 36); Waldmeier (1955: Fig. 75); Abetti (1957: Plates 47, 48, 50–55, 99–106).

Several points illustrated by these photographs deserve special mention. In the first place, any *light-bridges* present on the Hα or K spectroheliograms appear just as bright as the chromospheric faculae outside the spot (see, for example, Abetti, 1929: Figs. 62, 63, 120). This relationship is in striking agreement with the relationship between light-bridges and *photospheric* faculae discussed in the previous section.

Secondly, the distribution of the K faculae is related to the magnetic classification of the associated spot group (see Section 6.2.6). For example, when the following components of a bipolar group disappear the faculae in their immediate neighbourhood nevertheless persist: the facular region as a whole is then unsymmetrically disposed about the remaining spots, giving the group a magnetic classification of αp.

Finally, the appearance of the K faculae changes as the spectroheliograph is tuned from the wing (K_1) through one of the emission peaks (K_2) to the line centre (K_3). The K_1 spectroheliograms differ little from white-light photographs, but in K_2 the faculae become denser, larger, and brighter. Moreover in K_2 some of the smaller spots become completely covered by the faculae. Finally, when K_3 (which shows the highest layers) is reached, even the large spots often become completely or almost completely covered by the faculae. This result may be connected with the peculiar behaviour of the profile of the core of the K line over spot umbrae, discussed in Section 4.3.5. In Hα spectroheliograms, the spots usually remain well visible and even the distinction between umbra and penumbra is frequently well maintained in the larger spots.

7.3.2 CHROMOSPHERIC PATTERNS IN THE NEIGHBOURHOOD OF SUNSPOT GROUPS

In an undisturbed region of the Sun the dark elements of the chromospheric fine structure are round, or perhaps slightly oval, and more or less randomly distributed. However, in the neighbourhood of sunspot groups, the dark elements undergo two changes (Kiepenheuer, 1953: p. 394). In the first place, they become *elongated*, their characteristic dimensions then being some 1500 × 20,000 km (de Jager, 1959: p. 179, quotes the figures 2000 × 12,000 km). Secondly, they show a tendency to line up and form characteristic patterns. For convenience, the patterns may be classified into four types: (a) a very confused structure, in which the small dark filaments show no particular orientation; (b) an 'iron filing' structure, in which the filaments are orientated in a way reminiscent of iron filings in the field of a bar magnet; (c) a vortex structure, in which the filaments spiral around a spot; (d) a radial structure, in which the filaments are directed radially outward from a spot.

The structural rearrangement of the chromospheric fine structure associated with a new spot group begins several days before the appearance of the group and survives it by several rotations (Kiepenheuer, 1953: p. 394). The particular pattern occurring in any given case appears to depend on the complexity and the stage of evolution of the spot group. Thus the confused structure is characteristic of a group at an early stage of its development containing numerous spots (Abetti, 1957: p. 187). The 'iron filing' structure, on the other hand, is characteristic of a well-developed bipolar group; beautiful photographs illustrating this structure have been published by Abetti (1929: Fig. 69) and Kiepenheuer (1953: Fig. 33). The vortex structure (Abetti, 1929: Fig. 68) is typically associated with isolated spots, while the radial structure can be regarded as a special case of the vortex structure.

According to Richardson (1941), under perfect conditions about one-third of all spots with areas greater than 200 millionths of the visible hemisphere might be expected to show Hα vortices. However, well-marked vortices are actually an exceedingly rare phenomenon; the small dark filaments are generally so indefinite and confused that it is difficult even to decide on the presence of a whirl, let alone determine its direction. The vortex structure was systematically studied by Richardson, who concluded that the direction taken by the vortex is determined by purely *hydrodynamic* effects due to solar rotation. However, although this conclusion has been accepted by a number of subsequent authors, a careful examination of Richardson's data (see particularly his Table 2) shows that a *magnetic* interpretation cannot be entirely ruled out.

7.3.3 PROMINENCES

For the purpose of discussing the morphological relationship between sunspots and prominences it is convenient to take as our starting point the scheme of prominence classification proposed by Menzel and Evans (1953). In this scheme the prominences are first classified according to whether they originate in the corona, on the one hand, or in the underlying chromosphere on the other. They are then sub-classified according to whether or not they are associated with sunspot groups. Finally, they are further sub-classified according to their shape and appearance. The complete Menzel–Evans classification is as follows:

A. Prominences originating in the corona

   S. Spot prominences
      f. Funnels
      l. Loops

    N. Nonspot prominences
      a. Coronal rain
      b. Tree trunk
      c. Tree
      d. Hedgerow
      e. Mound

  B. Prominences originating in the chromosphere

    S. Spot prominences
      s. Surges
      p. Puffs
    N. Nonspot prominences
      s. Spicules

Photographs illustrating most of these types are given in Figs. 1–10 of Menzel and Evans's paper.[3]

The above classification is based on the appearance of prominences seen in emission at the *limb*, whereas in investigating the relationship with sunspots we are obviously more concerned with their appearance seen in absorption on the *disk*. However, experience indicates that in most cases a prominence seen on the disk can be given its correct designation with little or no ambiguity. We are, of course, concerned only with the sunspot prominences and we shall now discuss each of these types in turn:

(1) *ASf: funnel prominences*. Disk observations with a spectrohelioscope made by Ellison (1944) have shown that there is a very close relationship between these prominences and sunspots. The prominence material descends directly towards the penumbra of a spot in streamers some 60,000 km in length, which originate in one or more 'coronal clouds'. The streamers impinge almost vertically onto *the outer boundary of the penumbra*, a black dot being observed at the point of entry. The velocity of inflow is ~ 48 km/sec, and the inflow is not confined to spots of one polarity. When the spot group is surrounded by a well-marked vortex structure (see Section 7.3.2), the inward-moving prominence material often seems to travel along paths parallel to the small dark filaments outlining the structure. Several drawings illustrating the relationship between funnel prominences and sunspots are given in Ellison's paper (see his Fig. 1); a similar drawing has been published by Waldmeier (1955: Fig. 99 *c*).

When seen at the limb, the bright clouds which are the source of the inflowing material are conspicuous features which are often brighter than the streamers themselves. Excellent limb photographs of funnel prominences have been published by Waldmeier (1955: Fig. 97) and de Jager (1959: Fig. 72). However, these photographs do not, of course, show the associated spots.

(2) *ASl: loop prominences*. Ellison's disk observations also demonstrate a very

---

[3] For a general account of the properties of prominences, including the statistics of their occurrence, latitude distribution, etc., see Kiepenheuer (1953).

close connection between loop prominences and sunspots. According to Ellison, loops occur only during the most active stages of very complex groups.[4] Single or double arches are quite common, but complex formations of loops are rare. The material rises up from the region of a newly formed spot or pore, loops over, and descends into a neighbouring spot of the group. The mean outward velocity is 28 km/sec, while the inward velocity along the other leg of the arch is 39 km/sec. On occasion the motion along the loop may continue for up to several hours. The mean distance between the points of inflow and outflow is 43,000 km, and the height of the arch is 40,000–100,000 km. The trajectories are usually orientated approximately east–west, e.g. between the leader and follower components of a bipolar group. The direction of flow is independent of the magnetic polarities of the spots.

In the majority of the cases observed by Ellison the loops joined neighbouring spots of the group, but in a few cases there was no visible spot at the point of inflow. No case was seen of a loop re-entering the same spot from which it emerged. Only one case was observed in which matter descended from the crown of the arch along both legs, although limb observations by Menzel and Evans (1953) and McMath suggest that this is the general rule.

Figure 2 of Ellison's paper gives six drawings demonstrating the close relationship between loop prominences seen on the disk and sunspots. A similar drawing has been published by Waldmeier (1955: Fig. 99 d), who comments that in the case observed by him the point of inflow lay on the edge of the penumbra, as in the funnel prominences discussed above. More recently, excellent disk photographs of loop prominences have been published by Bruzek (1962: Fig. 1). In contradiction to Ellison, this author observed downward motions at *both* ends of the loop, the velocities being $\sim$ 100 km/sec.

When seen at the limb, a system of loop prominences presents a striking appearance: a particularly impressive photograph obtained by R. B. Dunn has been reproduced by de Jager (1959: Fig. 71). On the other hand, by themselves such photographs do not reveal the positions of the associated sunspots. However, Bumba and Kleczek (1961) not only obtained a limb photograph of a system of loop prominences but in addition carefully established the positions of the associated spots. The result is shown in Plate 7.3, which is taken from their paper. In agreement with Ellison's observations, this plate demonstrates the close connection between loop prominences and sunspots, most or all of the loops being anchored to umbrae of spots in the group.

(3) *BSs and BSp: surges and puffs.* Seen on the disk, a surge is a very dark scimitar-shaped jet of outward-moving material, whose velocity often reaches several hundred km/sec. Surges are closely associated with flares: according to Giovanelli and McCabe (1958), 18% of flares are accompanied by surges and all surges are associated with flares, mostly of class 1–. These authors (see their Fig. 2) and independently, Shaposhnikova (1958), found that with few exceptions

---

[4] However, Ramseyer, Warwick, and Zirin (1960) have observed a fine set of loops over a single large sunspot (Zürich class H). See also Dodson (1961).

the initial paths of surges are directed more or less radially away from the nearest large sunspot. The trajectory as a whole is often gently curved, and after the surge matter is often observed to return along exactly the same path. Moreover, surges—like flares—show a strong tendency to recur at the same place. Good photographs illustrating the surge–spot relationship have been published by Giovanelli and McCabe (1958: Plate 1, Fig. 2; Plate 2, Fig. 1) and Kiepenheuer (1960: Fig. 13).

Disk observations of *puffs* have not been described in the literature, as far as the authors are aware. At the limb they appear as small spherical balls whose velocities and trajectories are indistinguishable from those of surges (Menzel and Evans, 1953).

This completes our discussion of observations of the relationship between sunspots and sunspot prominences. However, the relationship between sunspots and dark chromospheric *filaments*[5] also deserves mention. When a filament is in the neighbourhood of a bipolar sunspot group, it is found that in 80% of the cases the filament is directed towards the leader component and only in 20% of the cases towards the follower (see Kiepenheuer, 1953: p. 405). Moreover, on the average the distance between the end of the filament and the nearest spot is only 10,000–20,000 km. Kiepenheuer remarks that the influence of the spot evidently extends along the full length of the filament, a distance of some 100,000 km. The development of filaments in relation to the activity centre as a whole is outlined in Table 7.1.

7.3.4 FLARES

Chromospheric flares, like photospheric and chromospheric faculae, show a very close statistical relationship with sunspots.[6] This is well illustrated by the relationship between flare frequency and sunspot number found by Waldmeier (1955: p. 241): if E is the mean number of flares per day and R the sunspot number, then approximately $E = 0.061R$. Thus near the peak of a solar cycle, when R might be, say, 150, the number of flares per day would be ~ 9. Waldmeier has remarked that more than 50 flares can occur in a single spot group during the course of its life.

However, crude statistics of this kind fail to reveal the important fact that the flare activity in a given spot group is very strongly dependent on *the stage of development of the group*. This is shown in Fig. 7.1, which is taken from Waldmeier (1955: Fig. 46). In this figure both the mean daily flare number and the area of the group are plotted against the time in days from the appearance of the first spot; the curves are representative of a large spot group which passes through all stages of the Zürich classification (cf. Section 6.2.2) during the course of its development and decline. It can be seen that during the first few days of the group's life, when its area is rapidly increasing, the flare activity

---

[5] ANd on the Menzel–Evans classification; also called 'quiescent prominences'.
[6] For general accounts of the properties of flares, see Ellison (1949); Kiepenheuer (1953); Waldmeier (1955); de Jager (1959).

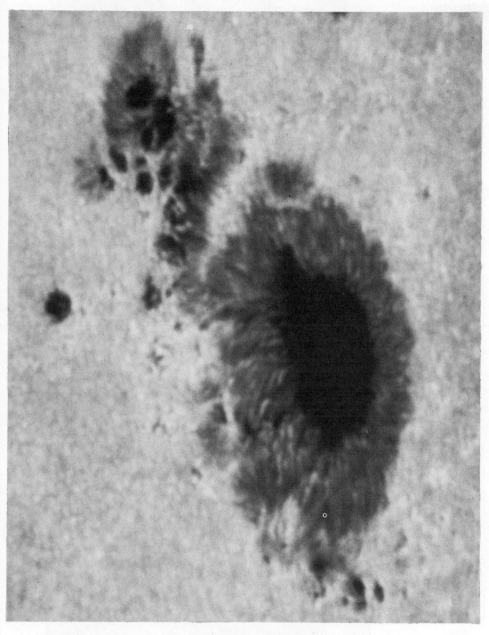

0″   10″

PLATE 7.1. Large sunspot group fairly near limb (February 25, 1959). Facular material
penetrates into the penumbra at top of main spot.

TO LIMB ←

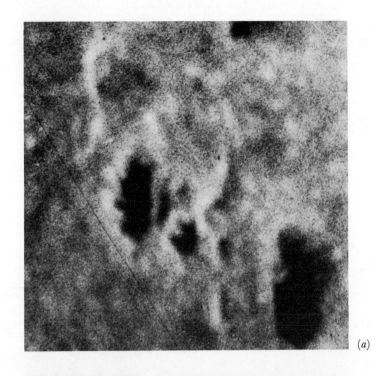

(a)

0″ ⌊⌄⌄⌄⌄⌄⌄⌄⌄⌄⌄⌄⌄⌄⌄⌄⌄ 5″

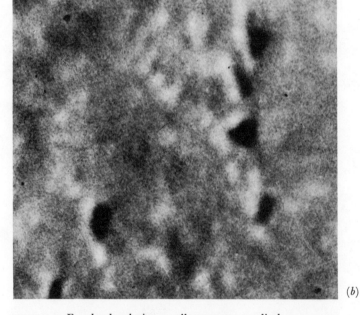

(b)

PLATE 7.2. Faculae bordering small sunspots near limb.

(a) Portion of an isolated group of small spots (January 10, 1958); heliocentric angle, 58°. (b) Small spots associated with a bipolar group (January 13, 1958); heliocentric angle, 64°.

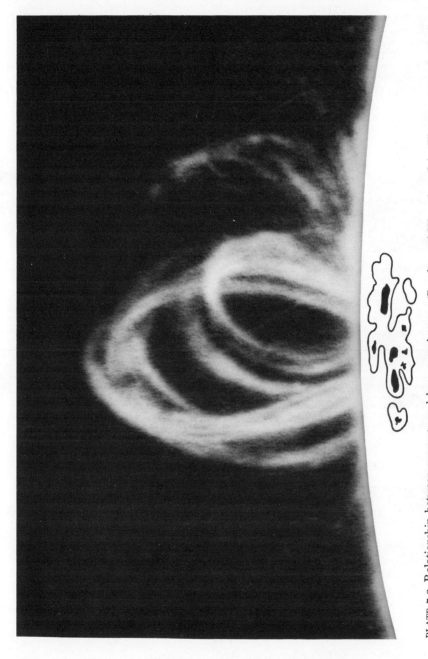

PLATE 7.3. Relationship between sunspots and loop prominences (Bumba and Kleczek, 1961). The spot group has been drawn underneath the associated loop system. Most or all of the loops are anchored to spot umbrae.

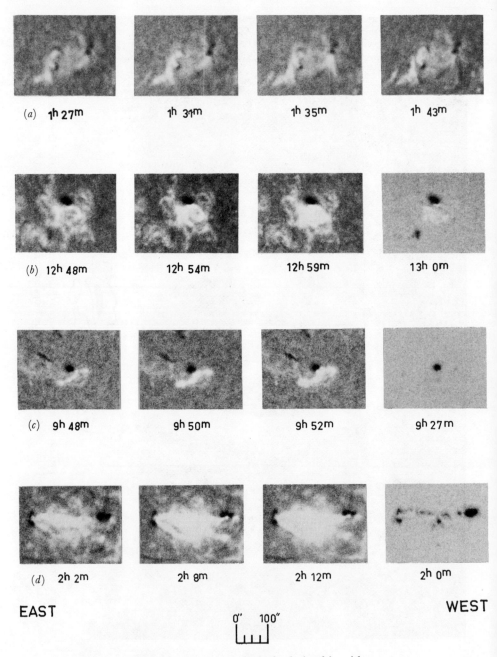

(a) 1ʰ 27ᵐ   1ʰ 31ᵐ   1ʰ 35ᵐ   1ʰ 43ᵐ

(b) 12ʰ 48ᵐ   12ʰ 54ᵐ   12ʰ 59ᵐ   13ʰ 0ᵐ

(c) 9ʰ 48ᵐ   9ʰ 50ᵐ   9ʰ 52ᵐ   9ʰ 27ᵐ

(d) 2ʰ 2ᵐ   2ʰ 8ᵐ   2ʰ 12ᵐ   2ʰ 0ᵐ

EAST   0″   100″   WEST

PLATE 7.4. Flares showing a distinct morphological relationship with sunspots.
  (a) Flare filaments terminating at spots (August 23, 1956).
  (b) Flare tangential to spot (January 24, 1957).
  (c) Flare with two 'nodes' at spot (May 27, 1959).
  (d) Flare extending between components of a bipolar group (September 19, 1957).
The final photographs in (b), (c), and (d) were taken with the filter de-tuned. For a general description of the development of flares in relation to the associated spots, see text.

rises sharply. Maximum activity is attained on or about the 10th day, when the area of the group is also at its maximum. Thereafter, the flare activity declines steeply; very few flares occur after the third or fourth week although the area of the group at 4 weeks is still 60% of its maximum value. It is evident from the figure that the flare activity of a spot group is confined almost completely to the first third of the group's lifetime.

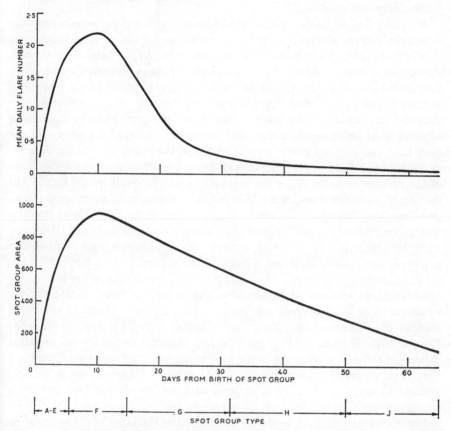

Fig. 7.1. Dependence of flare activity on the stage of development of the associated sunspot group (Waldmeier, 1955). Ordinates: mean daily flare number (upper curve), spot group area in millionths of the visible hemisphere (lower curve). Abscissa: time in days from appearance of first spot. The Zürich classification is shown at the bottom.

At the bottom of the figure are given the Zürich classes corresponding to the various stages of development of the group. It is evident that the flare frequency is greatest during stages D, E, and F, whereas stages H and J are nearly completely flare-free. According to Waldmeier, stages D, E, and F also correspond to the largest flares, a result confirmed by Kleczek (1953) and Künzel (1960).

There is some evidence that the flare activity of a spot group depends on its *magnetic* classification (Section 6.2.6), increasing as we pass from the simple unipolar α-type through the bipolar types β and βγ to the magnetically complex γ-type (Giovanelli, 1939; Künzel, 1960). However, the figures published by Giovanelli (see his Table 2) indicate that the differences in flare productivity (per unit area of the group) for groups of different magnetic class are not particularly well-marked.

About 95% of all flares are associated with spot groups; the remainder occur in activity centres which either previously contained spots or will subsequently do so (Waldmeier, 1955: p. 241). In fact, never has a flare been observed outside the spot zone or outside a region containing chromospheric faculae. With rare exceptions, according to Ellison (1949), flares are located within a radius of 100,000 km of a visible spot. Most of the time, however, they are much closer than this. The smaller flares (classes 1 and 2) show a preference for the regions of small spots and pores in the central and eastern parts of a spot group; the most favoured location is the central region of the group at a mean distance ≃ 50,000 km behind the leading spot (Ellison, 1949: Fig. 4). Some authors are of the opinion that the *larger* flares usually begin as small bright specks just outside of, or actually on, the penumbra of a major spot; these specks later grow into long bright ribbons which sometimes run right across the spot group and have a tendency to touch the umbrae of major spots (Dodson and Hedeman, 1949; Ellison, 1949; Kiepenheuer, 1953: p. 381; Waldmeier, 1955: Fig. 83 and p. 243; Bruzek, 1960). However, a study by the present authors (see below) has failed to confirm that large flares usually originate as specks in the immediate neighbourhood of spots. A property of the larger flares not shared by the smaller ones is an occasional tendency to partially or completely cover the umbrae of major spots in the group[7] (Dodson and Hedeman, 1949, 1960; Ellison, 1949; Bruzek, 1958: Figs. 2 and 3). Apart from occasional so-called 'white-light' flares, no changes in the underlying photosphere or sunspots have ever been observed during the course of a flare. However, no systematic attempt to detect such changes appears to have been made (see, however, Gopasyuk, 1961, 1962).

Flares exhibit a striking tendency to recur after the lapse of a few hours or days at the same position relative to the spots in the associated group (see, for example, Dodson and Hedeman, 1949). Photographs illustrating this tendency have been published by Kiepenheuer (1953: Fig. 48) and Waldmeier (1955: Fig. 84).

In order to elucidate more fully the detailed morphological relationship between sunspots and flares the present authors undertook a special study of the question, using as observational material Sydney flare patrol films obtained during the period 1956–59. The study was restricted to *large* flares (area greater

---

[7] Martres and Pick (1962) have found that unusually great centimetre and continuum radio emission is associated with such events, in agreement with an earlier suggestion of Dodson and Hedeman (1960).

than 500 millionths of the visible hemisphere) photographed under good observing conditions. Moreover, a flare was included only when both its start and its time of maximum brightness had been recorded. Limb flares were excluded since foreshortening would cause any sunspots in the vicinity to be invisible on the small-scale solar image (diameter = 16 mm). The final list contained 35 flares which, as far as their morphological relationship with spots is concerned, can be divided into four distinct classes:

Class (A), containing no less than 22 members, comprises those flares for which *no* morphological relationship with spots could be established; also included in this class are flares for which no spots were visible on the photographs examined and, in addition, flares which completely covered some of the spots present, thus obscuring any possible relationship;

Class (B) (6 members) includes flares in which a flare filament or a converging portion of the flare *terminated* at a spot umbra or penumbra: an example is given in Plate 7.4 (*a*);

Class (C) (5 members) includes flares which were tangent to a spot umbra or penumbra, or had one or more 'nodes' there, but which did not terminate there: good examples are given in Plates 7.4 (*b*) and (*c*);

Class (D) (2 members) includes flares in which one or more flare filaments arched between the components of a bipolar group: an unusually spectacular example of this type is given in Plate 7.4 (*d*).

It is evident that only about one-third of large flares show a distinct morphological relationship with nearby sunspots. On the other hand, the relationship, when present, is quite a striking one, as Plate 7.4 demonstrates. In the case of the 13 flares showing a distinct relationship, some idea of the exact process of development in relation to the associated spot or spots was obtained by carefully examining the films of the events. The results of this analysis can be embodied in the following four rules which, although based on a modest number of examples, give some idea of the development of large, spot-associated flares:

(1) in almost all cases, the flare commences as a brightening of a pre-existing vein in the chromospheric faculae, which may or may not touch spots in the vicinity;

(2) the brightening occurs simultaneously along the entire length of the facular vein: it does *not*, for example, proceed outward from a spot;

(3) sometimes during the course of a flare's development, portion expands towards a spot and then stops there; in such cases the region of the flare immediately adjacent to the spot is often the brightest (cf. Plates 7.4 *b* and *c*);

(4) only seldom (2 out of 13 cases) does a flare *originate* immediately adjacent to a spot, in contradiction to the remarks of certain other workers (see above).

In view of the close statistical association found between sunspots and flares, it is rather surprising that only a minority of large flares shows a distinct morphological relationship with nearby spots. Further systematic studies of this aspect of the relationship between spots and flares are clearly very desirable.

7.4 The Role Played by Magnetic Fields in Activity Centres

To conclude this chapter we shall briefly summarize the available information concerning the role played by magnetic fields in three of the phenomena associated with activity centres, namely chromospheric faculae, sunspot prominences, and flares.

(1) *Chromospheric faculae.* According to Leighton (1959), the distribution of the longitudinal component of the magnetic field throughout an activity centre is closely correlated with the distribution of the K faculae (Section 7.3.1). In fact, the correspondence is much closer than that found between the field and the visible structure of the associated sunspot group. Leighton observed fields of up to 100–200 gauss in the faculae and identified every region showing a measurable longitudinal field with a prominent K emission area having the same general shape.

Leighton's results have been confirmed by Howard (1959) and Stepanov and Petrova (1959). Like Leighton, Howard found a nearly one-to-one correspondence between the magnetic features and the faculae, not only with respect to their generally similar location but also in most small structural details. He found that the outlines of the K emission regions correspond well with a magnetic contour line of about 10 gauss. Stepanov and Petrova found that the lines of equal brightness closely follow lines of equal field strength.

The closeness of the relation between the magnetic field and the K faculae would seem to imply, as Leighton has emphasized, that the field is not merely an accompanying feature of the faculae but rather an indispensable factor in the underlying physical mechanism. A number of authors have in fact suggested various mechanisms for the magnetic heating of facular regions, but as yet there is no general agreement concerning the exact nature of the process.

(2) *Prominences.* Kiepenheuer (1960) has remarked: 'There can be no doubt that *all* prominences, whether in the vicinity of sunspots or far away in high latitudes, are related intimately to photospheric magnetic fields. They define their form and they screen them against the heavy bombardment of coronal ions. Only by this protection can a prominence survive as a cool body imbedded in the much hotter corona.'

In the case of loop prominences the connection with sunspots is particularly close, as we have seen in Section 7.3.3. Bumba and Kleczek (1961) have suggested that the observed loops of prominence material (see Plate 7.3) actually represent *magnetic flux loops* anchored to the spots. They go on to remark that since the cross-section of an individual loop prominence does not change appreciably along its length, the field strength at the top of the flux loop cannot differ too much from that at its anchorage points, i.e. the sunspot umbrae (cf. Section 5.4.6). Bumba and Kleczek's suggestion could be tested by making detailed measurements of the magnetic field distribution in a system of loop prominences at the limb. However, although magnetic fields have recently been detected in prominences, measurements having the necessary high spatial resolution are lacking at the present time. If the suggestion is correct it follows, as Kiepenheuer (1953: p. 416) has remarked, that a study of these prominences could yield data about the form of spot group fields in the corona.

(3) *Flares.* The energy liberated in chromospheric flares is so large that the only possible source seems to be the magnetic fields associated with neighbouring sunspots. However, at the present time, although various mechanisms involving the field have been suggested (see, for example, Gold and Hoyle, 1960; Severny, 1961), the exact nature of the process remains unknown. If the energy is derived from the field, then we would expect the field to show a significant alteration during the course of a flare. Unfortunately, the various observations which have been made to detect such changes fail to agree (see Severny, 1958 a, b, 1960, 1962; Howard, Cragg, and Babcock, 1959; Howard and Babcock, 1960; Gopasyuk, 1961, 1962; Michard, Mouradian, and Semel, 1961; Leroy, 1962). For this reason, it would be premature to discuss this topic any further at the present time. However, one may expect that in the years to come the application of novel magnetic techniques, such as those described in Section 5.3.8, will improve our insight not only into the mechanism of flares but also into the nature of solar activity as a whole.

REFERENCES

ABETTI, G. [1929] 'Solar physics', *Handbuch der Astrophysik*, ed. G. EBERHARD, A. KOHL-SCHÜTTER, and H. LUDENDORFF, vol. 4, p. 57. Berlin, Springer.

ABETTI, G. [1957] *The Sun.* London, Faber and Faber.

ATHAY, R. G., and WARWICK, C. S. [1961] 'Indices of solar activity', *Advances in Geophysics* **8**, 1.

BABCOCK, H. W., and BABCOCK, H. D. [1958] 'Photospheric magnetic fields', *Electromagnetic Phenomena in Cosmical Physics*, ed. B. LEHNERT, p. 239. Cambridge Univ. Press.

BRUZEK, A. [1958] 'Die Filamente und Eruptionen eines Aktivitätszentrums', *Z. Astrophys.* **44**, 183.

BRUZEK, A. [1960] 'Beobachtungen über die Beziehungen zwischen Eruptionen und Fleckenfeldern', *Z. Astrophys.* **50**, 110.

BRUZEK, A. [1962] 'Über Fleckenprotuberanzen vor der Sonnenschcibe', *Z. Astrophys.* **54**, 225.

BUMBA, V., and KLECZEK, J. [1961] 'On a sunspot group with an outstanding loop activity', *Observatory* **81**, 141.

CHRISTIANSEN, W. N., MATHEWSON, D. S., PAWSEY, J. L., SMERD, S. F., BOISCHOT, A., DENISSE, J. F., SIMON, P., KAKINUMA, T., DODSON-PRINCE, H., and FIROR, J. [1960] 'A study of a solar active region using combined optical and radio techniques', *Ann. Astrophys.* **23**, 75.

DODSON, H. W. [1961] 'Observation of loop-type prominences in projection against the disk at the time of certain solar flares', *Proc. Nat. Acad. Sci.* **47**, 901.

DODSON, H. W., and HEDEMAN, E. R. [1949] 'The frequency and positions of flares within three active sunspot areas', *Astrophys. J.* **110**, 242.

DODSON, H. W., and HEDEMAN, E. R. [1956] 'Detailed study of the development of an active solar region, 1954 August 20–27', *Mon. Not. R.A.S.* **116**, 428.

DODSON, H. W., and HEDEMAN, E. R. [1960] 'Flares of July 16, 1959', *Astron. J.* **65**, 51.

ELLISON, M. A. [1944] 'Sunspot prominences—some comparisons between limb and disk appearances', *Mon. Not. R.A.S.* **104**, 22.

ELLISON, M. A. [1949] 'Characteristic properties of chromospheric flares', *Mon. Not. R.A.S.* **109**, 3.

GIOVANELLI, R. G. [1939] 'The relations between eruptions and sunspots', *Astrophys. J.* **89**, 555.

GIOVANELLI, R. G., and MCCABE, M. K. [1958] 'The flare–surge event', *Aust. J. Phys.* **11**, 191.

GOLD, T., and HOYLE, F. [1960] 'On the origin of solar flares', *Mon. Not. R.A.S.* **120**, 89.

GOPASYUK, S. I. [1961] 'A study of the configuration variations of the magnetic field and spot groups in connection with solar flares. A determination of the total energy of flares', *Astron. J. U.S.S.R.* **38**, 209 (*Sov. Astron. AJ* **5**, 158).

GOPASYUK, S. I. [1962] 'The motion of spots connected with solar flares and the possible character of energy exit from flare regions', *Izv. Crim. Astrophys. Obs.* **27**, 110.

HOWARD, R. [1959] 'Observations of solar magnetic fields', *Astrophys. J.* **130**, 193.

HOWARD, R., and BABCOCK, H. W. [1960] 'Magnetic fields associated with the solar flare of July 16, 1959', *Astrophys. J.* **132**, 218.

HOWARD, R., CRAGG, T., and BABCOCK, H. W. [1959] 'Magnetic field associated with a great solar flare', *Nature* **184**, 351.

JAGER, C. DE [1959] 'Structure and dynamics of the solar atmosphere', *Handbuch der Physik*, ed. S. FLÜGGE, vol. 52, p. 80. Berlin, Springer.

JAGER, C. DE [1961] 'The development of a solar centre of activity', *Vistas in Astronomy*, ed. A. BEER, vol. 4, p. 143. London, Pergamon.

KIEPENHEUER, K. O. [1953] 'Solar activity', *The Sun*, ed. G. KUIPER, p. 322. Univ. Chicago Press.

KIEPENHEUER, K. O. [1960] 'The optical phenomena forming a solar centre of activity', *Soc. Ital. Fisica, Rendiconti Scuola Internazionale di Fisica 'Enrico Fermi'* (Director: G. RIGHINI), p. 39.

KLECZEK, J. [1953] 'Relations between flares and sunspots', *Bull. Astron. Inst. Czech.* **4**, 9.

KRIVSKY, L., and LETFUS, V. [1960] 'The active region on the Sun on May 5, 1958', *Bull. Astron. Inst. Czech.* **11**, 53.

KÜNZEL, H. [1960] 'Die Flare-Häufigkeit in Fleckengruppen unterschiedlicher Klasse und magnetischer Struktur', *Astron. Nach.* **285**, 271.

LEIGHTON, R. B. [1959] 'Observations of solar magnetic fields in plage regions', *Astrophys. J.* **130**, 366.

LEROY, J.-L. [1962] 'Contributions à l'étude de la polarisation de la lumière solaire', *Ann. Astrophys.* **25**, 127.

LOUGHHEAD, R. È., and BRAY, R. J. [1959] 'Observations of faculae bordering small sunspots near the limb', *Aust. J. Phys.* **12**, 97.

MARTRES, M.-J., and PICK, M. [1962] 'Caractères propres aux éruptions chromosphériques associées à des émissions radioélectriques', *Ann. Astrophys.* **25**, 293.

MENZEL, D. H., and EVANS, J. W. [1953] 'The behaviour and classification of solar prominences', ('Problemi della fisica solare'), *Acc. Naz. Lincei, Convegno* **11**, p. 119.

MICHARD, R., MOURADIAN, Z., and SEMEL, M. [1961] 'Champs magnétiques dans un centre d'activité solaire avant et pendant une éruption', *Ann. Astrophys.* **24**, 54.

RAMSEYER, H., WARWICK, J. W., and ZIRIN, H. [1960] 'Connection of a loop prominence with a sunspot', *Pub. Astron. Soc. Pac.* **72**, 509.

RICHARDSON, R. S. [1941] 'The nature of solar hydrogen vortices', *Astrophys. J.* **93**, 24.

SEVERNY, A. B. [1958a] 'Solar flares as a pinch effect', *Astron. J. U.S.S.R.* **35**, 335 (*Sov. Astron. AJ* **2**, 310).

SEVERNY, A. B. [1958b] 'The appearance of flares in neutral points of the solar magnetic field and the pinch effect', *Izv. Crim. Astrophys. Obs.* **20**, 22.

SEVERNY, A. B. [1960] 'An investigation of magnetic fields connected with solar flares', *Izv. Crim. Astrophys. Obs.* **22**, 12.

SEVERNY, A. B. [1961] 'On the generation of flares with the growth of the solar magnetic field', *Astron. J. U.S.S.R.* **38**, 402 (*Sov. Astron. AJ* **5**, 299).

SEVERNY, A. B. [1962] 'Some peculiarities of magnetic fields connected with solar flares', *Astron. J. U.S.S.R.* **39**, 961 (*Sov. Astron. AJ* **6**, 747).

SHAPOSHNIKOVA, E. F. [1958] 'Filaments directly connected with sunspots', *Izv. Crim. Astrophys. Obs.* **18**, 151.

STEPANOV, V. E., and PETROVA, N. N. [1959] 'The brightness of flocculi, magnetic fields and mechanisms of heating', *Izv. Crim. Astrophys. Obs.* **21**, 152.

WALDMEIER, M. [1955] *Ergebnisse und Probleme der Sonnenforschung*, 2nd ed. Leipzig, Geest u. Portig.

CHAPTER 8

Magnetohydrodynamic Theories of the Origin of Sunspots and the Solar Cycle

8.1 Introduction

As we have seen in Chapter 1, sunspots have been observed for a much longer time than any other solar feature. During this period a great deal of observational knowledge of their properties and behaviour has been accumulated, the record of which fills the preceding five chapters of this book. In this chapter we shall turn to the question of the theoretical explanation of their physical mechanism and mode of origin, but here we shall find that our knowledge is much less complete. There are several reasons why progress on the theoretical side has lagged behind that on the observational side. The most important is that the theory of sunspots and the solar cycle lies in the domain of *magnetohydrodynamics*, a subject which began to assume the status of a separate discipline only in the early 1940's and which is still in the process of development. Another reason has been our ignorance until quite recently of certain basic observational facts, such as the existence of granulation in the umbra (cf. Section 3.6), which have since proved to be of crucial importance.

Attempts were made to explain the origin of sunspots long before the advent of magnetohydrodynamics, but today these theories are either forgotten or remembered only for their historical interest. Nowadays a knowledge of the basic principles of magnetohydrodynamics is an essential pre-requisite to an understanding of theories of sunspots and the solar cycle. These principles are therefore summarized in Section 8.2: the basic equations are set out in Section 8.2.2, while the fundamental question of the growth and decay of sunspot magnetic fields is discussed in Section 8.2.3. It is found that the natural decay time of a spot field due to ohmic dissipation exceeds the average lifetime of a spot. This leads to the important conclusion that the spot field can neither grow nor decay *in situ*.

The problem of the origin of sunspots may, for convenience, be divided into two parts: in the first place, one has to elucidate the mechanism responsible for the coolness and continued existence of individual spots in a medium hotter than themselves and, in the second place, one has to explain the cause of the solar cycle as a whole. The first aspect of the problem is dealt with in Section 8.3, while the second is discussed in Section 8.4. As we have seen in Section 7.1, it is now generally accepted that sunspots owe their origin to the presence of internal solar magnetic fields which from time to time push their way through the photospheric convection zone. The role of the magnetic field is fundamental to the whole sunspot phenomenon, and therefore the explanation of the coolness of spots must be sought in terms of the field itself. The first magnetic theories of sunspot cooling were based on the suggestions that convection within the spot umbra is suppressed (Biermann) or diluted (Hoyle) by the field (cf. Section 8.3.2). However, observations of granulation in sunspot umbrae, described in Section 3.6, show that the convection is not suppressed; for this and other reasons both Biermann's and Hoyle's theories are no longer tenable. Nevertheless, it is almost certain that the coolness of a spot is due to a reduction in the amount of energy convected upwards, resulting in some way from the influence of the magnetic field.

Accepting this explanation for the coolness of a sunspot, we also require an explanation for the fate of the energy which, in the presence of the spot field, is no longer carried upwards by convection. Various ideas have been advanced, the most promising of which, in the authors' opinion, is the suggestion that the energy deficit is made good by an increase in *magnetic* energy. A possible mechanism for the energy conversion is discussed at the end of Section 8.3.2. It is emphasized that in order to make progress in elucidating the energy balance of sunspots, we shall have to enlarge our point of view to include not merely the photospheric layers but also the sub-photospheric magnetic field region and the active chromosphere above the spot.

The role played by magnetic forces in maintaining the equilibrium of sunspots is discussed in Section 8.3.3. Here the fundamental question arises of whether or not the spot field is 'force-free' in character. If it were, it could not contribute to the maintenance of pressure equilibrium between the spot and its surroundings, an assumption which is in fact embodied in Mattig's sunspot model, discussed in Section 4.2.4. Possible configurations of such force-free spot fields have been investigated by Schatzman, but a more general approach to the whole problem was introduced by Schlüter and Temesváry, who have attempted to obtain numerical solutions of the equations of *magnetohydrostatic* equilibrium for a spot. However, the solutions show a marked instability; this results from the assumption of purely radiative energy transport and indicates, in agreement with observation, that in order to obtain a realistic sunspot model it is also necessary to take account of the transport of energy by convection.

The question of the origin of the solar cycle is discussed in Section 8.4. Most modern theories assume the existence of a weak *general* magnetic field and seek to find ways whereby portions of this field can be brought to the surface and amplified to give intensities comparable to those observed in sunspots (cf. Section 8.4.1). The most detailed theoretical attack on the problem to date is that due to E. N. Parker, two aspects of which are described in Section 8.4.2.

This author introduced the valuable concept of the 'magnetic buoyancy' of flux tubes; this concept has since been incorporated by H. W. Babcock into the most elaborate synthesis yet attempted of existing observational data and modern theoretical ideas on the origin of the solar cycle. Babcock's theory is briefly outlined in Section 8.4.3. Most probably neither of the theories contains the final answers to the many difficult questions associated with the problem of the solar cycle, but they do in large measure reflect the present state of knowledge and thinking on the subject.

Magnetic activity is by no means a phenomenon unique to the Sun: thanks to the continuing efforts of H. W. Babcock since 1947, magnetic fields have been detected on stars covering a wide range of spectral types and the study, both observational and theoretical, of these so-called 'magnetic' stars has become a subject of rapidly increasing importance in astrophysics. In practically all cases the fields of magnetic stars are found to be variable, and the question thus arises of whether the solar cycle is related to some general form of magnetic activity occurring on a faster and more spectacular scale on other stars. At the moment no definite answer can be given to this question. However, in order to round off the discussion of the origin of sunspots, we conclude by summarizing the main observational facts relating to magnetic stars (Section 8.5.2) and by briefly considering possible explanations of their behaviour (Section 8.5.3).

8.2 Basic Magnetohydrodynamic Principles

8.2.1 INTRODUCTION

The discussion of the physical conditions in sunspots given in Chapter 4 has shown that the material in the visible layers of a spot umbra is relatively lowly ionized. For example, at optical depth unity the degree of ionization of the most abundant element, hydrogen, is only of the order of 10^{-6} (Table 4.14). The degree of ionization of the metals is much higher, namely about 0·6, but since the metals are only 10^{-4} times as abundant as hydrogen, the effective degree of ionization of the umbral material is only of the order of 10^{-4} at optical depth unity. (Further down, of course, the degree of ionization of the material greatly increases as a result of increasing hydrogen ionization.) Only the charged particles are directly affected by the spot magnetic field. However, although in the visible layers charged particles are greatly outnumbered by neutral particles, owing to the high frequency of collisions between charged and neutral particles, the gas as a whole must move as a single entity—at least within the time-scale of any macroscopic motions likely to be encountered (cf. Lehnert, 1959: p. 75). Moreover, unless the field in the spot is substantially 'force-free' (see Section 8.3.3), magnetic forces will generally predominate over any purely hydrodynamic forces: for example, taking a gas density of, say, $\rho = 1\cdot5 \times 10^{-7}$ gm cm^{-3} and a representative velocity of, say, $v = 2$ km sec^{-1} (cf. Table 4.14), the kinetic energy per unit volume is only $\frac{1}{2}\rho v^2 = 3 \times 10^3$ erg cm^{-3} compared to a magnetic energy density for a field of 1000 gauss of $H^2/8\pi = 4 \times 10^4$ erg cm^{-3}. Consequently, the study of the dynamical behaviour of sunspots falls within the domain of *magnetohydrodynamics*.

8.2.2 MAGNETOHYDRODYNAMIC EQUATIONS

Using the notation and units set out in Table 8.1 below, the basic equations of magnetohydrodynamics may be stated as follows.[1]
Conservation of mass:

$$\frac{d\rho}{dt} + \rho \operatorname{div} \mathbf{v} = 0 \tag{8.1}$$

Conservation of momentum:

$$\rho \frac{d\mathbf{v}}{dt} = \rho\mathbf{g} + \frac{1}{c}\mathbf{j} \times \mathbf{H} - \operatorname{grad} p \tag{8.2}$$

Conservation of thermal energy:

$$\frac{dU}{dt} + U \operatorname{div} \mathbf{v} = \mathbf{j} \cdot \left(\mathbf{E} + \frac{1}{c}\mathbf{v} \times \mathbf{H} \right) - p \operatorname{div} \mathbf{v} - \operatorname{div} \mathbf{F} \tag{8.3}$$

Equation of state:

$$p = \frac{k\rho T}{\mu} \tag{8.4}$$

Current equation:

$$\frac{m_e}{n_e e^2} \frac{\partial \mathbf{j}}{\partial t} + \frac{1}{n_e e c}\mathbf{j} \times \mathbf{H} + \frac{1}{\sigma}\mathbf{j} = \mathbf{E} + \frac{1}{c}\mathbf{v} \times \mathbf{H} + \frac{1}{n_e e} \operatorname{grad} p_e \tag{8.5}$$

Maxwell's equations:

$$\frac{\partial \mathbf{H}}{\partial t} = -c \operatorname{curl} \mathbf{E} \tag{8.6}$$

$$\mathbf{j} = \frac{c}{4\pi} \operatorname{curl} \mathbf{H} \tag{8.7}$$

$$\operatorname{div} \mathbf{E} = 4\pi q \tag{8.8}$$

$$\operatorname{div} \mathbf{H} = 0. \tag{8.9}$$

In these equations the operator

$$\frac{d}{dt} = \frac{\partial}{\partial t} + \mathbf{v} \cdot \operatorname{grad}$$

indicates differentiation following the motion of the gas.

[1] For detailed discussions of the derivation of the equations, and of the assumptions embodied therein, the reader is referred to the following sources: Cowling (1953, 1957, 1962); Westfold (1953); Spitzer (1956); Dungey (1958); Chu (1959); Lehnert (1959); Ferraro and Plumpton (1961).

Since any deviation from electrical neutrality inside a sunspot can be shown to be negligibly small (cf. Lehnert, 1959: Table 1), terms involving the electric charge density q as a factor have been neglected, although q must be retained on the R.H.S. of (8.8) (Dungey, 1958: p. 9). Moreover, the hydrodynamical pressure tensor has been approximated by a scalar hydrostatic pressure, thus neglecting the effect of viscous forces in (8.2) and of viscous dissipation in (8.3) (see Cowling, 1957: p. 38; Lehnert, 1959: p. 61).

TABLE 8.1

Principal Symbols used in Chapter Eight[1]

a radiation constant $(7.57 \times 10^{-15} \text{ erg cm}^{-3} \text{ deg}^{-4})$

c velocity of light $(2.998 \times 10^{10} \text{ cm sec}^{-1})$

$\mathbf{E}$ electric field (in e.s.u.)

e electron charge $(4.802 \times 10^{-10} \text{ e.s.u.})$

$\mathbf{F}$ thermal flux vector

$\mathbf{g}$ acceleration due to gravity

$\mathbf{H}$ magnetic field (in e.m.u.)

I ionization energy per unit volume

$\mathbf{j}$ electric current (in e.s.u.)

k Boltzmann constant $(1.380 \times 10^{-16} \text{ erg deg}^{-1})$

m_e electron mass $(9.107 \times 10^{-28} \text{ gm})$

n_e density of electrons

p gas pressure

p_e electron pressure

q electric charge density (in e.s.u.)

T gas temperature

t_D decay time of a magnetic field due to ohmic dissipation

U internal energy of gas per unit volume

$\mathbf{v}$ gas velocity

γ ratio of specific heats of gas

$\bar{\kappa}$ Rosseland mean mass absorption coefficient

μ mean molecular weight of gas

ρ gas density

σ electrical conductivity (in e.s.u.)

τ optical depth in continuum

[1] C.g.s. units are used throughout.

To complete the system of equations (8.1)–(8.9) we need expressions for the electrical conductivity σ, the total internal energy U, and the thermal flux vector $\mathbf{F}$. A formula for the electrical conductivity, valid under sunspot conditions, has already been given in Section 4.2.5, viz:

$$\sigma = 0.26 \, \frac{e^2}{(3km_e)^{\frac{1}{2}}} \cdot \frac{1}{S} \cdot \frac{p_e}{pT^{\frac{1}{2}}}. \tag{8.10}$$

Values of σ calculated from (8.10) on the basis of Michard's sunspot model are found to range from 3.4×10^{10} e.s.u. at optical depth $\tau = 0.02$ to 2.5×10^{11} e.s.u. at $\tau = 4.00$ (Table 4.7). The figure 1×10^{11} e.s.u. may therefore be adopted as a representative value for approximate calculations.

The internal energy U of the gas is simply the sum of the thermal and ionization energies of the particles. Consequently, we can write

$$U = \frac{1}{\gamma - 1} p + I, \qquad (8.11)$$

where $\dfrac{1}{\gamma - 1} p$ is the thermal energy per unit volume (Chapman and Cowling, 1952: p. 42) and I is the ionization energy. The relevant value of γ, the ratio of the specific heats, is that for a monatomic gas, namely $5/3$.

Finally, it can be shown that the contribution to the thermal flux vector $\mathbf{F}$ due to radiation far exceeds that due to thermal conduction (Stepanov, 1949). Consequently, following Ledoux and Walraven (1958: p. 444), we may write

$$\mathbf{F} = -\frac{c}{\bar{\kappa}\rho}\,\mathrm{grad}\,(\tfrac{1}{3}\,aT^4). \qquad (8.12)$$

The values of the Rosseland mean mass absorption coefficient $\bar{\kappa}$ have been conveniently tabulated by Allen (1955: p. 94).

In conclusion, it should be remarked that in practice the inertia term $\dfrac{m_e}{n_e e^2}\dfrac{\partial \mathbf{j}}{\partial t}$, the Hall electric field $\dfrac{1}{n_e ec}\mathbf{j} \times \mathbf{H}$, and the electron pressure gradient term $\dfrac{1}{n_e e}\,\mathrm{grad}\,p_e$ in the current equation (8.5) are frequently neglected (cf. Dungey, 1958: p. 25). The current equation then assumes the familiar simple form

$$\mathbf{j} = \sigma\left(\mathbf{E} + \frac{1}{c}\mathbf{v} \times \mathbf{H}\right). \qquad (8.13)$$

However, this approximation may not be justifiable in all circumstances and, in fact, Lehnert (1959: p. 74) has concluded that the Hall field may be of some importance in sunspots.

8.2.3   GROWTH AND DECAY OF SUNSPOT MAGNETIC FIELDS

An equation determining the variation of the magnetic field with time can be derived from the basic equations given in the previous section. If we take the electrical conductivity to be constant, then it can be shown from (8.6), (8.7), (8.9), and (8.13) that

$$\frac{\partial \mathbf{H}}{\partial t} = \mathrm{curl}\,(\mathbf{v} \times \mathbf{H}) + \frac{c^2}{4\pi\sigma}\nabla^2\mathbf{H}. \qquad (8.14)$$

In a medium at rest ($\mathbf{v} = 0$) this reduces to

$$\frac{\partial \mathbf{H}}{\partial t} = \frac{c^2}{4\pi\sigma}\nabla^2\mathbf{H}, \qquad (8.15)$$

which is the equation governing the natural decay of the field due to ohmic dissipation. Following Chandrasekhar (1961: p. 151), we may explore the physical consequences of this equation quite simply by considering the case of an infinite, homogeneous medium in which the field is a periodic function of position, varying as $\exp(i\mathbf{k} \cdot \mathbf{r})$. We then have

$$\frac{\partial \mathbf{H}}{\partial t} = -\frac{c^2 k^2}{4\pi\sigma} \cdot \mathbf{H},$$

whose solution is

$$\mathbf{H}(t) = \mathbf{H}(0) \exp\left(-\frac{c^2 k^2}{4\pi\sigma} \cdot t\right),$$

where $\mathbf{H}(0)$ is the value of the field at time $t = 0$. It follows that $\mathbf{H}$ decays exponentially with a time constant, t_0, given by

$$t_0 = \frac{4\pi\sigma}{c^2 k^2}.$$

This result can be written in terms of the linear scale $l = 2\pi/k$ (or 'spatial wavelength') of the field variations as

$$t_0 = \frac{1}{\pi} \cdot \frac{\sigma l^2}{c^2}, \tag{8.16}$$

thus establishing the important conclusion that *the natural decay time of a magnetic field possessing spatial variations of scale l is of the order of* $t_D = \dfrac{\sigma l^2}{c^2}$. Physically, t_D is simply the time required for the ohmic losses $\dfrac{1}{\sigma} j^2$ to dissipate an amount of energy comparable with the magnetic energy $H^2/8\pi$ (cf. Spitzer, 1956: p. 38).

In the case of a sunspot, let us take $l = 1500$ km (roughly the scale of the fine structure of the magnetic field described in Section 5.4.4) and $\sigma = 1 \times 10^{11}$ e.s.u.; then t_D is about one month. This figure is several times greater than the lifetime of most spots (cf. Section 6.2.3). It follows that the dissolution of a sunspot field cannot be due to ohmic decay in material at rest; instead, the field must be carried away by material motions (cf. Cowling, 1953: p. 545).[2] This conclusion is supported by the fact that, although the conductivity is smallest (and hence the ohmic dissipation greatest) in the central part of a spot, the field strength there is observed to remain remarkably constant over much of a spot's lifetime (cf. Section 5.4.7).

[2] Cowling estimated a decay time of about 1000 years. The discrepancy between this figure and the figure given above is due mainly to the higher value of σ used by Cowling: he takes $\sigma = 2 \cdot 7 \times 10^{13}$ e.s.u., although in the same article (*loc. cit.*, p. 539) he quotes the much smaller value $\sigma = 9 \times 10^{11}$ e.s.u. for the solar photosphere. In addition, he takes $l = 3000$ km. If the scale of the magnetic field variations should turn out to be only 1″ of arc (i.e., $l = 750$ km), then t_D is reduced to one week.

As shown by Cowling (1953: cf. p. 546), t_D also gives the time required for the field, if generated *in situ* by electric currents, to increase to roughly its steady-state value against the inertial effect of self-induction. The actual period of growth for a long-lived spot is about 10 days (cf. Section 5.4.7) and this is short compared with a t_D value of one month. Consequently, one can conclude that the spot field neither grows nor decays *in situ*.

8.3 Origin and Stability of Individual Sunspots

8.3.1   INTRODUCTION

The problem of the origin of sunspots may, for convenience, be divided into two parts: in the first place, one has to elucidate the mechanism responsible for the coolness and continued existence of individual spots in a medium hotter than themselves and, in the second place, one has to explain the cause of the solar cycle as a whole. In this section we deal only with the first aspect of the problem, leaving the question of the origin of the solar cycle to be discussed in Section 8.4.

Early theories of the origin of sunspots sought to explain the darkness of a spot in terms of the adiabatic cooling of gas moving upwards in a region of stable radiative equilibrium. Noteworthy among these were the theories proposed by Russell in 1921 and by Rosseland in 1926, and the vortex theory put forward by Bjerknes in 1926; in all cases the magnetic field was assumed to be merely a by-product of the cooling mechanism. However, the subsequent recognition of the existence of an outer hydrogen convection zone in the Sun raised insurmountable difficulties in the way of theories based on adiabatic cooling, with the result that they are now only of historical interest.[3] Moreover, the calculations in Section 8.2.3 show that the field cannot be produced or dissolved *in situ*, as the times required exceed the observed lifetimes of sunspots. Accordingly, it is now generally accepted that spots owe their origin to the presence of *internal* solar magnetic fields which from time to time push their way through the photospheric convection zone (cf. Section 7.1). On the modern view, the role of the magnetic field is fundamental and the explanation of the coolness must be sought in terms of the field itself.

8.3.2   MAGNETIC THEORIES OF SUNSPOT COOLING

The first attempt to explain the coolness of a sunspot in terms of its magnetic field was made by Biermann in 1941 (cf. Cowling, 1953: p. 570). He suggested that the field *suppressed* convection within the spot umbra, thereby decreasing the upward flow of energy. Valuable as this suggestion was for stimulating further theoretical work (see, for example, Cowling, 1957: p. 74), it is no longer tenable in view of the discovery of granulation in sunspot umbrae (cf. Section

---

[3] The reasons for discarding the adiabatic upflow theories have been discussed in some detail by Cowling (1953: p. 565) and Sweet (1955: p. 682), who also give references to the early literature. Additional references are given by Waldmeier (1955: p. 194).

3.6). The existence of umbral granulation shows that even in the largest spots the convection currents are *not* suppressed. On the other hand, the convection may well be retarded by the field and, indeed, the observed difference in properties between the umbral and photospheric granulation (see Section 3.6.4) supplies direct evidence of an interaction between the field and the convection currents. Further evidence that the convection is not entirely suppressed by the field is provided by the sunspot curve of growth (cf. Section 4.3.2): this indicates the existence of relatively large *non-thermal velocities* which, if the observational estimates are to be believed, are actually greater than those derived from the photospheric curve of growth.

An alternative explanation for the coolness of a sunspot was proposed by Hoyle (1949: p. 11), who suggested that the energy convected up from a given area at the base of a spot is spread over a larger area at the surface. This suggestion rests on two assumptions: (*a*) that motion can take place only along lines of force and (*b*) that these fan out near the surface in a way consistent with the classical picture of sunspot magnetic fields, described in Section 5.4. However, there are good reasons for supposing that both these assumptions are false. In regard to the first, Sweet (1955: p. 683) has pointed out that the upward movement of material must automatically cause the sideways displacement of material in its path. As far as the second assumption is concerned, we need only recall that modern observations suggest a configuration of the lines of force in a spot very different from the classical one (cf. Section 5.4.9). For these reasons Hoyle's hypothesis that the convection in a spot is diluted by the field must also be discarded.

At the present time, although Biermann's and Hoyle's suggestions are untenable in the light of modern observations, it does seem almost certain that the coolness of a spot is due to a reduction in the amount of energy convected upwards, resulting in some way from the influence of the spot magnetic field. However, almost nothing is known about the exact nature of the interaction between the field and the convection currents, although observations of the birth and development of sunspot pores do suggest that it may take place on the scale of the *individual* convection columns (cf. Section 3.4.4). Since a theory of established convection under solar conditions, even in the absence of a magnetic field, is entirely lacking (cf. Section 3.2.8), little help can be obtained from existing theoretical discussions of convection in the presence of a magnetic field, such as those of Thompson (1951), Malkus (1959), Chandrasekhar (1961), and Murgai and Khosla (1962). The work of Thompson, Chandrasekhar, and Murgai and Khosla, in particular, deals specifically with the *onset* of thermal instability but, as emphasized by Sweet (1955: p. 683), the conditions envisaged may bear little relation to the velocities and field distortions in a region of fully-developed convection.

It must be emphasized that in addition to an explanation of the origin of the coolness in a sunspot, some explanation is also required for the fate of the energy which, in the presence of the spot field, is no longer carried upwards by convection. As stated in Section 3.3.3, it seems unlikely on observational grounds that the energy deficit is made good by a slight enhancement of the intensity in the photosphere outside the spot. Moreover, explanations based on the Evershed effect (cf. Sweet, 1955: p. 680) appear equally improbable in view of the existence of sunspot pores, which are remarkably long-lived yet possess no penumbral structure (cf. Section 3.4.3). Gurevich and Lebedinsky (1945) sought to explain the flux deficit by the conversion of radiant energy into magnetic energy, but failed to advance any specific mechanism for carrying out the conversion.

However, the idea that the observed energy deficit is made good by an increase in *magnetic* energy is of fundamental importance. Let us see if we can gain some understanding of how this might occur. Observations reveal the presence of convection currents in sunspot umbrae. If the magnetic field is predominantly vertical, then the upward- and downward-moving material in an individual convection cell travels parallel to the field; no electric currents are induced in these portions of the cell, because the electric force $\frac{1}{c} \mathbf{v} \times \mathbf{H}$ vanishes there. However, the closing of the stream lines requires the material to move *perpendicular* to the field during part of its journey. The field will induce currents in these portions of the cell, since $\frac{1}{c} \mathbf{v} \times \mathbf{H}$ no longer vanishes. Although these induced currents $\mathbf{j}$ will give rise to an electromagnetic force $\frac{1}{c} \mathbf{j} \times \mathbf{H}$ which will oppose motion across the field, the motion will nevertheless be maintained by the hydrodynamical forces arising from thermal instability—indeed, the observed granulation shows that the motion is not suppressed. The net result, therefore, is a continuous conversion of heat into magnetic energy, the latter being the energy of the currents induced in material moving perpendicular to the field.

Some external 'sink' for the magnetic energy must be present; the observations described in Section 7.4 strongly suggest that the site of the sink is the activity centre associated with the spot group.

If the new ideas described above should prove correct, then it is clear that in order to make progress in elucidating the energy balance of sunspots, we shall have to enlarge our point of view to include not merely the photospheric layers but also the sub-photospheric field region and the active chromosphere above the spot.

8.3.3   ROLE PLAYED BY MAGNETIC FORCES IN MAINTAINING THE EQUILIBRIUM OF SUNSPOTS

As we have seen in Section 4.2.4, it is possible on the basis of radiation measurements alone to make considerable progress towards the formulation of a satisfactory sunspot model. However, this approach leaves undecided the very important question of the relative roles played by hydrodynamic and magnetic forces in maintaining the equilibrium of spots. For example, Mattig's model is based on the assumption of *hydrostatic* equilibrium, which implies that the magnetic field in the spot exerts no pressure, being essentially 'force-free' in character (see below). On the other hand, Michard's *empirical* model leads to gas pressures in the umbra some two to three times less than that in the surrounding photosphere, thereby suggesting that the field is in fact instrumental in maintaining equilibrium (cf. Cowling, 1957: p. 21).

Most attempts to investigate the mechanical stability of sunspots have been based on the equations of *magnetohydrostatic* equilibrium. These follow at once from the basic relations (8.1)–(8.4), (8.6)–(8.9), and (8.13) on putting $\mathbf{v} = 0$ and omitting all terms containing time derivatives except $\dfrac{\partial \mathbf{H}}{\partial t}$ (cf. Section 8.2.3).

Eliminating the electric field $\mathbf{E}$ one can write the resulting equations in the form

$$\operatorname{grad} p = \rho \mathbf{g} + \frac{1}{c} \mathbf{j} \times \mathbf{H} \tag{8.17}$$

$$\operatorname{div} \mathbf{F} = \frac{1}{\sigma} j^2 \tag{8.18}$$

$$p = \frac{k\rho T}{\mu} \tag{8.4}$$

$$\frac{\partial \mathbf{H}}{\partial t} = -\frac{c^2}{4\pi} \operatorname{curl}\left(\frac{1}{\sigma} \operatorname{curl} \mathbf{H}\right) \tag{8.19}$$

$$\mathbf{j} = \frac{c}{4\pi} \operatorname{curl} \mathbf{H} \tag{8.7}$$

$$\operatorname{div} \mathbf{H} = 0. \tag{8.9}$$

The term on the R.H.S. of (8.18) represents the rate of ohmic heating, which is so small that it can generally be neglected. Equation (8.18) then reduces to the equation of radiative equilibrium

$$\operatorname{div} \mathbf{F} = 0. \tag{8.20}$$

The role played by magnetic forces in maintaining the equilibrium of sunspots is determined by the term $\dfrac{1}{c} \mathbf{j} \times \mathbf{H}$ in (8.17). In the case of Mattig's model it is assumed that

$$\frac{1}{c} \mathbf{j} \times \mathbf{H} = 0, \tag{8.21}$$

so that (8.17) reduces to the ordinary equation of hydrostatic equilibrium. A magnetic field satisfying the condition (8.21) is said to be *force-free*. If we rewrite (8.21) with the aid of (8.7) in the form

$$(\text{curl } \mathbf{H}) \times \mathbf{H} = 0, \tag{8.21A}$$

then it is evident that a force-free field must satisfy the relation

$$\text{curl } \mathbf{H} = \alpha \mathbf{H}, \tag{8.22}$$

where α is a scalar function of position and time. The parameter α is not entirely arbitrary but must satisfy the conditions

$$\mathbf{H} \cdot \text{grad } \alpha = 0$$

and

$$\frac{\partial \mathbf{H}}{\partial t} = -\frac{c^2 \alpha^2}{4\pi\sigma} \mathbf{H} + \frac{c^2}{4\pi} \mathbf{H} \times \text{grad } \frac{\alpha}{\sigma}.$$

The first condition is derived by taking the divergence of (8.22) and substituting from (8.9), while the second is obtained by substituting (8.22) into (8.19).

Solutions of (8.22) for the case in which α is a constant and the field strength decreases exponentially with height have been obtained by Schatzman (1961), assuming axial symmetry. On the basis of these solutions he has proposed a possible model of a force-free field in a unipolar spot: the field configuration bears a crude resemblance to the classical picture of a sunspot field (see Section 5.4.1), although the lines of force are twisted and, in addition, are perpendicular to the solar surface at the outer boundary of the spot. However, Schatzman's analysis, as it stands, supplies no evidence to justify the original assumption that the field is force-free.

A more general approach to the problem of sunspot equilibrium was introduced by Schlüter and Temesváry (1958), who attempted to solve the magneto-hydrostatic equations (8.17) and (8.20) numerically, taking full account of the magnetic force term $\frac{1}{c} \mathbf{j} \times \mathbf{H}$. To simplify the calculations these authors considered the case of a circularly symmetrical unipolar spot and introduced cylindrical coordinates r, ϕ, and z with the positive direction of z directed downwards along the solar radius through the centre of the spot. It was assumed that the magnetic field strength vanished for $z \rightarrow -\infty$ and for $r \rightarrow \infty$; the azimuthal field component H_ϕ was put equal to zero. In addition, it was assumed that the relative distribution of magnetic flux was similar across any horizontal cross-section of the spot. The solutions of the magnetohydrostatic equations were found to show a marked instability; this resulted from the assumption of purely radiative equilibrium (equation 8.20). One can conclude, in agreement with the observational evidence, that to obtain a realistic sunspot model it is necessary also to take account of the transport of energy by *convection*.

When convection is taken into account, one cannot of course put $\mathbf{v} = 0$ but must seek steady-state solutions of the basic magnetohydrodynamic equations (8.1)–(8.9). Future theoretical studies along these lines, such as that now being undertaken by Deinzer at the Max-Planck Institut in Munich (cf. *Phys. Verh.* **11,** 211), will be of fundamental importance in obtaining a proper understanding of the mechanism and equilibrium of the individual sunspot.

8.3.4 ORIGIN OF SUNSPOT PENUMBRAE

It must be admitted that neither the mode of origin of the penumbra nor the role it plays in the sunspot phenomenon as a whole is yet properly understood. On the observational side (cf. Section 3.4.3) it is found that, while pores not exceeding about 5″ of arc in diameter exist as stable structures devoid of penumbrae, those above this size show an increasing tendency to develop into small spots having at least some rudimentary penumbral structure. The actual process of formation and dissolution of a rudimentary penumbra is described in Section 3.5.2: the observations suggest that the formation of a penumbra is due to the penetration of part of the umbral magnetic field into the surrounding photosphere, resulting first in a darkening of the intergranular material and then in the replacement of the granules by thin bright filaments directed more or less radially outwards from the spot.

Attempts have been made to explain the properties of the penumbra in terms of the influence of the spot magnetic field on convection in the penumbra (see, for example, Danielson, 1961). Any adequate theory must, of course, take full account of the Evershed effect (Section 4.4.2), which at the photospheric level takes the form of a *horizontal* outflow of matter across the penumbra with a velocity of a few kilometres per second. It has usually been assumed that this motion must take place along the lines of force, since theory indicates that any motion across the field should be strongly resisted by induction drag. However, modern observations indicate that the field in the penumbra is predominantly *vertical* (cf. Section 5.4.8), thus implying that the Evershed flow is, in fact, directed across the lines of force.[4] At the moment there seems no way of resolving this apparent paradox and the whole question of the theoretical interpretation of the penumbra remains obscure.

8.4 Origin of the Solar Cycle

8.4.1 MODERN THEORIES OF THE ORIGIN OF THE SOLAR CYCLE

Accepting the conclusion of the previous section that the coolness of a sunspot results from the influence of a rising magnetic field on the overlying convection zone, one must now explain the origin of the field and the various regularities associated with the solar cycle (cf. Section 6.3). Most modern theories assume the existence of a weak *general* solar magnetic field and seek to find ways whereby

---

[4] It is noteworthy that observational evidence of a similar situation in the case of photospheric velocity fields has recently been obtained by Stepanov (1961).

portions of this field can be brought to the surface and amplified to give intensities comparable to those observed in spots. Work along these lines has been stimulated by the recent observation (Babcock, 1959) that the Sun's polar field weakened and then reversed its polarity during the maximum of the last solar cycle. Owing to the high electrical conductivity within the Sun and the consequent very long decay time (cf. Section 8.2.3), such a reversal suggests that the field cannot permeate the whole interior of the Sun.[5] Instead, it suggests that the general field must be fairly superficial in character; if so, it may be closely related to sunspot magnetic fields (cf. Allen, 1960). This conclusion is strongly supported by the fact that the polarity change actually took place at different times in the two hemispheres, as did the maximum of the sunspot cycle: both occurred about a year earlier in the south than in the north (Waldmeier, 1960).

The first attempt to explain the origin of sunspot fields from a general internal field was made by Alfvén in a series of papers beginning in 1942. This theory was further developed by Walén (1944), whose paper (though subsequently repudiated by its author) remains the most complete account available of Alfvén's theory (cf. Cowling, 1958: p. 107). Although successful in explaining many of the features of the solar cycle, the theory is subject to a number of serious objections (Cowling, 1946; 1953: p. 574) and is now mainly of historical interest. A second attempt to explain the origin of sunspot fields was made by Walén himself, based on the concept of torsional magnetic oscillations inside the Sun. However, as pointed out by Cowling (1957: p. 49), any torsional oscillation around the Sun's axis should produce changes in the observed rotation at the surface but, in fact, no such variations have been detected (cf. Goldberg, 1953: p. 20).[6]

A magnetic field inside a conducting fluid sphere may be either poloidal or toroidal in character, or a combination of both. In the former case, the lines of force lie in meridional planes while, in the latter, they are parallel to the equator. Alfvén's theory rests on the assumption that the field is essentially poloidal in character, whereas most of the more recent theories embody the idea that sunspots arise from internal toroidal fields. There are two schools of thought regarding the generation of toroidal fields: some authors (Cowling, 1953: p. 575; Bullard, 1955: p. 686; Babcock, 1961) suggest that the toroidal fields are derived from a general poloidal field as a result of differential solar rotation, while others (Parker, 1955a, b; Elsasser, 1956) go further and postulate a magnetohydrodynamic 'dynamo' mechanism for the maintenance of a solar field with both toroidal and poloidal components.

To explain the progression of the sunspot zone towards the solar equator

[5] It seems safe to assume that the polarity of the polar field is reversed at each solar maximum, although magnetograph observations extending over a sufficiently long period are not yet available for confirmation.

[6] Although both Alfvén's and Walén's theories assume the existence of a convective core in the Sun, this is not now believed to be the case (see, for example, Schwarzschild, 1958: p. 132). The absence of a convective core would, by itself, be a fatal objection to Alfvén's theory.

during the course of a solar cycle (cf. Section 6.3.5), Bullard (1955: p. 688) suggested that the toroidal field is transported by a meridional current of moving material which, when it reaches the equator, sinks down and moves back towards the pole beneath the current carrying the field of the next solar cycle. The first current is supposed to eventually emerge again near latitude 25° to start the next cycle but one (see Bullard's Fig. 5). Such a system of meridional circulation was first proposed by Bjerknes in 1926, but with purely hydro-dynamic 'ring vortices' in place of the rings of magnetic field postulated by Bullard.[7] The concept has recently been extended by Allen (1960) to include in each hemisphere a circulation in the opposite direction between the mid-latitudes and the pole (see also Waldmeier, 1955: p. 197). Allen invokes the poleward movement of magnetic regions originating in the sunspot zone to explain the reversal of the polar field around the time of sunspot maximum. However, the idea that the poleward migration of disintegrating magnetic regions might be responsible for the formation and reversal of the polar field was actually first put forward by Babcock and Babcock (1955) and has since been elaborated by Babcock (1961).

Given internal toroidal fields and some system of mass circulation to explain the existence and progression of the sunspot zones, it is still necessary to find a specific dynamical mechanism for raising occasional strands of magnetic flux up to the surface. Furthermore, one must also explain the exact manner in which the rising flux tubes generate the observed sunspot fields. Both problems have been attacked theoretically by Parker (1955a). He introduced the valuable concept of the 'magnetic buoyancy' of flux tubes, which has since been incorporated by Babcock (1961) into the most elaborate synthesis yet attempted of existing observational data and modern theoretical ideas on the origin of the solar cycle. Although neither paper contains the final answers to the many difficult questions associated with the problem of the solar cycle, they do in large measure reflect the present state of knowledge and thinking on the subject. For this reason, the concept of magnetic buoyancy is discussed in more detail in the next section, while Babcock's synthesis of existing ideas and information is outlined in Section 8.4.3.

8.4.2 MAGNETIC BUOYANCY AND THE GENERATION OF SUNSPOT FIELDS

Following Parker (1955a) let us consider the magnetohydrostatic equilibrium of a horizontal tube of magnetic flux embedded in the sub-photospheric layers of the Sun. Using (8.7) one can write the equation of mechanical equilibrium (8.17) in the form

$$\operatorname{grad} p = \rho \mathbf{g} + \frac{1}{4\pi} (\operatorname{curl} \mathbf{H}) \times \mathbf{H}. \qquad (8.23)$$

---

[7] For a discussion of modern theoretical ideas concerning the existence of meridional circulation in the solar convection zone the reader is referred to Biermann (1958) and Rubashev (1958).

If we introduce a cartesian reference system OX, Y, Z, such that OZ is vertical and the field lines are parallel to OX, then it follows from (8.23) that

$$\frac{\partial}{\partial y}\left(p + \frac{H^2}{8\pi}\right) = 0.$$

Hence, integrating with respect to y, we obtain

$$p_{ex} = p_{in} + H^2/8\pi, \tag{8.24}$$

where p_{ex} is the gas pressure outside the flux tube and p_{in} the pressure inside. If we suppose that the material in the flux tube is in thermal equilibrium with its surroundings, then (8.24) can be rewritten with the aid of (8.4) in the form

$$\rho_{ex} = \rho_{in} + \frac{\mu}{kT} \cdot \frac{H^2}{8\pi}. \tag{8.25}$$

Since the magnetic field term on the R.H.S. of (8.25) is necessarily positive, it follows that $\rho_{ex} > \rho_{in}$ and hence the flux tube is subject to an upward force

$$(\rho_{ex} - \rho_{in})\,g = \frac{\mu g}{kT} \cdot \frac{H^2}{8\pi} \tag{8.26}$$

per unit volume, called by Parker *magnetic buoyancy*.

This buoyancy force will of course be opposed by a magnetic tension acting along the lines of force (cf. Dungey, 1958: p. 44); however, Parker (1955a) has shown that the buoyancy force is dominant over any length L of the flux tube exceeding twice the local scale height in the medium, i.e. for $L > \dfrac{2kT}{\mu g}$. Consequently, a long horizontal flux tube can never, in fact, be in static equilibrium. Moreover, as Parker has shown, raising a portion of a long flux tube results in a flow of fluid from the raised section, thus enhancing its buoyancy. Hence, once the tube has begun to rise, it will not stop unless swept down again by some motion in the surrounding medium.

To obtain a quantitative estimate of the buoyancy force acting on a horizontal flux tube in the sub-photospheric layers, Parker considers first the case of a flux tube of field strength 100 gauss at a depth of 20,000 km. Taking $\rho_{ex} \simeq 2\cdot5 \times 10^{-4}$ gm cm^{-3} and $T_{ex} \simeq 2\cdot5 \times 10^5$ °K, he obtains from (8.25) the result $\rho_{ex} - \rho_{in} \simeq 10^{-7}\,\rho_{ex}$; this density change is equivalent to that which would occur were the temperature to be increased by a mere $0\cdot02$°K, and is accordingly quite negligible. Now let us consider a flux tube of field strength 1000 gauss at a depth of only 1000 km. Taking $\rho_{ex} \simeq 0\cdot8 \times 10^{-8}$ gm cm^{-3} and $T_{ex} \simeq 1\cdot5 \times 10^4$ °K, Parker finds that $\rho_{ex} - \rho_{in} \simeq 0\cdot04\,\rho_{ex}$; hence in this case we have a strong buoyancy force equivalent to that which would be produced by raising the temperature of the flux tube region some 600°K higher than its surroundings. It follows, therefore, that in the *upper* parts of the convection zone, strong magnetic buoyancy forces will occasionally bring to the surface strands of the toroidal field. At depths of 20,000 km or greater, on the other hand, the buoyancy force is quite negligible and there is no tendency for the general internal field as a whole to rise to the surface. This means that, on Parker's theory, the depth of origin of sunspot fields must be of the order of 10,000 km, implying that a spot is a relatively shallow phenomenon in the solar convection zone.

Parker next considers the configuration which a rising strand of toroidal field would assume on reaching the solar surface. He supposes that the field brought up into the upper part of the convection zone ultimately becomes so steeply inclined to the horizontal that the problem becomes one of the magnetohydrostatic equilibrium of a *vertical* flux tube (cf. Parker, 1955a: Fig. 2). Then, taking the z-axis directed vertically upwards, Parker finds that, subject to certain simplifying assumptions, the magnetic field strength decreases with height according to the relation

$$H(z) = H(o) \exp\left(-\int_0^z \frac{\mu g}{2kT_{ex}} \cdot dz\right). \qquad (8.27)$$

At the same time it follows that the diameter of the flux tube must increase with height. Consequently, starting with an initial field of 100 gauss, Parker predicts the emergence in the photosphere of a vertical field of only about 10 gauss.

However, the relation (8.27) is derived on the assumption that the material in the flux tube remains in thermal equilibrium with its surroundings (i.e. $T_{ex} = T_{in}$). Parker now shows that, if this assumption is dropped and we suppose instead that the material in the flux tube is cooled as a result of the presence of the field, an actual *increase* of the field intensity with height can be obtained. The actual mechanism of magnetic cooling is not critical to Parker's theory and he assumes, for simplicity, that the temperature difference $T_{ex} - T_{in}$ is proportional to the magnetic field energy $H^2/8\pi$, viz:

$$T_{ex} - T_{in} = \eta \cdot \frac{H^2}{8\pi}. \qquad (8.28)$$

Using (8.28) he then obtains the result

$$\frac{[H(z)]^2}{8\pi} = \frac{[H(o)]^2}{8\pi} \exp\left\{\int_0^z \frac{\mu g}{kT_{in}}\left(\frac{k\rho_{ex}\eta}{\mu} - 1\right) dz\right\} \qquad (8.29)$$

for the variation of field strength with height. It can easily be seen that (8.29) reduces to (8.27) in the case of thermal equality ($\eta = 0$). According to (8.29) the field strength will increase with height (and hence the width of the flux tube will decrease) provided that, for each value of z, η exceeds the value η_0 given by

$$\frac{k\rho_{ex}\eta_0}{\mu} = 1.$$

As Parker points out, only a very modest decrease in temperature is required to satisfy this condition, at least in the deeper layers of the convection zone. For example, following Parker let us suppose that at the base of the flux tube the field intensity is 100 gauss, $\rho_{ex} = 10^{-5}\,\mathrm{gm\,cm^{-3}}$, and $T_{ex} = 4 \times 10^4\,°\mathrm{K}$. Then, taking $\mu \simeq 0.7\,m_H$ (cf. Unsöld, 1955: Table 23), we find

$$\eta_0 \simeq 0.85 \times 10^{-3},$$

and hence the field intensity will increase with height provided the temperature drop exceeds the value

$$T_{ex} - T_{in} \simeq 0.3\,°\mathrm{K}.$$

In view of the exponential factor involved in (8.29) it is evident that even a slight cooling of the interior of the flux tube can result in a large increase in the field intensity at the upper end, together with a corresponding decrease in the width of the tube. This is the principle of Parker's mechanism for the generation of sunspot fields.

8.4.3   H. W. BABCOCK'S THEORY OF THE SOLAR MAGNETIC CYCLE

The concept of magnetic buoyancy explained in the preceding section has recently been incorporated by Babcock (1961, 1962a) into the most elaborate synthesis yet attempted of existing observational data and modern theoretical ideas on the origin of the solar cycle. In formulating his theory Babcock has paid special regard to two important observational facts:

(1) the Sun has a weak *poloidal field* of strength 1–2 gauss which shows an irregular structure and is usually limited to high latitudes. The poloidal field was observed to reverse its polarity during the maximum of the last solar cycle and it is assumed, though not yet proven, that the polarity is reversed at the maximum of each solar cycle;

(2) extensive magnetograph observations have shown that the emergence of *bipolar magnetic regions* (called BM regions for short) in the photosphere underlies the development of sunspot groups and activity centres. Although some BM regions of low field strength may show little or no optical activity,' others increase in field strength and become identified with solar activity centres, whose subsequent development is summarized in Table 7.1. Like spots, BM regions are found to obey Hale's laws of sunspot polarities, given in Section 6.3.4. It is most significant that with increasing age BM regions tend to expand and diminish in field strength: as Babcock (1962a) has emphasized, this indicates that BM regions eventually disappear by expanding, *not* by contracting and sinking below the solar surface.

The essential content of the new theory is well summarized by Babcock (1962a) himself in the following terms: 'In brief outline, the model proposed here begins with a poloidal field whose lines of force lie in meridian planes. The submerged lines of force are shallow and are distorted by the differential rotation of the Sun, so that kinetic energy of rotation is converted to magnetic energy at an increasing rate as the lines of force are drawn out in longitude to form rather tight spirals—essentially toroidal fields—on opposite sides of the equator. This continues until the magnetic energy of the submerged toroidal fields attains locally a critical limit where instability sets in, forming concentrated flux loops that are brought to the surface by magnetic buoyancy.[8] Each such loop produces a BM region on the surface, with related sunspots. Thus we see that the poloidal and toroidal components are simply parts of the same general magnetic field. By following the development of the lines of force, one finds that, as the BM regions expand and migrate, the initial poloidal field is neutralized and then supplanted by another of reversed polarity. Of perhaps fully as much interest, most of the magnetic flux that is created in the formation of the toroidal fields eventually finds its way through the photosphere and is liberated in the form of large, detached loops, high in the corona.'

To explain the progression of the sunspot zone during the course of a solar cycle, Babcock points out that, as a consequence of differential solar rotation, the spiral winding of the toroidal fields is tighter in moderate latitudes than near the equator. The field may therefore be expected to become unstable (and so give rise to sunspots), first at moderate latitudes and then at progressively lower latitudes. Babcock explains Hale's laws of sunspot polarities in terms of the reversal of the poloidal field during the maximum of each successive solar cycle.

Babcock's theory has the special advantage of being closely tied to all the relevant observational material at present available, but at the moment it would be premature to attempt a critical evaluation. One obvious difficulty is to find the source of energy required to maintain the differential solar rotation which, in turn, supplies the energy for the generation of the toroidal fields. Babcock (1961) has estimated the amount of magnetic energy dissipated during the course of a solar cycle and has compared it with the energy of differential rotation, with the result that the latter turns out to be sufficient to maintain the solar cycle for only a few thousand years.

8.5 Magnetic Activity on Stars Other than the Sun

8.5.1 INTRODUCTION

The Sun is in no sense a peculiar star: of spectral class dGo-2, it is—as far as we know—a typical member of the lower main sequence of the Hertzsprung–Russell diagram. The existence of a magnetic cycle on the Sun thus suggests that magnetic activity of some sort may also be present on other stars, although in view of their greater distance it would have to be far more extensive in order to be detectable.[9] The first observational confirmation of the existence of stellar magnetic fields came with H. W. Babcock's discovery of a fluctuating field in

[8] A similar process of field amplification is outlined by Alfvén (1961).
[9] At the peak of the solar cycle, sunspots occupy only a few tenths of one per cent of the total area of the visible hemisphere (see Section 6.3.1).

78 *Virginis* in 1947. Since then, mainly as a result of Babcock's continued work, magnetic fields have been detected on stars covering a wide range of spectral types and the study, both observational and theoretical, of these so-called 'magnetic' stars has become a subject of rapidly increasing importance in astrophysics.

In practically all cases which have so far been adequately observed, the fields of magnetic stars have been found to be variable. In some cases the time-scale of the variations is only of the order of a week, whereas the period of the solar magnetic cycle is some 22 years (cf. Section 6.3.4). Nevertheless, the question does arise of whether the solar cycle is related to some general form of magnetic activity occurring on a faster and more spectacular scale on other stars. For example, do these stars possess large 'sunspots', occupying a substantial fraction of their surface (cf. Babcock, 1960a: p. 288)? And if so, what insight into the nature and origin of solar activity can be obtained from observations of magnetic stars? At the moment no definite answers can be given to these questions. However, in order to round off our discussion of the origin of sunspots, it seems appropriate to summarize the main observational facts relating to magnetic stars (Section 8.5.2) and to briefly consider possible explanations of their behaviour (Section 8.5.3).

8.5.2   OBSERVATIONS OF MAGNETIC STARS

Nearly all the existing observations of the Zeeman effect in stellar spectra have been obtained by H. W. Babcock, using a double image quarter-wave plate analyser for circularly polarized light (cf. Section 5.3.2).[10] His results are expressed in terms of the effective field H_{eff}, assumed uniform and longitudinal, required to produce the observed Zeeman displacements; the sign of H_{eff} is taken as positive when the field is directed outwards from the star (i.e., towards the observer). However, the actual field will not in general be purely longitudinal; consequently, as a result of blending of the σ and π components in the Zeeman patterns of the lines employed, the values of the field intensities may be under-estimated (cf. Section 5.2.4). In fact, measurements made by Babcock (1960b) on lines of different magnetic sensitivities in the spectrum of the Aop star HD 215441, photographed at a time when its effective field strength attained the extraordinarily high value of some 34,000 gauss, suggest that the true intensities for stars with significantly weaker fiedls may be several times greater than the H_{eff} values actually derived.

In contrast to HD 215441, the measured field strengths of most magnetic stars are only of the order of a thousand gauss (see below). Consequently, the Zeeman displacements normally to be expected are relatively small (cf. Section 5.2.5) and in these cases conditions on the star must be rather specialized for the field to be detectable at all. The stellar field must be mainly longitudinal and

---

[10] For a detailed account of all the varied aspects of magnetic star observations the reader is referred to review articles by Babcock and Cowling (1953), Deutsch (1958a), and Babcock (1960a, 1962b).

one polarity must predominate; moreover, line broadening from other sources, such as Doppler broadening due to axial rotation, must not be so large as to mask the Zeeman displacements (Babcock, 1958a). In practice, measurement of the Zeeman effect becomes difficult when the line widths exceed about 0·3Å. However, sufficiently strong fields can sometimes be detected even if the line widths are as great as 0·6 or 0·8Å (Babcock, 1958b).[11]

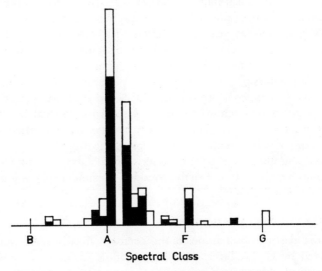

Fig. 8.1. Distribution of magnetic and probably magnetic stars according to spectral class (Schatzman, 1962). The stars included all show 'peculiar' spectral characteristics (see text). The shaded areas represent magnetic stars, the unshaded areas probably magnetic stars.

Most of the existing data on magnetic stars is contained in a paper published by Babcock (1958b), who examined the spectra of 338 stars for the presence of the Zeeman effect. Among the 150 stars whose lines are sufficiently sharp to permit the ready detection of effective fields larger than, say, 500 gauss (Deutsch, 1958a: p. 692), 89 show definite evidence of magnetic fields and 61 do not. Among the stars with broader lines, 66 are listed as probably showing some evidence of the Zeeman effect, and 122 are listed as having lines too broad for measurement. The distribution of the magnetic and probably magnetic stars showing 'peculiar' spectral characteristics (see below) according to spectral class is shown in Fig. 8.1, which is taken from Schatzman (1962). It is evident that magnetic fields are observed most frequently in A-type stars. This fact is explained by Babcock (1958a) on the hypothesis that strong coherent fields

---

[11] In the spectra of some magnetic stars the line widths in one state of polarization are at times systematically wider than those in the other state of polarization. The phenomenon is usually seen in stars showing strong reversing fields near the times of reversal, and is known as the 'crossover effect' (cf. Deutsch, 1958a: p. 700).

might be expected to occur in those stars which have both rapid axial rotation and an outer convection zone. Rapid rotation is characteristic of the hotter stars (types B, A, and early F) while convection zones prevail in the cooler stars, becoming thin and finally absent as we proceed from the F to the hotter B types. Consequently, on Babcock's hypothesis the most prominent magnetic fields are to be expected for stars of types A and early F, in agreement with observation. However, it must be stressed that the stars observed by Babcock do not form a random sample and, in fact, a large part of the observing programme was devoted to stars of type A.

The 89 stars which definitely show the Zeeman effect comprise 70 A-type stars, 1 cluster-type variable (RR *Lyrae*), 1 sub-dwarf, 2 S-type stars, 3 M-type giants, 7 'metallic-line' stars, and a few unusual objects (Babcock, 1960a: cf. Table 1). The magnetic fields of all these stars appear to vary with time; moreover, in the great majority of cases, the variations are irregular in character. The measured effective field strengths range from a few hundred to over 5000 gauss, with an average value of the order of 1000 gauss.

Nearly all the 70 A-type magnetic stars mentioned above have 'peculiar' spectra (spectral designation Ap), indicating anomalous abundances of certain elements (Deutsch, 1958a: p. 694). On the other hand, no Ap stars are found among the small number of A stars known *not* to show the Zeeman effect. This suggests that the abundance anomalies in Ap stars are closely related to the presence of strong magnetic fields, and attempts have been made to explain the abundance anomalies in terms of nuclear reactions occurring at the surface of the stars. It has been suggested that the nuclear reactions are initiated by energetic protons accelerated as the result of changes in the magnetic field, and an analogy has been drawn with possible acceleration mechanisms operating at the times of very large solar flares (cf. Deutsch, 1958a: p. 710). However, at the present time the suggestion that the abundance anomalies are the result of surface nuclear reactions can by no means be regarded as proven (Babcock, 1960a: p. 303).

Babcock (1960a: cf. Tables 3–5, 10) has divided the A-type magnetic variables into four magnetic classes α, β, γ, and δ, whose properties may be briefly summarized as follows:

(1) the α variables show essentially periodic magnetic variations, but the amplitude often shows considerable deviations from the mean value. Most of these stars are also spectrum variables, showing changes in line intensity in synchronism with the magnetic variations. Their periods range from about 5 to 9 days, although one (HD 188041) is known to have a period of 226 days. The typical α variable has a large magnetic field which undergoes a nearly symmetrical reversal of polarity, and it shows the crossover effect in the line widths (see above);

(2) the β variables show irregular magnetic fluctuations, including reversal of polarity. These stars usually have smaller magnetic fields than the α variables. They do not show spectrum variations or the crossover effect;

(3) the γ variables show irregular magnetic fluctuations, but without reversal of polarity. In other respects they are similar to the β variables;

(4) the δ variables show irregular magnetic fluctuations, together with erratic and apparently uncorrelated spectrum variations. The range of the field variations in these stars is much less than that in the α variables.

8.5.3 POSSIBLE EXPLANATIONS OF THE BEHAVIOUR OF α-TYPE MAGNETIC VARIABLES

Theoretical work on the problem of magnetic stars has been concentrated to date on the possible explanation of the properties of Babcock's α-type magnetic variables. These stars are of great intrinsic interest for a number of reasons: they possess large magnetic fields which undergo periodic and nearly symmetrical reversals of polarity with periods normally in the range 5–9 days. They show the crossover effect and, in addition, they show changes in line intensity in synchronism with the magnetic fluctuations. Finally, like other Ap-type magnetic stars (cf. Abt and Golson, 1962), they display small changes in luminosity which in some cases occur in phase with the field variations, the epoch of minimum light coinciding with that of greatest positive magnetic field (Rakos, 1962).

Two different models have so far been advanced to explain the α-type variables, one based on the concept of an oblique rotator and the other on the concept of a greatly amplified and accelerated magnetic cycle similar in character to the 22-year cycle observed on the Sun. The simplest form of *oblique rotator* is a star rotating as a rigid body and possessing a dipole field whose axis is inclined both to the axis of rotation and to the observer's line-of-sight. To the observer, the effective field of such a star would evidently appear to undergo periodic variations. However, in this simple form the oblique rotator model is inadequate to explain all the observed features of α variables. More elaborate versions have therefore been developed, based on the idea that large magnetic areas with associated abundance anomalies are distributed irregularly over the surface of a uniformly rotating star (cf. Deutsch, 1958b). However, the assumption of rigid body rotation appears to be highly artificial, particularly in view of the known non-uniform rotation of the Sun. In any event, completely uniform rotation cannot be admitted as most α variables do, in fact, show random deviations in amplitude or phase. Moreover, as Rakos (1962) has pointed out, the oblique rotator model fails to explain the observed luminosity variations.

On the '*solar cycle*' model, the behaviour of the α-type magnetic variables is attributed to a greatly amplified and accelerated magnetic cycle similar in character to the 22-year cycle observed on the Sun. Some possible features of such a model have been discussed by Babcock (1960a: p. 314), whose ideas are closely related to the various known aspects of solar activity. However, as we have seen in Section 8.4, no complete theory is yet available of the solar cycle, much less of its possible presence in magnetic stars in an enhanced form.

In Babcock's view, the typical, rapidly-rotating A star possesses both a strong coherent magnetic field and a peculiar spectrum. However, unless the star is observed *pole-on*, the presence of both the Zeeman effect and the peculiar spectrum characteristics is masked by the large Doppler broadening of the spectral lines due to the rapid rotation. Consequently, if the solar cycle model is valid, the A-type magnetic variables represent those A-type stars, out of the hundreds observable, whose rotation axes are inclined at only *small* angles to the line-of-sight. On the other hand, if the oblique rotator model is correct, the A-type magnetic variables represent A stars with rotational velocities much smaller than average, seen with their rotation axes inclined at *large* angles to the line-of-sight. In principle, therefore, it should be possible to discriminate between the two theories from the observed distribution of rotational velocities among the A-type stars (Sweet, 1961); however, owing to the numerous uncertainties involved, no definite decision is at present possible.

8.6 Concluding Remarks

It is now generally accepted that sunspots and the other diverse and often spectacular phenomena associated with activity centres owe their origin to the presence of magnetic fields which have risen from the solar interior and pushed their way through the photospheric layers into the chromosphere and corona above (cf. Section 7.1). Although there are reasons for believing that the internal fields responsible for the formation of spots may be fairly superficial in character (cf. Section 8.4), the possibility of large fields being present in the deeper layers cannot be excluded (see, for example, Wentzel, 1961). In the final analysis, therefore, as Dungey (1958: p. 130) has remarked, the problem of explaining the origin of spots is one involving the theory of the internal constitution of the Sun. On the other hand, the classical theory of the structure and evolution of the stars takes no account of the possible role played by internal magnetic fields (see, for example, Schwarzschild, 1958), although recent work (Burbidge and Burbidge, 1958: p. 284; Sweet, 1961; Mestel, 1962; Schatzman, 1962) has raised the question of whether this neglect is entirely justified.

As we have seen in Section 8.5.2, there is some evidence to support Babcock's view that strong magnetic activity is to be expected on those stars having both rapid axial rotation and an outer convection zone. This suggests that magnetic fields may play a significant role in stellar evolution. For example, on the assumption that surface activity of the solar type is a normal feature of a star possessing both a magnetic field and an outer convection zone, Schatzman (1962) has suggested an explanation of the distribution of rotational velocities along the main sequence in terms of the angular momentum carried away from the star by material ejected under the influence of the field. Whether this conjecture be right or wrong, it is evident that the study of stellar magnetism constitutes one of the most important and exciting problems confronting present-day astrophysics. In this connection the study of sunspots and other magnetic phenomena on the Sun can be expected to play a fundamental role.

REFERENCES

ABT, H. A., and GOLSON, J. C. [1962] 'Colors and variability of magnetic stars', *Astrophys. J.* **136**, 35.

ALFVÉN, H. [1961] 'On the origin of cosmic magnetic fields', *Astrophys. J.* **133**, 1049.

ALLEN, C. W. [1955] *Astrophysical Quantities*. Univ. London Press.

ALLEN, C. W. [1960] 'A sunspot cycle model', *Observatory* **80**, 94.

BABCOCK, H. D. [1959] 'The Sun's polar magnetic field', *Astrophys. J.* **130**, 364.

BABCOCK, H. W. [1958a] 'Stellar magnetic fields', *Electromagnetic Phenomena in Cosmical Physics*, ed. B. LEHNERT, p. 161. Camb. Univ. Press.

BABCOCK, H. W. [1958b] 'A catalog of magnetic stars', *Astrophys. J. Supp.* **3**, 141.

BABCOCK, H. W. [1960a] 'Stellar magnetic fields', *Stellar Atmospheres*, ed. J. L. GREENSTEIN, p. 282. Univ. Chicago Press.

BABCOCK, H. W. [1960b] 'The 34-kilogauss magnetic field of HD 215441', *Astrophys. J.* **132**, 521.

BABCOCK, H. W. [1961] 'The topology of the Sun's magnetic field and the 22-year cycle', *Astrophys. J.* **133**, 572.

BABCOCK, H. W. [1962a] 'The solar magnetic cycle', *Trans. I.A.U.* **11B**, 419.

BABCOCK, H. W. [1962b] 'Measurement of stellar magnetic fields', *Astronomical Techniques*, ed. W. A. HILTNER, p. 107. Univ. Chicago Press.

BABCOCK, H. W., and BABCOCK, H. D. [1955] 'The Sun's magnetic field, 1952–1954', *Astrophys. J.* **121**, 349.

BABCOCK, H. W., and COWLING, T. G. [1953] 'General magnetic fields in the Sun and stars', *Mon. Not. R.A.S.* **113**, 357.

BIERMANN, L. [1958] 'On meridional circulations in stellar convective zones', *Electromagnetic Phenomena in Cosmical Physics*, ed. B. LEHNERT, p. 248. Camb. Univ. Press.

BULLARD, E. C. [1955] 'The magnetic fields of sunspots', *Vistas in Astronomy*, ed. A. BEER, vol. 1, p. 685. London, Pergamon.

BURBIDGE, G. R., and BURBIDGE, E. M. [1958] 'Stellar evolution', *Handbuch der Physik*, ed. S. FLÜGGE, vol. 51, p. 134. Berlin, Springer.

CHANDRASEKHAR, S. [1961] *Hydrodynamic and Hydromagnetic Stability*. Oxford Univ. Press.

CHAPMAN, S., and COWLING, T. G. [1952] *The Mathematical Theory of Non-uniform Gases*. Camb. Univ. Press.

CHU, B.-T. [1959] 'Thermodynamics of electrically conducting fluids', *Phys. Fluids* **2**, 473.

COWLING, T. G. [1946] 'Alfvén's theory of sunspots', *Mon. Not. R.A.S.* **106**, 446.

COWLING, T. G. [1953] 'Solar electrodynamics', *The Sun*, ed. G. KUIPER, p. 532. Univ. Chicago Press.

COWLING, T. G. [1957] *Magnetohydrodynamics*. New York, Interscience.

COWLING, T. G. [1958] 'Solar electrodynamics', *Electromagnetic Phenomena in Cosmical Physics*, ed. B. LEHNERT, p. 105. Camb. Univ. Press.

COWLING, T. G. [1962] 'Magnetohydrodynamics', *Rep. Progress Phys.* **25**, 244.

DANIELSON, R. E. [1961] 'The structure of sunspot penumbras. II. Theoretical', *Astrophys. J.* **134**, 289.

DEUTSCH, A. J. [1958a] 'Magnetic fields of stars', *Handbuch der Physik*, ed. S. FLÜGGE, vol. 51, p. 689. Berlin, Springer.

DEUTSCH, A. J. [1958b] 'Harmonic analysis of the periodic spectrum variables', *Electromagnetic Phenomena in Cosmical Physics*, ed. B. LEHNERT, p. 209. Camb. Univ. Press.

DUNGEY, J. W. [1958] *Cosmic Electrodynamics*. Camb. Univ. Press.

ELSASSER, W. M. [1956] 'Hydromagnetic dynamo theory', *Rev. Mod. Phys.* **28**, 135.

FERRARO, V. C. A., and PLUMPTON, C. [1961] *Magneto-Fluid Mechanics*. Oxford Univ. Press.

GOLDBERG, L. [1953] *Introduction to 'The Sun'*, ed. G. KUIPER. Univ. Chicago Press.

GUREVICH, L. E., and LEBEDINSKY, A. I. [1945] 'The magnetic field of the sun spots', *C.R. Acad. Sci. U.S.S.R.* **49,** 92.

HOYLE, F. [1949] *Some Recent Researches in Solar Physics.* Camb. Univ. Press.

LEDOUX, P., and WALRAVEN, T. [1958] 'Variable stars', *Handbuch der Physik*, ed. s. FLÜGGE, vol. 5¹, p. 353. Berlin, Springer.

LEHNERT, B. [1959] 'Plasma physics on cosmical and laboratory scale', *Nuovo Cimento Supp.* **13,** 59.

MALKUS, W. V. R. [1959] 'Magnetoconvection in a viscous fluid of infinite electrical conductivity', *Astrophys. J.* **130,** 259.

MESTEL, L. [1962] 'Rotation and magnetism in stellar evolution', *Symposium on Stellar Evolution*, p. 251. Argentina, National Univ. of La Plata.

MURGAI, M. P., and KHOSLA, P. K. [1962] 'A study of the combined effect of thermal radiative transfer and a magnetic field on the gravitational convection of an ionized fluid', *J. Fluid Mech.* **14,** 433.

PARKER, E. N. [1955a] 'The formation of sunspots from the solar toroidal field', *Astrophys. J.* **121,** 491.

PARKER, E. N. [1955b] 'Hydromagnetic dynamo models', *Astrophys. J.* **122,** 293.

RAKOS, K. D. [1962] 'Photoelectric investigation of magnetic and spectrum variable stars', *Lowell Obs. Bull.* **5,** 227.

RUBASHEV, B. M. [1958] 'Some problems of subphotospheric stratification and circulation on the Sun', *Izv. Astron. Obs. Pulkovo* **21,** No. 162, p. 39.

SCHATZMAN, E. [1961] 'Modèles de taches solaires et protubérances de bord de taches', *Ann. Astrophys.* **24,** 251.

SCHATZMAN, E. [1962] 'A theory of the role of magnetic activity during star formation', *Ann. Astrophys.* **25,** 18.

SCHLÜTER, A., and TEMESVÁRY, S. [1958] 'The internal constitution of sunspots', *Electromagnetic Phenomena in Cosmical Physics*, ed. B. LEHNERT, p. 263. Camb. Univ. Press.

SCHWARZSCHILD, M. [1958] *Structure and Evolution of the Stars.* Princeton Univ. Press.

SPITZER, L. [1956] *Physics of Fully Ionized Gases.* New York, Interscience.

STEPANOV, V. E. [1949] 'On the problem of the electromagnetic nature of sunspots', *Sci. Trans. Lvov State Univ.: Series Phys.-Math.* **15,** 45.

STEPANOV, V. E. [1961] 'On motions in different levels of the solar atmosphere', *Izv. Crim. Astrophys. Obs.* **25,** 154.

SWEET, P. A. [1955] 'The structure of sunspots', *Vistas in Astronomy*, ed. A. BEER, vol. 1, p. 675. London, Pergamon.

SWEET, P. A. [1961] 'A survey of stellar magnetic problems and of the mixing of material in stars', *Proc. Roy. Soc.* **260,** 160.

THOMPSON, W. B. [1951] 'Thermal convection in a magnetic field', *Phil. Mag.* **42,** 1417.

UNSÖLD, A. [1955] *Physik der Sternatmosphären*, 2nd ed. Berlin, Springer.

WALDMEIER, M. [1955] *Ergebnisse und Probleme der Sonnenforschung*, 2nd ed. Leipzig, Geest u. Portig.

WALDMEIER, M. [1960] 'Zirkulation und Magnetfeld der solaren Polarzone', *Z. Astrophys.* **49,** 176.

WALÉN, C. [1944] 'On the theory of sun-spots', *Ark. Mat. Astron. Fys.* **30A,** No. 15.

WENTZEL, D. G. [1961] 'On the shape of magnetic stars', *Astrophys. J.* **133,** 170.

WESTFOLD, K. C. [1953] 'Collisional effects and the conduction current in an ionized gas', *Phil. Mag.* **44,** 712.

INDEXES

Name Index

Subject Index

A CATALOGUE OF SELECTED DOVER BOOKS
IN ALL FIELDS OF INTEREST

A CATALOGUE OF SELECTED DOVER BOOKS IN ALL FIELDS OF INTEREST

CELESTIAL OBJECTS FOR COMMON TELESCOPES, T. W. Webb. The most used book in amateur astronomy: inestimable aid for locating and identifying nearly 4,000 celestial objects. Edited, updated by Margaret W. Mayall. 77 illustrations. Total of 645pp. 5⅜ x 8½.
20917-2, 20918-0 Pa., Two-vol. set $8.00

HISTORICAL STUDIES IN THE LANGUAGE OF CHEMISTRY, M. P. Crosland. The important part language has played in the development of chemistry from the symbolism of alchemy to the adoption of systematic nomenclature in 1892. ". . . wholeheartedly recommended,"—Science. 15 illustrations. 416pp. of text. 5⅝ x 8¼. 63702-6 Pa. $6.00

BURNHAM'S CELESTIAL HANDBOOK, Robert Burnham, Jr. Thorough, readable guide to the stars beyond our solar system. Exhaustive treatment, fully illustrated. Breakdown is alphabetical by constellation: Andromeda to Cetus in Vol. 1; Chamaeleon to Orion in Vol. 2; and Pavo to Vulpecula in Vol. 3. Hundreds of illustrations. Total of about 2000pp. 6⅛ x 9¼.
23567-X, 23568-8, 23673-0 Pa., Three-vol. set $26.85

THEORY OF WING SECTIONS: INCLUDING A SUMMARY OF AIR-FOIL DATA, Ira H. Abbott and A. E. von Doenhoff. Concise compilation of subatomic aerodynamic characteristics of modern NASA wing sections, plus description of theory. 350pp. of tables. 693pp. 5⅜ x 8½.
60586-8 Pa. $6.50

DE RE METALLICA, Georgius Agricola. Translated by Herbert C. Hoover and Lou H. Hoover. The famous Hoover translation of greatest treatise on technological chemistry, engineering, geology, mining of early modern times (1556). All 289 original woodcuts. 638pp. 6¾ x 11.
60006-8 Clothbd. $17.50

THE ORIGIN OF CONTINENTS AND OCEANS, Alfred Wegener. One of the most influential, most controversial books in science, the classic statement for continental drift. Full 1966 translation of Wegener's final (1929) version. 64 illustrations. 246pp. 5⅜ x 8½. 61708-4 Pa. $3.00

THE PRINCIPLES OF PSYCHOLOGY, William James. Famous long course complete, unabridged. Stream of thought, time perception, memory, experimental methods; great work decades ahead of its time. Still valid, useful; read in many classes. 94 figures. Total of 1391pp. 5⅜ x 8½.
20381-6, 20382-4 Pa., Two-vol. set $13.00

THE PHILOSOPHY OF HISTORY, Georg W. Hegel. Great classic of Western thought develops concept that history is not chance but a rational process, the evolution of freedom. 457pp. 5⅜ x 8½.   20112-0 Pa. $4.50

LANGUAGE, TRUTH AND LOGIC, Alfred J. Ayer. Famous, clear introduction to Vienna, Cambridge schools of Logical Positivism. Role of philosophy, elimination of metaphysics, nature of analysis, etc. 160pp. 5⅜ x 8½. (Available in U.S. only)   20010-8 Pa. $1.75

A PREFACE TO LOGIC, Morris R. Cohen. Great City College teacher in renowned, easily followed exposition of formal logic, probability, values, logic and world order and similar topics; no previous background needed. 209pp. 5⅜ x 8½.   23517-3 Pa. $3.50

REASON AND NATURE, Morris R. Cohen. Brilliant analysis of reason and its multitudinous ramifications by charismatic teacher. Interdisciplinary, synthesizing work widely praised when it first appeared in 1931. Second (1953) edition. Indexes. 496pp. 5⅜ x 8½.   23633-1 Pa. $6.00

AN ESSAY CONCERNING HUMAN UNDERSTANDING, John Locke. The only complete edition of enormously important classic, with authoritative editorial material by A. C. Fraser. Total of 1176pp. 5⅜ x 8½.
20530-4, 20531-2 Pa., Two-vol. set $14.00

HANDBOOK OF MATHEMATICAL FUNCTIONS WITH FORMULAS, GRAPHS, AND MATHEMATICAL TABLES, edited by Milton Abramowitz and Irene A. Stegun. Vast compendium: 29 sets of tables, some to as high as 20 places. 1,046pp. 8 x 10½.   61272-4 Pa. $12.50

MATHEMATICS FOR THE PHYSICAL SCIENCES, Herbert S. Wilf. Highly acclaimed work offers clear presentations of vector spaces and matrices, orthogonal functions, roots of polynomial equations, conformal mapping, calculus of variations, etc. Knowledge of theory of functions of real and complex variables is assumed. Exercises and solutions. Index. 284pp. 5⅝ x 8¼.   63635-6 Pa. $4.50

THE PRINCIPLE OF RELATIVITY, Albert Einstein et al. Eleven most important original papers on special and general theories. Seven by Einstein, two by Lorentz, one each by Minkowski and Weyl. All translated, unabridged. 216pp. 5⅜ x 8½.   60081-5 Pa. $3.00

THERMODYNAMICS, Enrico Fermi. A classic of modern science. Clear, organized treatment of systems, first and second laws, entropy, thermodynamic potentials, gaseous reactions, dilute solutions, entropy constant. No math beyond calculus required. Problems. 160pp. 5⅜ x 8½.
60361-X Pa. $2.75

ELEMENTARY MECHANICS OF FLUIDS, Hunter Rouse. Classic undergraduate text widely considered to be far better than many later books. Ranges from fluid velocity and acceleration to role of compressibility in fluid motion. Numerous examples, questions, problems. 224 illustrations. 376pp. 5⅝ x 8¼.   63699-2 Pa. $5.00

TONE POEMS, SERIES II: TILL EULENSPIEGELS LUSTIGE STREICHE, ALSO SPRACH ZARATHUSTRA, AND EIN HELDENLEBEN, Richard Strauss. Three important orchestral works, including very popular *Till Eulenspiegel's Marry Pranks,* reproduced in full score from original editions. Study score. 315pp. 9⅜ x 12¼. (Available in U.S. only)
23755-9 Pa. $7.50

TONE POEMS, SERIES I: DON JUAN, TOD UND VERKLARUNG AND DON QUIXOTE, Richard Strauss. Three of the most often performed and recorded works in entire orchestral repertoire, reproduced in full score from original editions. Study score. 286pp. 9⅜ x 12¼. (Available in U.S. only)
23754-0 Pa. $7.50

11 LATE STRING QUARTETS, Franz Joseph Haydn. The form which Haydn defined and "brought to perfection." *(Grove's).* 11 string quartets in complete score, his last and his best. The first in a projected series of the complete Haydn string quartets. Reliable modern Eulenberg edition, otherwise difficult to obtain. 320pp. 8⅜ x 11¼. (Available in U.S. only)
23753-2 Pa. $6.95

FOURTH, FIFTH AND SIXTH SYMPHONIES IN FULL SCORE, Peter Ilyitch Tchaikovsky. Complete orchestral scores of Symphony No. 4 in F Minor, Op. 36; Symphony No. 5 in E Minor, Op. 64; Symphony No. 6 in B Minor, "Pathetique," Op. 74. Bretikopf & Hartel eds. Study score. 480pp. 9⅜ x 12¼. 23861-X Pa. $10.95

THE MARRIAGE OF FIGARO: COMPLETE SCORE, Wolfgang A. Mozart. Finest comic opera ever written. Full score, not to be confused with piano renderings. Peters edition. Study score. 448pp. 9⅜ x 12¼. (Available in U.S. only) 23751-6 Pa. $11.95

"IMAGE" ON THE ART AND EVOLUTION OF THE FILM, edited by Marshall Deutelbaum. Pioneering book brings together for first time 38 groundbreaking articles on early silent films from *Image* and 263 illustrations newly shot from rare prints in the collection of the International Museum of Photography. A landmark work. Index. 256pp. 8¼ x 11.
23777-X Pa. $8.95

AROUND-THE-WORLD COOKY BOOK, Lois Lintner Sumption and Marguerite Lintner Ashbrook. 373 cooky and frosting recipes from 28 countries (America, Austria, China, Russia, Italy, etc.) include Viennese kisses, rice wafers, London strips, lady fingers, hony, sugar spice, maple cookies, etc. Clear instructions. All tested. 38 drawings. 182pp. 5⅜ x 8.
23802-4 Pa. $2.50

THE ART NOUVEAU STYLE, edited by Roberta Waddell. 579 rare photographs, not available elsewhere, of works in jewelry, metalwork, glass, ceramics, textiles, architecture and furniture by 175 artists—Mucha, Seguy, Lalique, Tiffany, Gaudin, Hohlwein, Saarinen, and many others. 288pp. 8⅜ x 11¼. 23515-7 Pa. $6.95

HISTORY OF BACTERIOLOGY, William Bulloch. The only comprehensive history of bacteriology from the beginnings through the 19th century. Special emphasis is given to biography-Leeuwenhoek, etc. Brief accounts of 350 bacteriologists form a separate section. No clearer, fuller study, suitable to scientists and general readers, has yet been written. 52 illustrations. 448pp. 5⅝ x 8¼.   23761-3 Pa. $6.50

THE COMPLETE NONSENSE OF EDWARD LEAR, Edward Lear. All nonsense limericks, zany alphabets, Owl and Pussycat, songs, nonsense botany, etc., illustrated by Lear. Total of 321pp. 5⅜ x 8½. (Available in U.S. only)   20167-8 Pa. $3.00

INGENIOUS MATHEMATICAL PROBLEMS AND METHODS, Louis A. Graham. Sophisticated material from Graham *Dial*, applied and pure; stresses solution methods. Logic, number theory, networks, inversions, etc. 237pp. 5⅜ x 8½.   20545-2 Pa. $3.50

BEST MATHEMATICAL PUZZLES OF SAM LOYD, edited by Martin Gardner. Bizarre, original, whimsical puzzles by America's greatest puzzler. From fabulously rare *Cyclopedia*, including famous 14-15 puzzles, the Horse of a Different Color, 115 more. Elementary math. 150 illustrations. 167pp. 5⅜ x 8½.   20498-7 Pa. $2.50

THE BASIS OF COMBINATION IN CHESS, J. du Mont. Easy-to-follow, instructive book on elements of combination play, with chapters on each piece and every powerful combination team—two knights, bishop and knight, rook and bishop, etc. 250 diagrams. 218pp. 5⅜ x 8½. (Available in U.S. only)   23644-7 Pa. $3.50

MODERN CHESS STRATEGY, Ludek Pachman. The use of the queen, the active king, exchanges, pawn play, the center, weak squares, etc. Section on rook alone worth price of the book. Stress on the moderns. Often considered the most important book on strategy. 314pp. 5⅜ x 8½.   20290-9 Pa. $3.50

LASKER'S MANUAL OF CHESS, Dr. Emanuel Lasker. Great world champion offers very thorough coverage of all aspects of chess. Combinations, position play, openings, end game, aesthetics of chess, philosophy of struggle, much more. Filled with analyzed games. 390pp. 5⅜ x 8½.   20640-8 Pa. $4.00

500 MASTER GAMES OF CHESS, S. Tartakower, J. du Mont. Vast collection of great chess games from 1798-1938, with much material nowhere else readily available. Fully annotated, arranged by opening for easier study. 664pp. 5⅜ x 8½.   23208-5 Pa. $6.00

A GUIDE TO CHESS ENDINGS, Dr. Max Euwe, David Hooper. One of the finest modern works on chess endings. Thorough analysis of the most frequently encountered endings by former world champion. 331 examples, each with diagram. 248pp. 5⅜ x 8½.   23332-4 Pa. $3.50

"OSCAR" OF THE WALDORF'S COOKBOOK, Oscar Tschirky. Famous American chef reveals 3455 recipes that made Waldorf great; cream of French, German, American cooking, in all categories. Full instructions, easy home use. 1896 edition. 907pp. 6⅝ x 9⅜.   20790-0 Clothbd. $15.00

COOKING WITH BEER, Carole Fahy. Beer has as superb an effect on food as wine, and at fraction of cost. Over 250 recipes for appetizers, soups, main dishes, desserts, breads, etc. Index. 144pp. 5⅜ x 8½. (Available in U.S. only)   23661-7 Pa. $2.50

STEWS AND RAGOUTS, Kay Shaw Nelson. This international cookbook offers wide range of 108 recipes perfect for everyday, special occasions, meals-in-themselves, main dishes. Economical, nutritious, easy-to-prepare: goulash, Irish stew, boeuf bourguignon, etc. Index. 134pp. 5⅜ x 8½.
23662-5 Pa. $2.50

DELICIOUS MAIN COURSE DISHES, Marian Tracy. Main courses are the most important part of any meal. These 200 nutritious, economical recipes from around the world make every meal a delight. "I . . . have found it so useful in my own household,"—N.Y. Times. Index. 219pp. 5⅜ x 8½.   23664-1 Pa. $3.00

FIVE ACRES AND INDEPENDENCE, Maurice G. Kains. Great back-to-the-land classic explains basics of self-sufficient farming: economics, plants, crops, animals, orchards, soils, land selection, host of other necessary things. Do not confuse with skimpy faddist literature; Kains was one of America's greatest agriculturalists. 95 illustrations. 397pp. 5⅜ x 8½.
20974-1 Pa. $3.50

A PRACTICAL GUIDE FOR THE BEGINNING FARMER, Herbert Jacobs. Basic, extremely useful first book for anyone thinking about moving to the country and starting a farm. Simpler than Kains, with greater emphasis on country living in general. 246pp. 5⅜ x 8½.
23675-7 Pa. $3.50

HARDY BULBS, Louise Beebe Wilder. Fullest, most thorough book on plants grown from bulbs, corms, rhizomes and tubers. 40 genera and 335 species covered: selecting, cultivating, naturalizing; name, origins, blooming season, when to plant, special requirements. 127 illustrations. 432pp. 5⅜ x 8½.   23102-X Pa. $4.50

A GARDEN OF PLEASANT FLOWERS (PARADISI IN SOLE: PARADISUS TERRESTRIS), John Parkinson. Complete, unabridged reprint of first (1629) edition of earliest great English book on gardens and gardening. More than 1000 plants & flowers of Elizabethan, Jacobean garden fully described, most with woodcut illustrations. Botanically very reliable, a "speaking garden" of exceeding charm. 812 illustrations. 628pp. 8½ x 12¼.   23392-8 Clothbd. $25.00

THE COMPLETE WOODCUTS OF ALBRECHT DURER, edited by Dr. W. Kurth. 346 in all: "Old Testament," "St. Jerome," "Passion," "Life of Virgin," Apocalypse," many others. Introduction by Campbell Dodgson. 285pp. 8½ x 12¼. 21097-9 Pa. $6.95

DRAWINGS OF ALBRECHT DURER, edited by Heinrich Wolfflin. 81 plates show development from youth to full style. Many favorites; many new. Introduction by Alfred Werner. 96pp. 8⅛ x 11. 22352-3 Pa. $4.00

THE HUMAN FIGURE, Albrecht Dürer. Experiments in various techniques—stereometric, progressive proportional, and others. Also life studies that rank among finest ever done. Complete reprinting of *Dresden Sketchbook*. 170 plates. 355pp. 8⅜ x 11¼. 21042-1 Pa. $6.95

OF THE JUST SHAPING OF LETTERS, Albrecht Dürer. Renaissance artist explains design of Roman majuscules by geometry, also Gothic lower and capitals. Grolier Club edition. 43pp. 7⅞ x 10¾ 21306-4 Pa. $2.50

TEN BOOKS ON ARCHITECTURE, Vitruvius. The most important book ever written on architecture. Early Roman aesthetics, technology, classical orders, site selection, all other aspects. Stands behind everything since. Morgan translation. 331pp. 5⅜ x 8½. 20645-9 Pa. $3.75

THE FOUR BOOKS OF ARCHITECTURE, Andrea Palladio. 16th-century classic responsible for Palladian movement and style. Covers classical architectural remains, Renaissance revivals, classical orders, etc. 1738 Ware English edition. Introduction by A. Placzek. 216 plates. 110pp. of text. 9½ x 12¾. 21308-0 Pa. $7.50

HORIZONS, Norman Bel Geddes. Great industrialist stage designer, "father of streamlining," on application of aesthetics to transportation, amusement, architecture, etc. 1932 prophetic account; function, theory, specific projects. 222 illustrations. 312pp. 7⅞ x 10¾. 23514-9 Pa. $6.95

FRANK LLOYD WRIGHT'S FALLINGWATER, Donald Hoffmann. Full, illustrated story of conception and building of Wright's masterwork at Bear Run, Pa. 100 photographs of site, construction, and details of completed structure. 112pp. 9¼ x 10. 23671-4 Pa. $5.00

THE ELEMENTS OF DRAWING, John Ruskin. Timeless classic by great Viltorian; starts with basic ideas, works through more difficult. Many practical exercises. 48 illustrations. Introduction by Lawrence Campbell. 228pp. 5⅜ x 8½. 22730-8 Pa. $2.75

GIST OF ART, John Sloan. Greatest modern American teacher, Art Students League, offers innumerable hints, instructions, guided comments to help you in painting. Not a formal course. 46 illustrations. Introduction by Helen Sloan. 200pp. 5⅜ x 8½. 23435-5 Pa. $3.50

THE EARLY WORK OF AUBREY BEARDSLEY, Aubrey Beardsley. 157 plates, 2 in color: *Manon Lescaut, Madame Bovary, Morte Darthur, Salome,* other. Introduction by H. Marillier. 182pp. 8⅛ x 11. 21816-3 Pa. $4.50

THE LATER WORK OF AUBREY BEARDSLEY, Aubrey Beardsley. Exotic masterpieces of full maturity: *Venus and Tannhauser, Lysistrata, Rape of the Lock, Volpone,* Savoy material, etc. 174 plates, 2 in color. 186pp. 8⅛ x 11. 21817-1 Pa. $4.50

THOMAS NAST'S CHRISTMAS DRAWINGS, Thomas Nast. Almost all Christmas drawings by creator of image of Santa Claus as we know it, and one of America's foremost illustrators and political cartoonists. 66 illustrations. 3 illustrations in color on covers. 96pp. 8⅜ x 11¼. 23660-9 Pa. $3.50

THE DORÉ ILLUSTRATIONS FOR DANTE'S DIVINE COMEDY, Gustave Doré. All 135 plates from Inferno, Purgatory, Paradise; fantastic tortures, infernal landscapes, celestial wonders. Each plate with appropriate (translated) verses. 141pp. 9 x 12. 23231-X Pa. $4.50

DORÉ'S ILLUSTRATIONS FOR RABELAIS, Gustave Doré. 252 striking illustrations of *Gargantua and Pantagruel* books by foremost 19th-century illustrator. Including 60 plates, 192 delightful smaller illustrations. 153pp. 9 x 12. 23656-0 Pa. $5.00

LONDON: A PILGRIMAGE, Gustave Doré, Blanchard Jerrold. Squalor, riches, misery, beauty of mid-Victorian metropolis; 55 wonderful plates, 125 other illustrations, full social, cultural text by Jerrold. 191pp. of text. 9⅜ x 12¼. 22306-X Pa. $6.00

THE RIME OF THE ANCIENT MARINER, Gustave Doré, S. T. Coleridge. Dore's finest work, 34 plates capture moods, subtleties of poem. Full text. Introduction by Millicent Rose. 77pp. 9¼ x 12. 22305-1 Pa. $3.00

THE DORE BIBLE ILLUSTRATIONS, Gustave Doré. All wonderful, detailed plates: Adam and Eve, Flood, Babylon, Life of Jesus, etc. Brief King James text with each plate. Introduction by Millicent Rose. 241 plates. 241pp. 9 x 12. 23004-X Pa. $5.00

THE COMPLETE ENGRAVINGS, ETCHINGS AND DRYPOINTS OF ALBRECHT DURER. "Knight, Death and Devil"; "Melencolia," and more—all Dürer's known works in all three media, including 6 works formerly attributed to him. 120 plates. 235pp. 8⅜ x 11¼. 22851-7 Pa. $6.50

MAXIMILIAN'S TRIUMPHAL ARCH, Albrecht Dürer and others. Incredible monument of woodcut art: 8 foot high elaborate arch—heraldic figures, humans, battle scenes, fantastic elements—that you can assemble yourself. Printed on one side, layout for assembly. 143pp. 11 x 16. 21451-6 Pa. $5.00

DRAWINGS OF WILLIAM BLAKE, William Blake. 92 plates from Book of Job, *Divine Comedy, Paradise Lost,* visionary heads, mythological figures, Laocoon, etc. Selection, introduction, commentary by Sir Geoffrey Keynes. 178pp. 8⅛ x 11. 22303-5 Pa. $4.00

ENGRAVINGS OF HOGARTH, William Hogarth. 101 of Hogarth's greatest works: *Rake's Progress, Harlot's Progress, Illustrations for Hudibras, Before and After, Beer Street and Gin Lane,* many more. Full commentary. 256pp. 11 x 13¾. 22479-1 Pa. $7.95

DAUMIER: 120 GREAT LITHOGRAPHS, Honore Daumier. Wide-ranging collection of lithographs by the greatest caricaturist of the 19th century. Concentrates on eternally popular series on lawyers, on married life, on liberated women, etc. Selection, introduction, and notes on plates by Charles F. Ramus. Total of 158pp. 9⅜ x 12¼. 23512-2 Pa. $5.50

DRAWINGS OF MUCHA, Alphonse Maria Mucha. Work reveals drafts-man of highest caliber: studies for famous posters and paintings, render-ings for book illustrations and ads, etc. 70 works, 9 in color; including 6 items not drawings. Introduction. List of illustrations. 72pp. 9⅜ x 12¼. (Available in U.S. only) 23672-2 Pa. $4.00

GIOVANNI BATTISTA PIRANESI: DRAWINGS IN THE PIERPONT MORGAN LIBRARY, Giovanni Battista Piranesi. For first time ever all of Morgan Library's collection, world's largest. 167 illustrations of rare Piranesi drawings—archeological, architectural, decorative and visionary. Essay, detailed list of drawings, chronology, captions. Edited by Felice Stampfle. 144pp. 9⅜ x 12¼. 23714-1 Pa. $7.50

NEW YORK ETCHINGS (1905-1949), John Sloan. All of important American artist's N.Y. life etchings. 67 works include some of his best art; also lively historical record—Greenwich Village, tenement scenes. Edited by Sloan's widow. Introduction and captions. 79pp. 8⅜ x 11¼. 23651-X Pa. $4.00

CHINESE PAINTING AND CALLIGRAPHY: A PICTORIAL SURVEY, Wan-go Weng. 69 fine examples from John M. Crawford's matchless private collection: landscapes, birds, flowers, human figures. etc., plus calligraphy. Every basic form included: hanging scrolls, handscrolls, album leaves, fans, etc. 109 illustrations. Introduction. Captions. 192pp. 8⅞ x 11¾. 23707-9 Pa. $7.95

DRAWINGS OF REMBRANDT, edited by Seymour Slive. Updated Lipp-mann, Hofstede de Groot edition, with definitive scholarly apparatus. All portraits, biblical sketches, landscapes, nudes, Oriental figures, classical studies, together with selection of work by followers. 550 illustrations. Total of 630pp. 9⅛ x 12¼. 21485-0, 21486-9 Pa., Two-vol. set $14.00

THE DISASTERS OF WAR, Francisco Goya. 83 etchings record horrors of Napoleonic wars in Spain and war in general. Reprint of 1st edition, plus 3 additional plates. Introduction by Philip Hofer. 97pp. 9⅜ x 8¼. 21872-4 Pa. $3.75

THE ANATOMY OF THE HORSE, George Stubbs. Often considered the great masterpiece of animal anatomy. Full reproduction of 1766 edition, plus prospectus; original text and modernized text. 36 plates. Introduction by Eleanor Garvey. 121pp. 11 x 14¾.   23402-9 Pa. $6.00

BRIDGMAN'S LIFE DRAWING, George B. Bridgman. More than 500 illustrative drawings and text teach you to abstract the body into its major masses, use light and shade, proportion; as well as specific areas of anatomy, of which Bridgman is master. 192pp. 6½ x 9¼. (Available in U.S. only)
22710-3 Pa. $2.50

ART NOUVEAU DESIGNS IN COLOR, Alphonse Mucha, Maurice Verneuil, Georges Auriol. Full-color reproduction of *Combinaisons ornementales* (c. 1900) by Art Nouveau masters. Floral, animal, geometric, interlacings, swashes—borders, frames, spots—all incredibly beautiful. 60 plates, hundreds of designs. 9⅜ x 8-1/16.   22885-1 Pa. $4.00

FULL-COLOR FLORAL DESIGNS IN THE ART NOUVEAU STYLE, E. A. Seguy. 166 motifs, on 40 plates, from *Les fleurs et leurs applications decoratives* (1902): borders, circular designs, repeats, allovers, "spots." All in authentic Art Nouveau colors. 48pp. 9⅜ x 12¼.
23439-8 Pa. $5.00

A DIDEROT PICTORIAL ENCYCLOPEDIA OF TRADES AND IN-DUSTRY, edited by Charles C. Gillispie. 485 most interesting plates from the great French Encyclopedia of the 18th century show hundreds of working figures, artifacts, process, land and cityscapes; glassmaking, paper-making, metal extraction, construction, weaving, making furniture, clothing, wigs, dozens of other activities. Plates fully explained. 920pp. 9 x 12.
22284-5, 22285-3 Clothbd., Two-vol. set $40.00

HANDBOOK OF EARLY ADVERTISING ART, Clarence P. Hornung. Largest collection of copyright-free early and antique advertising art ever compiled. Over 6,000 illustrations, from Franklin's time to the 1890's for special effects, novelty. Valuable source, almost inexhaustible.
Pictorial Volume. Agriculture, the zodiac, animals, autos, birds, Christmas, fire engines, flowers, trees, musical instruments, ships, games and sports, much more. Arranged by subject matter and use. 237 plates. 288pp. 9 x 12.
20122-8 Clothbd. $13.50

Typographical Volume. Roman and Gothic faces ranging from 10 point to 300 point, "Barnum" German and Old English faces, script, logotypes, scrolls and flourishes, 1115 ornamental initials, 67 complete alphabets, more. 310 plates. 320pp. 9 x 12.   20123-6 Clothbd. $13.50

CALLIGRAPHY (CALLIGRAPHIA LATINA), J. G. Schwandner. High point of 18th-century ornamental calligraphy. Very ornate initials, scrolls, borders, cherubs, birds, lettered examples. 172pp. 9 x 13.
20475-8 Pa. $6.00

ART FORMS IN NATURE, Ernst Haeckel. Multitude of strangely beautiful natural forms: Radiolaria, Foraminifera, jellyfishes, fungi, turtles, bats, etc. All 100 plates of the 19th-century evolutionist's *Kunstformen der Natur* (1904). 100pp. 9⅜ x 12¼. 22987-4 Pa. $4.50

CHILDREN: A PICTORIAL ARCHIVE FROM NINETEENTH-CENTURY SOURCES, edited by Carol Belanger Grafton. 242 rare, copyright-free wood engravings for artists and designers. Widest such selection available. All illustrations in line. 119pp. 8⅜ x 11¼.
23694-3 Pa. $3.50

WOMEN: A PICTORIAL ARCHIVE FROM NINETEENTH-CENTURY SOURCES, edited by Jim Harter. 391 copyright-free wood engravings for artists and designers selected from rare periodicals. Most extensive such collection available. All illustrations in line. 128pp. 9 x 12.
23703-6 Pa. $4.00

ARABIC ART IN COLOR, Prisse d'Avennes. From the greatest ornamentalists of all time—50 plates in color, rarely seen outside the Near East, rich in suggestion and stimulus. Includes 4 plates on covers. 46pp. 9⅜ x 12¼. 23658-7 Pa. $6.00

AUTHENTIC ALGERIAN CARPET DESIGNS AND MOTIFS, edited by June Beveridge. Algerian carpets are world famous. Dozens of geometrical motifs are charted on grids, color-coded, for weavers, needleworkers, craftsmen, designers. 53 illustrations plus 4 in color. 48pp. 8¼ x 11. (Available in U.S. only) 23650-1 Pa. $1.75

DICTIONARY OF AMERICAN PORTRAITS, edited by Hayward and Blanche Cirker. 4000 important Americans, earliest times to 1905, mostly in clear line. Politicians, writers, soldiers, scientists, inventors, industrialists, Indians, Blacks, women, outlaws, etc. Identificatory information. 756pp. 9¼ x 12¾. 21823-6 Clothbd. $40.00

HOW THE OTHER HALF LIVES, Jacob A. Riis. Journalistic record of filth, degradation, upward drive in New York immigrant slums, shops, around 1900. New edition includes 100 original Riis photos, monuments of early photography. 233pp. 10 x 7⅞. 22012-5 Pa. $6.00

NEW YORK IN THE THIRTIES, Berenice Abbott. Noted photographer's fascinating study of city shows new buildings that have become famous and old sights that have disappeared forever. Insightful commentary. 97 photographs. 97pp. 11⅜ x 10. 22967-X Pa. $4.50

MEN AT WORK, Lewis W. Hine. Famous photographic studies of construction workers, railroad men, factory workers and coal miners. New supplement of 18 photos on Empire State building construction. New introduction by Jonathan L. Doherty. Total of 69 photos. 63pp. 8 x 10¾.
23475-4 Pa. $3.00

A MAYA GRAMMAR, Alfred M. Tozzer. Practical, useful English-language grammar by the Harvard anthropologist who was one of the three greatest American scholars in the area of Maya culture. Phonetics, grammatical processes, syntax, more. 301pp. 5⅜ x 8½.   23465-7 Pa. $4.00

THE JOURNAL OF HENRY D. THOREAU, edited by Bradford Torrey, F. H. Allen. Complete reprinting of 14 volumes, 1837-61, over two million words; the sourcebooks for *Walden*, etc. Definitive. All original sketches, plus 75 photographs. Introduction by Walter Harding. Total of 1804pp. 8½ x 12¼.   20312-3, 20313-1 Clothbd., Two-vol. set $50.00

CLASSIC GHOST STORIES, Charles Dickens and others. 18 wonderful stories you've wanted to reread: "The Monkey's Paw," "The House and the Brain," "The Upper Berth," "The Signalman," "Dracula's Guest," "The Tapestried Chamber," etc. Dickens, Scott, Mary Shelley, Stoker, etc. 330pp. 5⅜ x 8½.   20735-8 Pa. $3.50

SEVEN SCIENCE FICTION NOVELS, H. G. Wells. Full novels. *First Men in the Moon, Island of Dr. Moreau, War of the Worlds, Food of the Gods, Invisible Man, Time Machine, In the Days of the Comet.* A basic science-fiction library. 1015pp. 5⅜ x 8½. (Available in U.S. only)
                                                   20264-X Clothbd. $8.95

ARMADALE, Wilkie Collins. Third great mystery novel by the author of *The Woman in White* and *The Moonstone*. Ingeniously plotted narrative shows an exceptional command of character, incident and mood. Original magazine version with 40 illustrations. 597pp. 5⅜ x 8½.
                                                   23429-0 Pa. $5.00

MASTERS OF MYSTERY, H. Douglas Thomson. The first book in English (1931) devoted to history and aesthetics of detective story. Poe, Doyle, LeFanu, Dickens, many others, up to 1930. New introduction and notes by E. F. Bleiler. 288pp. 5⅜ x 8½. (Available in U.S. only)
                                                   23606-4 Pa. $4.00

FLATLAND, E. A. Abbott. Science-fiction classic explores life of 2-D being in 3-D world. Read also as introduction to thought about hyperspace. Introduction by Banesh Hoffmann. 16 illustrations. 103pp. 5⅜ x 8½.
                                                   20001-9 Pa. $1.50

THREE SUPERNATURAL NOVELS OF THE VICTORIAN PERIOD, edited, with an introduction, by E. F. Bleiler. Reprinted complete and unabridged, three great classics of the supernatural: *The Haunted Hotel* by Wilkie Collins, *The Haunted House at Latchford* by Mrs. J. H. Riddell, and *The Lost Stradivarius* by J. Meade Falkner. 325pp. 5⅜ x 8½.
                                                   22571-2 Pa. $4.00

AYESHA: THE RETURN OF "SHE," H. Rider Haggard. Virtuoso sequel featuring the great mythic creation, Ayesha, in an adventure that is fully as good as the first book, *She*. Original magazine version, with 47 original illustrations by Maurice Greiffenhagen. 189pp. 6½ x 9¼.
                                                   23649-8 Pa. $3.00

AN AUTOBIOGRAPHY, Margaret Sanger. Exciting personal account of hard-fought battle for woman's right to birth control, against prejudice, church, law. Foremost feminist document. 504pp. 5⅜ x 8½.
20470-7 Pa. $5.50

MY BONDAGE AND MY FREEDOM, Frederick Douglass. Born as a slave, Douglass became outspoken force in antislavery movement. The best of Douglass's autobiographies. Graphic description of slave life. Introduction by P. Foner. 464pp. 5⅜ x 8½. 22457-0 Pa. $5.00

LIVING MY LIFE, Emma Goldman. Candid, no holds barred account by foremost American anarchist: her own life, anarchist movement, famous contemporaries, ideas and their impact. Struggles and confrontations in America, plus deportation to U.S.S.R. Shocking inside account of persecution of anarchists under Lenin. 13 plates. Total of 944pp. 5⅜ x 8½.
22543-7, 22544-5 Pa., Two-vol. set $9.00

LETTERS AND NOTES ON THE MANNERS, CUSTOMS AND CONDITIONS OF THE NORTH AMERICAN INDIANS, George Catlin. Classic account of life among Plains Indians: ceremonies, hunt, warfare, etc. Dover edition reproduces for first time all original paintings. 312 plates. 572pp. of text. 6⅛ x 9¼. 22118-0, 22119-9 Pa.. Two-vol. set $10.00

THE MAYA AND THEIR NEIGHBORS, edited by Clarence L. Hay, others. Synoptic view of Maya civilization in broadest sense, together with Northern, Southern neighbors. Integrates much background, valuable detail not elsewhere. Prepared by greatest scholars: Kroeber, Morley, Thompson, Spinden, Vaillant, many others. Sometimes called Tozzer Memorial Volume. 60 illustrations, linguistic map. 634pp. 5⅜ x 8½.
23510-6 Pa. $7.50

HANDBOOK OF THE INDIANS OF CALIFORNIA, A. L. Kroeber. Foremost American anthropologist offers complete ethnographic study of each group. Monumental classic. 459 illustrations, maps. 995pp. 5⅜ x 8½.
23368-5 Pa. $10.00

SHAKTI AND SHAKTA, Arthur Avalon. First book to give clear, cohesive analysis of Shakta doctrine, Shakta ritual and Kundalini Shakti (yoga). Important work by one of world's foremost students of Shaktic and Tantric thought. 732pp. 5⅜ x 8½. (Available in U.S. only)
23645-5 Pa. $7.95

AN INTRODUCTION TO THE STUDY OF THE MAYA HIEROGLYPHS, Syvanus Griswold Morley. Classic study by one of the truly great figures in hieroglyph research. Still the best introduction for the student for reading Maya hieroglyphs. New introduction by J. Eric S. Thompson. 117 illustrations. 284pp. 5⅜ x 8½. 23108-9 Pa. $4.00

A STUDY OF MAYA ART, Herbert J. Spinden. Landmark classic interprets Maya symbolism, estimates styles, covers ceramics, architecture, murals, stone carvings as artforms. Still a basic book in area. New introduction by J. Eric Thompson. Over 750 illustrations. 341pp. 8⅜ x 11¼.
21235-1 Pa. $6.95

HOLLYWOOD GLAMOUR PORTRAITS, edited by John Kobal. 145 photos capture the stars from 1926-49, the high point in portrait photography. Gable, Harlow, Bogart, Bacall, Hedy Lamarr, Marlene Dietrich, Robert Montgomery, Marlon Brando, Veronica Lake; 94 stars in all. Full background on photographers, technical aspects, much more. Total of 160pp. 8⅜ x 11¼. 23352-9 Pa. $5.00

THE NEW YORK STAGE: FAMOUS PRODUCTIONS IN PHOTOGRAPHS, edited by Stanley Appelbaum. 148 photographs from Museum of City of New York show 142 plays, 1883-1939. *Peter Pan, The Front Page, Dead End, Our Town,* O'Neill, hundreds of actors and actresses, etc. Full indexes. 154pp. 9½ x 10. 23241-7 Pa. $4.50

MASTERS OF THE DRAMA, John Gassner. Most comprehensive history of the drama, every tradition from Greeks to modern Europe and America, including Orient. Covers 800 dramatists, 2000 plays; biography, plot summaries, criticism, theatre history, etc. 77 illustrations. 890pp. 5⅜ x 8½. 20100-7 Clothbd. $10.00

THE GREAT OPERA STARS IN HISTORIC PHOTOGRAPHS, edited by James Camner. 343 portraits from the 1850s to the 1940s: Tamburini, Mario, Caliapin, Jeritza, Melchior, Melba, Patti, Pinza, Schipa, Caruso, Farrar, Steber, Gobbi, and many more—270 performers in all. Index. 199pp. 8⅜ x 11¼. 23575-0 Pa. $6.50

J. S. BACH, Albert Schweitzer. Great full-length study of Bach, life, background to music, music, by foremost modern scholar. Ernest Newman translation. 650 musical examples. Total of 928pp. 5⅜ x 8½. (Available in U.S. only) 21631-4, 21632-2 Pa., Two-vol. set $9.00

COMPLETE PIANO SONATAS, Ludwig van Beethoven. All sonatas in the fine Schenker edition, with fingering, analytical material. One of best modern editions. Total of 615pp. 9 x 12. (Available in U.S. only) 23134-8, 23135-6 Pa., Two-vol. set $13.00

KEYBOARD MUSIC, J. S. Bach. Bach-Gesellschaft edition. For harpsichord, piano, other keyboard instruments. English Suites, French Suites, Six Partitas, Goldberg Variations, Two-Part Inventions, Three-Part Sinfonias. 312pp. 8⅛ x 11. (Available in U.S. only) 22360-4 Pa. $5.50

FOUR SYMPHONIES IN FULL SCORE, Franz Schubert. Schubert's four most popular symphonies: No. 4 in C Minor ("Tragic"); No. 5 in B-flat Major; No. 8 in B Minor ("Unfinished"); No. 9 in C Major ("Great"). Breitkopf & Hartel edition. Study score. 261pp. 9⅜ x 12¼. 23681-1 Pa. $6.50

THE AUTHENTIC GILBERT & SULLIVAN SONGBOOK, W. S. Gilbert, A. S. Sullivan. Largest selection available; 92 songs, uncut, original keys, in piano rendering approved by Sullivan. Favorites and lesser-known fine numbers. Edited with plot synopses by James Spero. 3 illustrations. 399pp. 9 x 12. 23482-7 Pa. $7.95

THE DEPRESSION YEARS AS PHOTOGRAPHED BY ARTHUR ROTH-STEIN, Arthur Rothstein. First collection devoted entirely to the work of outstanding 1930s photographer: famous dust storm photo, ragged children, unemployed, etc. 120 photographs. Captions. 119pp. 9¼ x 10¾.
23590-4 Pa. $5.00

CAMERA WORK: A PICTORIAL GUIDE, Alfred Stieglitz. All 559 illustrations and plates from the most important periodical in the history of art photography, Camera Work (1903-17). Presented four to a page, reduced in size but still clear, in strict chronological order, with complete captions. Three indexes. Glossary. Bibliography. 176pp. 8⅜ x 11¼.
23591-2 Pa. $6.95

ALVIN LANGDON COBURN, PHOTOGRAPHER, Alvin L. Coburn. Revealing autobiography by one of greatest photographers of 20th century gives insider's version of Photo-Secession, plus comments on his own work. 77 photographs by Coburn. Edited by Helmut and Alison Gernsheim. 160pp. 8⅛ x 11.
23685-4 Pa. $6.00

NEW YORK IN THE FORTIES, Andreas Feininger. 162 brilliant photographs by the well-known photographer, formerly with Life magazine, show commuters, shoppers, Times Square at night, Harlem nightclub, Lower East Side, etc. Introduction and full captions by John von Hartz. 181pp. 9¼ x 10¾.
23585-8 Pa. $6.00

GREAT NEWS PHOTOS AND THE STORIES BEHIND THEM, John Faber. Dramatic volume of 140 great news photos, 1855 through 1976, and revealing stories behind them, with both historical and technical information. Hindenburg disaster, shooting of Oswald, nomination of Jimmy Carter, etc. 160pp. 8¼ x 11.
23667-6 Pa. $5.00

THE ART OF THE CINEMATOGRAPHER, Leonard Maltin. Survey of American cinematography history and anecdotal interviews with 5 masters—Arthur Miller, Hal Mohr, Hal Rosson, Lucien Ballard, and Conrad Hall. Very large selection of behind-the-scenes production photos. 105 photographs. Filmographies. Index. Originally Behind the Camera. 144pp. 8¼ x 11.
23686-2 Pa. $5.00

DESIGNS FOR THE THREE-CORNERED HAT (LE TRICORNE), Pablo Picasso. 32 fabulously rare drawings—including 31 color illustrations of costumes and accessories—for 1919 production of famous ballet. Edited by Parmenia Migel, who has written new introduction. 48pp. 9⅜ x 12¼. (Available in U.S. only)
23709-5 Pa. $5.00

NOTES OF A FILM DIRECTOR, Sergei Eisenstein. Greatest Russian filmmaker explains montage, making of Alexander Nevsky, aesthetics; comments on self, associates, great rivals (Chaplin), similar material. 78 illustrations. 240pp. 5⅜ x 8½.
22392-2 Pa. $4.50

GEOMETRY, RELATIVITY AND THE FOURTH DIMENSION, Rudolf Rucker. Exposition of fourth dimension, means of visualization, concepts of relativity as Flatland characters continue adventures. Popular, easily followed yet accurate, profound. 141 illustrations. 133pp. 5⅜ x 8½.
23400-2 Pa. $2.75

THE ORIGIN OF LIFE, A. I. Oparin. Modern classic in biochemistry, the first rigorous examination of possible evolution of life from nitrocarbon compounds. Non-technical, easily followed. Total of 295pp. 5⅜ x 8½.
60213-3 Pa. $4.00

THE CURVES OF LIFE, Theodore A. Cook. Examination of shells, leaves, horns, human body, art, etc., in *"the* classic reference on how the golden ratio applies to spirals and helices in nature . . . . "—Martin Gardner. 426 illustrations. Total of 512pp. 5⅜ x 8½.
23701-X Pa. $5.95

PLANETS, STARS AND GALAXIES, A. E. Fanning. Comprehensive introductory survey: the sun, solar system, stars, galaxies, universe, cosmology; quasars, radio stars, etc. 24pp. of photographs. 189pp. 5⅜ x 8½. (Available in U.S. only)
21680-2 Pa. $3.00

THE THIRTEEN BOOKS OF EUCLID'S ELEMENTS, translated with introduction and commentary by Sir Thomas L. Heath. Definitive edition. Textual and linguistic notes, mathematical analysis, 2500 years of critical commentary. Do not confuse with abridged school editions. Total of 1414pp. 5⅜ x 8½.
60088-2, 60089-0, 60090-4 Pa., Three-vol. set $18.00

DIALOGUES CONCERNING TWO NEW SCIENCES, Galileo Galilei. Encompassing 30 years of experiment and thought, these dialogues deal with geometric demonstrations of fracture of solid bodies, cohesion, leverage, speed of light and sound, pendulums, falling bodies, accelerated motion, etc. 300pp. 5⅜ x 8½.
60099-8 Pa. $4.00

Prices subject to change without notice.

Available at your book dealer or write for free catalogue to Dept. GI, Dover Publications, Inc., 180 Varick St., N.Y., N.Y. 10014. Dover publishes more than 175 books each year on science, elementary and advanced mathematics, biology, music, art, literary history, social sciences and other areas.